Analysis of Qualitative Data

Volume 1   *INTRODUCTORY TOPICS*

SHELBY J. HABERMAN

DEPARTMENT OF STATISTICS
UNIVERSITY OF CHICAGO
CHICAGO, ILLINOIS

Analysis of Qualitative Data

Volume 1 *INTRODUCTORY TOPICS*

ACADEMIC PRESS   New York San Francisco London 1978
A Subsidiary of Harcourt Brace Jovanovich, Publishers

ACADEMIC PRESS, INC.
111 Fifth Avenue, New York, New York 10003

United Kingdom Edition published by
ACADEMIC PRESS, INC. (LONDON) LTD.
24/28 Oval Road, London NW1 7DX

Library of Congress Cataloging in Publication Data

Haberman, Shelby J
    Analysis of qualitative data.

    Includes bibliographical references and indexes.
    CONTENTS: v. 1. Introductory topics.
    1. Log–linear models. 2. Contingency tables.
3. Social sciences––Statistical methods.
I. Title.
HA33.H27   519.5   77–25731
ISBN 0–12–312501–4

Contents

Preface   vii
Acknowledgments   xi
Contents of Volume 2   xiii

1 POLYTOMOUS RESPONSES   1

1.1   Effects of Memory—An Introduction to Log–Linear Models   2
1.2   Self-Classification by Social Class   23
1.3   Suicide and the Poisson Distribution   43
1.4   Distribution of Reported Suicides by Region   53
1.5   General Techniques for Log–Linear Models   60
      Exercises   79

2 COMPLETE TWO-WAY TABLES   91

2.1   A Two-by-Two Example   92
2.2   A $2 \times s$ Table: Ethnicity and Women's Role Attitudes   109
2.3   An $r \times s$ Table: Psychiatric Therapy and the Psychoses   112
2.4   Comparison of Samples   119
2.5   Additive Log–Linear Models for Poisson Observations   124
2.6   Additive Log–Linear Models   131
      Exercises   151

3 COMPLETE THREE-WAY TABLES   160

3.1   Race, Sex, and Homicide Weapons—A $2 \times 2 \times 2$ Table   162
3.2   Woman's Place, Sex, and Education   183

3.3   Possible Hierarchical Models for Three-Way Tables   197
3.4   Computational Procedures   207
3.5   Selection of Models   223
      Exercises   234

4   *COMPLETE HIGHER-WAY TABLES*   247

4.1   Comparison of Three-Way Tables: Year of Survey, Sex,
      Education, and Place   248
4.2   Year of Survey, Religion, Education, and Attitudes Toward
      Nontherapeutic Abortions   262
      Exercises   285

5   *LOGIT MODELS*   292

5.1   Logits and Table 3.1   293
5.2   Logits and Women's Roles   302
5.3   Logits, Women's Roles, and Small Frequency Counts   311
5.4   Properties of Logit Models   331
5.5   Quantal-Response Models   344
      Exercises   346

References   354
Example Index   361
Subject Index   363

Preface

Qualitative data are encountered throughout the social sciences. Despite this widespread occurrence, statistical methods for analysis of these data remained quite primitive until the 1960s. Since the early 1960s, development of log–linear models has resulted in very rapid progress in the sophisticated analysis of qualitative and nominal data. Although much of the development of log–linear models has been brought about by statisticians such as Leo Goodman and Frederick Mosteller with close ties to the social sciences, few social scientists have had more than a limited acquaintance with this class of statistical models.

A social scientist interested in learning about log–linear models currently has available a welter of journal articles. These articles appear in the statistics, social science, and biological science literature, and varying levels of mathematical sophistication are required to read them. Three recent books, Plackett (1974), Bishop, Fienberg, and Holland (1975), and Haberman (1974a), are also available. None is specifically oriented toward the needs of social scientists, either in choice of examples and topics or in mathematical level.

This book attempts to provide an introduction to log–linear models that is oriented toward social scientists. Mathematical requirements are kept to a minimum. Some familiarity with matrix algebra sometimes will be helpful, but no explicit use will be made of calculus. Although the formal mathematical requirements are limited, the book does contain numerous algebraic formulas and summation notation is repeatedly used. Without such expressions it is not possible to present general computational algorithms, and assessment of the variability of parameter estimates is impossible. The book assumes sufficient prior knowledge of statistics so that concepts such as confidence intervals, hypothesis tests, and point estimates are familiar. Knowledge of analysis of variance and regression analysis is very helpful in gaining perspective on the relationships between the analysis of continuous and of discrete variables; however, this knowledge is not required for understanding the material in this book. General familiarity with maximum-likelihood estimation is also helpful but not necessary. These restrictions prevent inclusion of formal proofs of results; however, such proofs can be found in Haberman (1974a).

Examples used to illustrate the statistical techniques described in this book are obtained from real data of interest to social scientists. Some examples will involve basic problems in survey research, such as memory error. Some examples will consider topics of public interest such as variations in homicide rates related to variables such as the race and sex of victim. Still other examples will use the General Social Survey of the National Opinion Research Center to examine public opinion concerning abortion.

The first four chapters are ordered by the number of variables under study. Chapter 1 introduces some basic methods for study of the distribution of a single polytomous variable. Several simple examples provide an introduction to the basic estimation and testing procedures associated with log–linear models.

Chapter 2 considers contingency tables in which two polytomous variables are cross classified. Tests for independence and methods for describing departures from independence are discussed.

In Chapter 3 hierarchical log–linear models for three-way tables are introduced. In these tables, three polytomous variables are cross classified. Models for independence, conditional independence, and no three-factor interaction are described. The iterative proportional fitting algorithm of Deming and Stephen (1940) is introduced.

Chapter 4 extends procedures developed in Chapter 3 to contingency tables in which four or more variables are cross classified.

Chapter 5 provides an introduction to logit models. In these models discrete or continuous independent variables are used to predict a dichotomous dependent variable. The models considered are analogous to regression models used with continuous dependent variables.

In Chapter 6 multinomial-response models are developed for prediction of one or more polytomous dependent variables. These models are generalizations of the logit models of Chapter 5.

Chapters 7 and 8 examine contingency tables with special structures. In Chapter 7 tables are considered in which certain cells are unusual in their behavior. The quasi-independence model and other models that ignore these unusual cells are considered. In Chapter 8 models are considered for contingency tables in which several polytomous variables have the same categories. Symmetry models, quasi-symmetry models, and distance models are introduced.

Chapter 9 considers a classical problem of applied statistics, the adjustment of data. The methods developed by Deming and Stephan in the 1940s are illustrated, and some newer possibilities are discussed.

Chapter 10 examines the relationship of log–linear models to latent-structure models. The material in this chapter is quite new. It is intended to provide an indication of possible future developments.

When used as a textbook in a one-quarter or one-semester course, the first five chapters, which constitute Volume 1, can be regarded as a minimal goal.

Acknowledgments

Work on this book has been supported in part by National Science Foundation Grant No. SOC72-05228 A04, National Science Foundation Grant No. MCS72-04364 A04, and National Institutes of Health Grant No. GM22648.

Discussions with Professors Leo Goodman, William Mason, and Clifford Clogg have been particularly helpful in preparation of the manuscript. Much of the data used in this book has been provided by the National Opinion Research Center of the University of Chicago. Thanks are also due to the editors of *Sociological Analysis* for permission to use Table 2.2, to John Wiley & Sons for permission to use Tables 2.4 and 2.14, and to Harcourt Brace Jovanovich for permission to use Tables 2.1 and 2.12.

Contents of Volume 2

6 MULTINOMIAL-RESPONSE MODELS

7 INCOMPLETE TABLES

8 SYMMETRICAL TABLES

9 ADJUSTMENT OF DATA

10 LATENT-CLASS MODELS

APPENDIX: COMPUTER PROGRAMS FOR COMPUTATION OF MAXIMUM-LIKELIHOOD ESTIMATES

1 *Polytomous Responses*

The basic tools used to analyze frequency tables by means of log–linear models may be introduced by an examination of some one-way frequency tables which summarize observations on a single polytomous variable, examples for which are provided in Sections 1.1 to 1.4. In Section 1.5, some general observations are made concerning use of log–linear models with one-way tables; chapter references are provided here.

Sections 1.1 and 1.2 treat multinomial data. The example in Section 1.1 uses memory data to illustrate use of Pearson and likelihood-ratio chi-square statistics, adjusted residuals, and linear combinations of frequency counts to examine departures from a model of equal probabilities for each class. Departures from this model, which has many formal analogies to those of simple linear regression, are then described by a log–linear model. The analogies are used to describe the Newton–Raphson algorithm for computation of maximum-likelihood estimates in terms of a series of weighted regression analyses. Formulas used in weighted regression are also used to obtain normal approximations for the distribution of estimated parameters. Once maximum-likelihood estimates are obtained, chi-square tests and adjusted residuals are used to explore the adequacy of the proposed model.

In Section 1.2, which has a similar structure to Section 1.1, self-classification by social class is considered. Chi-square statistics and adjusted residuals are now used to determine whether the distribution of self-classifications is symmetrical. Departures from symmetry are examined by a log–linear model which is formally analogous to a multiple regression model, and shows that computational procedures for maximum-likelihood estimates correspond to weighted multiple regressions.

In Sections 1.3 and 1.4 attention shifts to Poisson data. Section 1.3 considers whether daily suicide rates are seasonally dependent and models of

1

constant daily rates and of constant daily rates for each season are tested with chi-square statistics and adjusted residuals. Regional variations in suicide rates are considered in Section 1.4 and simultaneous confidence intervals are introduced to determine which regional differences are clearly established.

Section 1.5 provides the general results on which the chapter is based. Maximum-likelihood equations and the Newton–Raphson algorithm are introduced for general log–linear models for both multinominal and Poisson data. General analogies with weighted regression analysis are used to describe the Newton–Raphson algorithm and to develop large-sample approximations to the distributions of parameter estimates, residuals and chi-square statistics.

1.1 Effects of Memory—An introduction to Log–Linear Models

Analysis of data by log–linear models involves several distinct stages. First, a plausible model is proposed for the data under study. Second, unknown parameters in the model are estimated from the data, generally by the method of maximum likelihood. This methods yields estimates available in closed form, as in the equiprobability model, or estimates computed by a version of the Newton–Raphson algorithm with analogies to weighted regression analysis, as in the log–linear time-trend model. Third, these parameter estimates are used in statistical tests of the model's adequacy. Pearson and likelihood-ratio chi-square tests provide overall measures of the compatibility of the model and the data. More specific insight into deviations between model and data are provided by analysis of adjusted residuals and of selected linear combinations of frequencies. Two possibilities exist at the fourth step. If the model appears adequate, then the parameter estimates are used to obtain quantitative implications concerning the data. Asymptotic standard deviations and approximate confidence intervals are major tools at this point. If the model appears inadequate, then the residual analyses and analyses of linear combinations of frequencies of the third stage are employed to suggest new models that are more consistent with the data, to which the new model are then applied. Thus, analysis will often be an iterative process in which the four-stage analysis is applied to several distinct models, many of which are suggested by previous exploration of the data.

As an introduction to the use of log–linear models, consider the data shown in Table 1.1. These data were compiled to study a problem observed during a study conducted to explore the relationship between life stresses and illnesses. The study, which is described in Uhlenhuth, Lipman, Balter,

Table 1.1

*Distribution by Months Prior to Interview of Stressful Events
Reported by Subjects: Subjects Reporting One Stressful Event
in the Period from 1 to 18 Months Prior to Interview[a]*

Months before interview	Number of subjects	Percentage of subjects
1	15	10.2
2	11	7.5
3	14	9.5
4	17	11.6
5	5	3.4
6	11	7.5
7	10	6.8
8	4	2.7
9	8	5.4
10	10	6.8
11	7	4.8
12	9	6.1
13	11	7.5
14	3	2.0
15	6	4.1
16	1	0.7
17	1	0.7
18	4	2.7
Total	147	100.0

[a] From data file used by Uhlenhuth, Balter, Lipman, and Haberman (1977) and Uhlenhuth, Lipman, Balter, and Stern (1974).

and Stern (1974), was conducted by interviewing randomly chosen household members from a probability sample of Oakland, California. Interviewees were required to be between 18 and 65 years old.

Part of the interview consisted of a questionnaire inquiring about 41 life events extracted from a large list of 61 events previously used by Paykel, Prusoff, and Uhlenhuth (1971). Respondents were asked to note which of these events had occurred within the last 18 months and to report the month of occurrence of these events.

The Equiprobability Model

An obvious difficulty associated with an interview of this type is the fallibility of human memory. Presumably some events are forgotten completely or at least misdated. Table 1.1 was compiled to help assess the magnitude of these memory problems. To see how this table can be used for this purpose, it is helpful to construct a simple probability model.

Let N be the number of subjects in the study who reported exactly one dated life event during the 18-month period in question. Number these subjects from 1 to N and let X_h be the number of months before the interview in which subject h's event was reported to have occurred. For integers i between 1 and 18, let n_i be the number of subjects h such that $X_h = i$. By referring to Table 1.1, one finds that $n_1 = 15, n_2 = 11$, etc.

The simplest conceivable probability model that might apply to these data would assume that the X_h are statistically independent random variables and that the probability p_i that X_h equals i is $\frac{1}{18}$ for each subject h and number of months i. This model assumes that no relationship exists between values of X_h associated with distinct subjects in the sample and that within the 18-month period under study, the probability of dating an event in a particular month i is constant for all i. This assumption will hold if reporting of events is accurate and if the actual distribution of events is such that an event is equally likely to occur in any of the 18 months prior to interview.

The most traditional approach to determine if this simple probability model is consistent with the data uses the Pearson chi-square statistic to provide an overall test of goodness of fit. A similar overall test can be made with the likelihood-ratio chi-square statistic. Neither of the statistics discussed here gives any specific information concerning the nature of departures from the model of equal probabilities, for which adjusted residuals and more general linear combinations of frequencies are helpful. Adjusted residuals provide indications of which individual frequency counts are larger or smaller than expected under the equiprobability model; linear combinations of frequencies can provide more subtle information. For example, in this section, the average of the X_h, $1 \leqslant h \leqslant N$, will be compared to the expected value of the X_h under equiprobability to find a systematic tendency to report more recent events rather than more distant events. The more specific tests of models are also considered in this section.

Chi-Square Tests

The chi-square test probably most familiar to readers is based on the Pearson (1900) chi-square test. In this test, the statistic

$$X^2 = \sum (n_i - m_i)^2 / m_i$$
$$= (n_1 - m_1)^2 / m_1 + \cdots + (n_{18} - m_{18})^2 / m_{18}$$

is computed, where m_i, the expected value of n_i under the proposed model, is given by

$$m_i = N p_i = \left(\frac{1}{18}\right)(147) = 8.17.$$

Under the hypothesis that the proposed probability model is correct, X^2 has an approximate chi-square distribution on $18 - 1 = 17$ degrees of freedom, where the approximation becomes increasingly accurate as N increases. In this example, $X^2 = 45.4$. This statistic provides quite strong evidence that the simple probability model is untenable as the probability that a chi-square random variable χ_{17}^2 with 17 degrees of freedom exceeds 45.4 is about 0.0002.

Likelihood-Ratio Chi-Square Statistic

An alternate test of the model of equal probabilities closely related to the X^2 statistic relies on the likelihood-ratio chi-square statistic

$$L^2 = 2\sum n_i \log(n_i/m_i)$$

of Fisher (1924), Neyman and Pearson (1928), and Wilks (1935, 1938). Here m_i is defined as before and log refers to natural logarithms. This statistic has large-sample properties similar to those of X^2. As is the case with X^2, under the simple probability model which has been proposed, L^2 is approximately distributed as a chi-square random variable with 17 degrees of freedom, where the approximation becomes increasingly accurate as N increases. Since the observed value of L^2 is 50.8, use of L^2 also provides considerable evidence that the model of equal probabilities for all months does not fit the observed data.

Residual Analysis

By themselves, the X^2 and L^2 statistics provide no information showing how the equiprobability model may fail to fit the data: they merely indicate whether the model is consistent with the data. To explore how the model fails, residual analysis and the exploration of linear combinations are both exceedingly helpful. To perform a residual analysis, the residuals

$$d_i = n_i - m_i$$

are computed and then adjusted by dividing by the respective standard errors $c_i^{1/2}$, where

$$c_i = Np_i(1 - p_i) = 147\left(\frac{1}{18}\right)\left(\frac{17}{18}\right) = 7.71.$$

One then obtains the adjusted residuals

$$r_i = d_i/c_i^{1/2}.$$

Under the equiprobability model, each r_i has expected value 0 and variance 1. As N becomes large, the distribution of r_i approaches the standardized normal distribution $N(0, 1)$. As noted in Section 1.5, alternate definitions of residuals are available. The definition used here has been chosen for its simplicity and for its standard normal approximation.

Adjusted residuals for the equiprobability model are listed in Table 1.2. The adjusted residuals display an obvious pattern, for the first four residuals are positive and the last five are negative. Of these nine adjusted residuals, five have magnitudes greater than 2. According to the normal approximation, if the equiprobability model holds, then the expected number of residuals of at least this magnitude is $18(0.05) = 0.90$. The natural conclusion from this analysis is that the deviation from the equiprobability model involves a tendency for an event to be reported in a recent month rather than in a more distant month. Such a tendency would result if subjects were more likely to remember recent events than distant events.

Individual residuals provide a suggestive analysis of the failure of the equiprobability model, but a more thorough analysis can be accomplished by a careful choice of linear combinations of the frequencies n_i. In this case, a reasonable linear combination to explore is the sample average

$$\bar{X} = \frac{1}{N} \sum X_h = \frac{1}{N} \sum i n_i.$$

Table 1.2

Adjusted Residuals for the Equiprobability Model as Applied to Table 1.1

Number of months before interview	Adjusted residual
1	2.46
2	1.02
3	2.10
4	3.18
5	−1.14
6	1.02
7	0.66
8	−1.50
9	−0.06
10	0.66
11	−0.42
12	0.30
13	1.02
14	−1.86
15	−0.78
16	−2.58
17	−2.58
18	−1.50

If subjects do in fact tend to remember recent events better than more distant events and if the actual distribution of events is uniform over the 18 months, then $\bar{X}$ should have a smaller expected value than is the case under the equiprobability model.

Under the equiprobability model, $\bar{X}$ is the average of N independent and identically distributed random variables X_h. By Stuart (1950), each X_h has mean

$$\theta = \sum i p_i = \tfrac{1}{2}(1 + 18) = 9.5$$

and variance

$$\sigma^2 = \sum i^2 p_i - \left(\sum i p_i\right)^2$$
$$= \tfrac{1}{12}(18 + 1)(18 - 1) = 26.9.$$

By the central limit theorem,

$$Z = N^{1/2}(\bar{X} - \theta)/\sigma$$

is then approximately distributed as an $N(0, 1)$ random variable. Since the observed value of $\bar{X}$ is 7.33 and the observed value of Z is -5.08, it is clear that $\bar{X}$ is much smaller than could be expected under the equiprobability model.

The Log–Linear Time-Trend Model

At this point, log–linear models become important. It is clear that an equiprobability model is unacceptable and that any acceptable model must be able to account for the observed value of $\bar{X}$. It is also desirable that any probability model appear physically plausible. One candidate for an acceptable model which meets all these criteria assumes that the X_h are independent and identically distributed random variables with

$$\log m_i = \alpha + \beta i$$

for some unknown constants α and β. Since the log of the probabilities is a linear function of the time before interview, the model is called a log–linear time-trend model. The equiprobability model is a special case of this model in which $\alpha = \log(N/18)$ and $\beta = 0$.

If $\beta > 0$, then the probability that an event is reported in a given month increases as the number of months i from interview increases. On the other hand, if $\beta < 0$, then as the number of months i from interview increases the probability of reporting decreases.

Many models for memory are consistent with log–linear time-trend model. The following exponential decay model, although simplistic, does provide some motivation for the log–linear time-trend model when $\beta < 0$. In the exponential decay model, E is an event occurring in month X, $1 \leqslant X \leqslant 18$, before interview. It is assumed that X has a uniform distribution on the integers 1 to 18. Associated with X is the last month $Z \leqslant X$ before interview in which the subject remembered E. If $Z = 0$, E is reported to have occurred in month X. If $Z > 0$, E is not reported. The distribution of Z is determined by a monthly rate q of forgetting, so that given $Z \leqslant i \leqslant X$, the probability is q that $Z = i$. In other words, given that E is remembered i months before interview, the probability is q that E is not remembered after $i - 1$ months before interview.

Under these assumptions,

$$\frac{p_{i+1}}{p_i} = 1 - q.$$

A simple induction shows that

$$\frac{p_i}{p_1} = (1 - q)^{i-1}.$$

For example, note that

$$\frac{p_4}{p_1} = \frac{p_4}{p_3} \frac{p_3}{p_2} \frac{p_2}{p_1}$$

$$= (1 - q)(1 - q)(1 - q) = (1 - q)^3.$$

To relate the memory decay model to the log–linear time-trend model, note that the natural logarithm has the properties that if x and y are positive and z is real, then

$$\log(xy) = \log x + \log y,$$

$$\log\left(\frac{x}{y}\right) = \log x - \log y,$$

and

$$\log(x^z) = z \log x.$$

Thus

$$\log\left(\frac{p_i}{p_1}\right) = (i - 1)\log(1 - q)$$

and

$$\log m_i = \log N + \log p_i$$
$$= [\log N + \log p_1 - \log(1 - q)] + [\log(1 - q)]i.$$

Thus the log–linear time-trend model holds with $\beta = \log(1 - q)$.

To express the probability q of forgetting an event in a given month in terms of β, it is helpful to use the exponential function exp. If y is positive and

$$x = \log y,$$

then $y = \exp x$. If $e = 2.718 \cdots$ is the base of the system of natural logarithms, then

$$\exp x = e^x.$$

It is important to note that for any x and y,

$$(\exp x)(\exp y) = \exp(x + y)$$

and

$$(\exp x)^y = \exp(xy).$$

Thus

$$1 - q = \exp \beta$$

or

$$q = 1 - \exp \beta.$$

Since

$$p_i = p_1 \exp \beta(i - 1),$$

the term exponential decay may be used. The exponent of e is $\beta(i - 1)$, a linear function of the time i before interview.

Maximum-Likelihood Estimates

To determine whether the log–linear time-trend model does fit the data, it is necessary to estimate α and β and then to use these estimates to compute goodness-of-fit test statistics. The most common technique for estimation of α and β is the method of maximum likelihood. Some general rules are discussed in Section 1.5; for now, the only results considered will be those required for the immediate purpose of analyzing the data in Table 1.1.

The method of maximum likelihood yields unique estimates $\hat{\alpha}$ of α and $\hat{\beta}$ of β such that

$$\sum \hat{m}_i = \sum \hat{n}_i = N \tag{1.1}$$

and

$$\sum i \hat{m}_i = \sum i n_i = N \bar{X}, \tag{1.2}$$

where

$$\hat{m}_i = N \hat{p}_i = \exp(\hat{\alpha} + \hat{\beta} i).$$

Division of (1.1) by N yields

$$\sum \hat{p}_i = 1,$$

so that the sum of the estimated probabilities must be 1. Division of (1.2) by N gives

$$\hat{\theta} = \sum i \hat{p}_i = \bar{X},$$

so that the estimated population mean $\hat{\theta}$ of the X_h is equal to the sample mean $\bar{X}$. Thus the method of maximum likelihood results in a model consistent with the observed sample mean $\bar{X}$.

The Newton–Raphson Algorithm

To compute $\hat{\alpha}$ and $\hat{\beta}$, an iterative procedure must be employed; however, the algorithm is relatively simple. The procedure is a version of the Newton–Raphson algorithm, a classical procedure for solution of nonlinear equations discussed in more detail in Section 1.5. A computer program in the Appendix in Volume 2 can be used to aid computations. In the case of log–linear models, the algorithm may be described in terms of a series of calculations analogous to those performed in regression analysis.

To begin the algorithm, an initial approximation m_{i0} is required for $\hat{m}_i$. Since the algorithm uses the logarithm y_{i0} of m_{i0}, each m_{i0} must be positive. In the example under study, it is reasonable to set m_{i0} equal to n_i for each i. The estimate n_i is an estimate of m_i and $\hat{m}_i$ is an estimate of m_i, so n_i should serve as an approximation to m_i. In this example, all n_i are positive, so that logarithms can be used. Since examples can arise in which some n_i are 0, it is common practice to let m_{i0} be $n_i + \frac{1}{2}$, which is always positive. This practice is adopted in the computer algorithm presented in the Appendix (see Vol. 2) and in Table 1.3.

To compute β_0, an initial approximation to $\hat{\beta}$, the formulas

$$\theta_0 = \frac{\sum_i i m_{i0}}{\sum_i m_{i0}}$$

and

$$\beta_0 = \frac{\sum_i (i - \theta_0) y_{i0} m_{i0}}{\sum_i (i - \theta_0)^2 m_{i0}}$$

are employed. Calculations are summarized in the first eight columns of Table 1.3.

An analogy to regression analysis should be noted. Consider a regression model

$$y_{i0} = \alpha + \beta i + \varepsilon_i,$$

Table 1.3

Calculation of Maximum-Likelihood Estimates[a] for the Log–Linear Time-Trend Model for Table 1.1

Month i before interview (1)	n_i (2)	in_i (3)	m_{i0} (4)	y_{i0} (5)	im_{i0} (6)	$(i - \theta_0)y_{i0}m_{i0}$ (7)	$(i - \theta_0)^2 m_{i0}$ (8)
1	15	15	15.5	2.7408	15.5	−100.00	645.22
2	11	22	11.5	2.4423	23.0	−62.697	341.82
3	14	42	14.5	2.6741	43.5	−64.553	287.38
4	17	68	17.5	2.8622	70.0	−60.409	208.53
5	5	25	5.5	1.7047	27.5	−13.486	33.066
6	11	66	11.5	2.4423	69.0	−16.697	24.243
7	10	70	10.5	2.3514	73.5	−4.7452	2.1445
8	4	32	4.5	1.5041	36.0	2.4664	1.3518
9	8	72	8.5	2.1401	76.5	13.159	20.371
10	10	100	10.5	2.3514	105.0	26.755	68.173
11	7	77	7.5	2.0149	82.5	26.611	94.416
12	9	108	9.5	2.2513	114.0	43.207	196.51
13	11	143	11.5	2.4423	149.5	63.803	353.98
14	3	42	3.5	1.2528	49.0	22.918	150.07
15	6	90	6.5	1.8718	97.5	49.062	370.33
16	1	16	1.1	0.4055	24.0	12.822	109.60
17	1	17	1.5	0.4055	25.5	14.322	136.75
18	4	72	4.5	1.5041	81.0	47.466	500.68
Total	147	1077	156.00		1162.5	−243.23	3544.6

$\bar{X} = 1077/147 = 7.3265$; $\theta_0 = 1162.5/156.0 = 7.4519$; $\beta_0 = -243.23/3544.6 = -0.068620$.

where the ε_i are independent random variables with zero means and respective variances m_{i0}^{-1}. The weighted-least-squares estimate of β is β_0, as in the Newton–Raphson algorithm.

Given β_0, an approximation α_0 for α may be obtained by solving the equation

$$\sum_i \exp(\alpha_0 + \beta_0 i) = N.$$

Let $g_0 = \exp(\alpha_0)$. Since

$$\sum_i \exp(\alpha_0 + \beta_0 i) = \sum_i \exp(\alpha_0)\exp(\beta_0 i)$$

$$= \exp(\alpha_0) \sum_i \exp(\beta_0 i),$$

it follows that

$$g_0 = N \Big/ \sum_i \exp(\beta_0 i).$$

Table 1.3(continued)

Month i before interview	$\exp(\beta_0 i)$ (9)	m_{i1} (10)	y_{i1} (11)[b]	im_{i1} (12)	$(i - \theta_1)y_{i1}m_{i1}$ (13)[b]	$(i - \theta_1)^2 m_{i1}$ (14)
1	0.93368	13.746	2.7120	13.746	−92.077	616.77
2	0.87176	12.834	2.4092	25.669	−73.136	416.76
3	0.81395	11.983	2.6518	35.950	−56.302	264.53
4	0.75997	11.189	2.9343	44.754	−41.380	153.04
5	0.70957	10.447	1.8249	52.233	−28.189	76.066
6	0.66251	9.7537	2.4054	58.522	−16.566	28.136
7	0.61857	9.1069	2.3071	63.748	−6.3605	4.4423
8	0.57755	8.5029	1.6108	68.023	2.5642	0.77331
9	0.53925	7.9390	2.0795	71.451	10.333	13.449
10	0.50348	7.4125	2.3522	74.125	17.060	39.266
11	0.47009	6.9209	1.9460	76.130	22.850	75.440
12	0.43892	6.4619	2.2587	77.543	27.796	119.57
13	0.40981	6.0334	2.6205	78.434	31.986	169.58
14	0.38263	5.6332	1.2612	78.865	35.498	223.70
15	0.35726	5.2597	1.8008	78.895	38.404	280.41
16	0.33556	4.9108	0.79508	78.573	40.768	338.44
17	0.31144	4.5852	0.74092	77.948	42.649	396.70
18	0.29079	4.2811	1.3885	77.059	44.102	454.32
Total	9.9850	147.00		1131.7	−306.60	3671.4

$g_0 = 147/9.9850 = 14.722$; $\alpha_0 = \log(14.722) = 2.6893$; $\theta_1 = 1131.7/147.00 = 7.6984$; $\beta_1 = -306.60/3671.4 = -0.068620 + (7.3265 - 7.6984)/3671.4 = -0.083511$.

Given g_0 and β_0, a new approximation m_{i1} for $\hat{m}_i$ is obtained from the equation

$$m_{i1} = \exp(\alpha_0 + \beta_0 i) = g_0 \exp(\beta_0 i).$$

Computations are summarized in columns 9 and 10 of Table 1.3. Note that

$$\sum_i m_{i1} = N.$$

At this point, a new cycle of the algorithm can begin. A *working logarithm* y_{i1} is defined by the equation

$$y_{i1} = \log m_{i1} + (n_i - m_{i1})/m_{i1},$$

as in column 11 of Table 1.3. A new approximation β_1 for $\hat{\beta}$ is computed from the formulas

$$\theta_1 = \frac{\sum_i im_{i1}}{\sum_i m_{i1}} = N^{-1} \sum_i im_{i1}$$

Table 1.3 (continued)

Month i before interview	$\exp(\beta_1 i)$ (15)	m_{i2} (16)	y_{i2} (17)[b]	im_{i2} (18)	$(i - \theta_2)y_{i2}m_{i2}$ (19)[b]	$(i - \theta_2)^2 m_{i2}$ (20)
1	0.91988	15.146	2.7081	15.146	−259.75	607.42
2	0.84618	13.933	2.4237	27.866	−180.08	396.22
3	0.77839	12.816	2.6431	38.449	−146.77	240.60
4	0.71602	11.790	2.9092	47.159	−114.31	130.95
5	0.65866	10.845	1.8447	54.225	−46.670	59.016
6	0.60589	9.9762	2.4028	59.857	−31.948	17.720
7	0.55734	9.1769	2.3064	64.238	−7.0430	1.0161
8	0.51269	8.4417	1.6070	67.533	9.0517	3.7583
9	0.47161	7.7653	2.0799	69.888	26.928	21.585
10	0.43383	7.1432	2.3661	71.432	45.080	50.818
11	0.39907	6.5709	1.9480	72.280	46.940	88.369
12	0.36769	6.0444	2.2881	72.533	64.549	131.67
13	0.33769	5.5602	2.6940	72.282	84.889	178.58
14	0.31063	5.1147	1.2187	71.606	41.557	227.36
15	0.28574	4.7049	1.8239	70.574	65.794	276.58
16	0.26285	4.3279	0.69615	69.247	26.113	325.12
17	0.24179	3.9812	0.63276	67.680	24.353	372.06
18	0.22242	3.6622	1.3903	65.920	54.313	416.73
Total	8.9278	147.00		1077.9	−297.01	3545.6

$g_1 = 147/8.9278 = 16.465$; $\alpha_1 = \log(16.465) = 2.8012$; $\theta_2 = 1077.9/147.00 = 7.3328$; $\beta_2 = 297.01/3545.6 = -0.083511 + (7.3265 - 7.3328)/3545.6 = -0.083769$.

and

$$\beta_1 = \frac{\sum_i (i - \theta_1)y_{i1}m_{i1}}{\sum_i (i - \theta_1)^2 m_{i1}} = \beta_0 + \frac{N(\bar{X} - \theta_1)}{\sum_i (i - \theta_1)^2 m_{i1}}.$$

The first expression for β_1 is the estimate of β in the regression model

$$y_{i1} = \alpha + \beta i + \varepsilon_i,$$

where the ε_i are independent random variables with respective means 0 and variances m_{i1}^{-1}. The second expression, which is algebraically equivalent to the first, is more convenient in numberical computations. The algebraic identity of these expressions may be illustrated by results of calculations summarized by columns 11 to 14 of Table 1.3. In practice, columns 11 and 13 are unnecessary. They are used here to illustrate the identity between the alternate approaches to computation of β_1.

A new approximation α_1 for $\hat{\alpha}$ is obtained by the equation

$$g_1 = N \bigg/ \sum_i \exp(\beta_1 i),$$

Table 1.3 (continued)

Month i before interview	$\exp(\beta_2 i)$ (21)	m_{i3} (22)	im_{i2} (23)	$(i - \theta_3)^2 m_{i2}$ (24)
1	0.91964	15.171	15.171	607.22
2	0.84574	13.952	27.904	395.84
3	0.77778	12.831	38.492	240.18
4	0.71528	11.800	47.199	130.57
5	0.65781	10.852	54.258	58.737
6	0.60495	9.9796	59.878	17.561
7	0.55634	9.1777	64.244	0.9786
8	0.51163	8.4402	67.522	3.8281
9	0.47052	7.7620	69.858	21.737
10	0.43271	7.1383	71.383	51.020
11	0.39794	6.5647	72.211	88.586
12	0.36596	6.0371	72.446	131.86
13	0.33655	5.5520	72.176	178.71
14	0.30951	5.1059	71.482	227.39
15	0.28464	4.6956	70.434	276.49
16	0.26177	4.3183	69.092	324.86
17	0.24073	3.9713	67.512	371.62
18	0.22139	3.6522	65.739	416.06
Total	8.9111	147.00	1077.0	3543.2

$g_2 = 147/8.9111 = 16.497$; $\alpha_2 = \log(16.497) = 2.8032$; $\theta_3 = 1077.0/147.00 = 7.3265$.

[a] Numbers are presented to five significant figures. Actual computations were performed on an IBM 370/168 computer, which carries at least six significant figures during computations. Consequently, rounding errors may cause some column totals obtained by adding the numbers listed in the column to differ slightly from the listed column total. Similar remarks apply to other tables presented in this book.

[b] These columns are presented for illustrative purposes. They are not actually needed for computations.

where $g_1 = \exp(\alpha_1)$. Given g_1 and β_1, a new approximation m_{i2} for $\hat{m}_i$ is obtained by the formula

$$m_{i2} = g_1 \exp(\beta_1 i).$$

Computations are summarized in columns 15 and 16 of Table 1.3.

Given m_{i2}, new working logarithms y_{i2} and new approximations θ_2, β_2, α_2, and g_2 may be computed. In general, given an integer $t \geqslant 1$ and approximations m_{it} for $\hat{m}_i$, one may define

$$y_{it} = \log m_{it} + \frac{(n_i - m_{it})}{m_{it}},$$

$$\theta_t = \frac{\sum_i i m_{it}}{\sum_i m_{it}} = N^{-1} \sum_i i m_{it},$$

$$\beta_t = \frac{\sum_i (i - \theta_t) y_{it} m_{it}}{\sum_i (i - \theta_t)^2 m_{it}} = \beta_{t-1} + \frac{N(\bar{X} - \theta_t)}{\sum_i (i - \theta_t)^2 m_{it}},$$

$$g_t = \frac{N}{\sum_i \exp(\beta_t i)},$$

and

$$m_{i(t+1)} = g_t \exp(\beta_t i).$$

Columns 17 through 22 summarize computations for $t = 2$.

Under normal circumstances, successive approximations β_t becomes closer and closer to the maximum-likelihood estimate $\hat{\beta}$. In the example under study, β_0 is -0.068620, β_1 is -0.083511, and $\beta_2 = -0.083769$. Further values of β_t agree with β_2 to five significant figures. Thus $\hat{\beta}$ can be taken to be -0.083769. Three indications that β_2 is a very good approximation to $\hat{\beta}$ can be seen: (1) the difference between β_2 and β_1 is only -0.000258. (2) The largest difference $|m_{i2} - m_{i3}|$ is only 0.0100. (3) θ_3 and $\bar{X}$ agree to five significant figures. Given that β_2 is a satisfactory approximation to $\hat{\beta}$ and m_{i3} is an approximation of $\hat{m}_i$ computed from β_2, m_{i3} may be regarded as a satisfactory approximation to $\hat{m}_i$.

Chi-Square Tests

Given estimates m_i, tests for goodness of fit may be made by means of the Pearson chi-square statistic

$$X^2 = \sum \frac{(n_i - \hat{m}_i)^2}{\hat{m}_i}$$

or the log-likelihood ratio chi-square statistic

$$L^2 = 2 \sum n_i \log\left(\frac{n_i}{\hat{m}_i}\right).$$

Since there are 18 entries in Table 1.2 and 2 parameters, α and β, have been estimated, there are $18 - 2 = 16$ degrees of freedom associated with these statistics. If the log–linear time-trend model holds, then both X^2 and

L^2 are approximately distributed as chi-square random variables with 16 degrees of freedom. In this example, X^2 is 22.7 and L^2 is 24.6, so that the model is consistent with the data, although the fit is not perfect. Note that the probability is 0.10 that a χ^2_{16} variable exceeds 23.5 and the probability is 0.05 that a χ^2_{16} variable exceeds 26.3.

Comparison of Models

The chi-square statistics of this model represent a considerable improvement over the equiprobability model. This can be seen in the reduction in L^2 from 50.8 to 24.6, a change of 26.2, resulting from removal of a single degree of freedom. If the equiprobability model fits the data, then this reduction in L^2 has an approximate chi-square distribution on 1 degree of freedom. The probability that $\chi_1{}^2$ exceeds 26.2 is less than 0.00001, so that further confirmation is obtained that the equiprobability model is inconsistent with the data. This comparison of chi-square statistics provides a better proof of the unsuitability of the equiprobability model than does the original L^2 statistic for that model, as may be seen by comparing their significance levels.

The foregoing comparison is made possible by the fact that the equiprobability model is a special case of the log–linear time-trend model in which one parameter, namely β, is assumed 0. Note that in the equiprobability model

$$\log m_i = \alpha,$$

while in the log–linear time-trend model

$$\log m_i = \alpha + \beta i.$$

Two measures of goodness-of-fit that are based on likelihood-ratio chi-square statistics may be used to compare the equiprobability model to the log–linear time-trend model. The first statistic is F, the ratio of the likelihood-ratio chi-square to the corresponding degrees of freedom. In the equiprobability model, F is

$$\left(\frac{1}{17}\right)(50.8) = 3.0,$$

while in the log–linear time-trend model, F is

$$\left(\frac{1}{16}\right)(24.6) = 1.5.$$

To interpret these two statistics, note that if a model holds, then the expected value of F is about 1. The second F statistic is much closer to 1 than the first; the suggestion is that the log–linear time-trend model provides a better fit.

The F statistic presented here is somewhat analogous to the F statistic of regression analysis or analysis of variance. For example, if the time-trend model holds, F has an approximate F distribution with 16 degrees of freedom in the numerator and infinite degrees of freedom in the denominator.

An alternate statistic is R^2, which is $(L_1{}^2 - L_2{}^2)/L_1{}^2$, where $L_1{}^2$ is the likelihood-ratio chi-square for the equiprobability model and $L_2{}^2$ is the likelihood-ratio chi-square for the log–linear time-trend model. The statistic is analogous to the R^2 statistic of regression analysis. It measures the success of the log–linear time-trend model in predicting the probabilities p_i. In this example,

$$R^2 = (50.8 - 24.6)/(50.8) = 0.52.$$

Residual Analysis

To examine the fitted table more more closely, residual analysis is again in order. The residual is now

$$d_i = n_i - \hat{m}_i.$$

As noted in Section 1.5, the variance c_i of d_i may be estimated by

$$\hat{c}_i = \hat{m}_i \left[1 - \frac{\hat{m}_i}{N} - \frac{(i - \bar{X})^2 \hat{m}_i}{\hat{S}} \right],$$

where

$$\hat{S} = \sum_i (i - \bar{X})^2 \hat{m}_i.$$

Division of d_i by $\hat{c}_i^{1/2}$ leads to the adjusted residual

$$r_i = (n_i - \hat{m}_i)/\hat{c}_i^{1/2}.$$

If the log–linear time-trend model holds, then each r_i has an approximate normal distribution with mean 0 and variance 1.

Computation of r_i is summarized in Table 1.4. Note that the program in the Appendix can compute the adjusted residuals. Intermediate numbers in computations are recorded to five significant figures as an aid for the reader who wishes to check calculations. Final values of r_i are approximated to three significant figures. Several totals are included to illustrate relationships that consistently hold. For example, the sum of the residuals d_i is always

Table 1.4

Computation of Adjusted Residuals for the Log–Linear Time-Trend Model
for Table 1.1

Months i before interview	Number of subjects n_i	Estimated expected number of subjects $\hat{m}_i$	Residual d_i	Estimated probability $\hat{p}_i = \hat{m}_i/N$
1	15	15.171	−0.17104	0.10320
2	11	13.952	−2.9519	0.094911
3	14	12.831	1.1692	0.087284
4	17	11.800	5.2002	0.080271
5	5	10.852	−5.8516	0.073820
6	11	9.9796	1.0204	0.067888
7	10	9.1777	0.82232	0.062433
8	4	8.4402	−4.4402	0.057416
9	8	7.7620	0.23802	0.052803
10	10	7.1383	2.8617	0.048560
11	7	6.5647	0.43535	0.044657
12	9	6.0371	2.9629	0.041069
13	11	5.5520	5.4480	0.037769
14	3	5.1059	−2.1059	0.034734
15	6	4.6956	1.3044	0.031943
16	1	4.3183	−3.3183	0.029376
17	1	3.9713	−2.9713	0.027015
18	4	3.6522	0.34784	0.024845
Total	147(N)	147.00	0.0000	1.0000

0 and the sum of the terms

$$1 - \frac{\hat{m}_i}{N} - \frac{(i - \bar{X})^2 \hat{m}_i}{\hat{S}}$$

is the number of degrees of freedom of the X^2 and L^2 statistics.

The adjusted residuals show no obvious patterns. This behavior can be discerned directly from Table 1.4 or from the graph in Figure 1.1 of the adjusted residual r_i by the month i of interview.

Two checks may be made of the sizes of the resulting adjusted residuals: Since they have approximate $N(0, 1)$ distributions, the probability is about 0.045 that a given value of $|r_i|$ exceeds 2 and the probability is about 0.012 that a given value of $|r_i|$ exceeds 2.5. Given 18 values of $|r_i|$, the expected number of $|r_i|$ to exceed 2 is about

$$18(0.045) = 0.82$$

and the expected number of $|r_i|$ to exceed 2.5 is about

$$18(0.012) = 0.22.$$

Table 1.4 (continued)

Months i before interview	$(i - \bar{X})^2 \hat{m}_i$	$(i - \bar{X})^2 \hat{m}_i / \hat{S}$	$1 - (\hat{m}_i / N) - \{(i - \bar{X})^2 \hat{m}_i / \hat{S}\}$	$\hat{c}_i^{1/2}$	r_i
1	607.22	0.17137	0.72542	3.3174	−0.05
2	395.84	0.11172	0.79337	3.3270	−0.89
3	240.18	0.067785	0.84493	3.2926	0.36
4	130.57	0.036852	0.88288	3.2277	1.61
5	58.737	0.016577	0.90960	3.1418	−1.86
6	17.561	0.0049562	0.92716	3.0418	0.34
7	0.97858	0.00027617	0.93729	2.9329	0.28
8	3.8281	0.0010804	0.94150	2.8190	−1.58
9	21.737	0.0061349	0.94106	2.7027	0.09
10	51.020	0.014399	0.93704	2.5863	1.11
11	88.586	0.025001	0.93034	2.4713	0.18
12	131.86	0.037214	0.92172	2.3589	1.26
13	178.71	0.050437	0.91179	2.2500	2.42
14	227.39	0.064176	0.90109	2.1450	−0.98
15	276.49	0.078032	0.89003	2.0443	0.64
16	324.86	0.091684	0.87894	1.9482	−1.70
17	371.62	0.10488	0.86810	1.8567	−1.60
18	416.06	0.11742	0.85773	1.7699	0.20
Total	3543.2($\hat{S}$)	1.0000	16.000		

$\bar{X} = 7.3265$

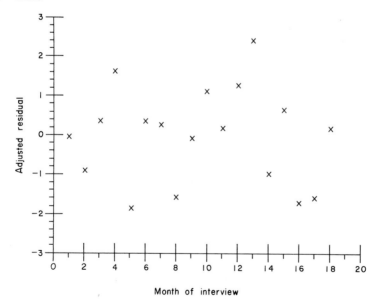

Figure 1.1 Adjusted residuals in Table 1.4 by month.

Given that the estimated expected values $\hat{m}_i$ are often less than 10, these approximations are somewhat crude. Nonetheless, they suggest that it is not surprising that one adjusted residual exceeds 2 in magnitude and that none exceeds 2.5 in magnitude.

A rankit plot or full-normal plot graph, discussed by Tukey (1962), also may be used to examine results. To construct the graph, order the residuals from smallest to largest, as in Table 1.5. Corresponding to the ith smallest residual is a score R_i called a rankit, which is an approximation for the expected value of the ith smallest value of r_i, assuming that the log-linear time-trend model holds. The approximation improves as the number of adjusted residuals becomes large and as the individual expected values m_i become large. In Table 1.5, R_i is chosen so that

$$(3i - 1)/(3s + 1)$$

is the probability that a standard normal deviate is less than R_i.

In Figure 1.2, the ith smallest adjusted residual is plotted against the rankit R_i. The points on the graph should resemble random deviations from the straight line with ordinate y and abscissa x equal. Roughly speaking,

Table 1.5

Preparation of a Rankit Plot for the Adjusted Residuals in Table 1.4

i	ith smallest adjusted residual	$\dfrac{3i-1}{3s+1}$ ($s = 18$)	Rankit R_i
1	−1.86	0.04	−1.79
2	−1.70	0.09	−1.34
3	−1.60	0.15	−1.06
4	−1.58	0.20	−0.84
5	−0.98	0.25	−0.66
6	−0.89	0.31	−0.50
7	−0.05	0.36	−0.35
8	0.09	0.42	−0.21
9	0.18	0.47	−0.07
10	0.20	0.53	0.07
11	0.28	0.58	0.21
12	0.34	0.64	0.35
13	0.36	0.69	0.50
14	0.64	0.75	0.66
15	1.11	0.80	0.84
16	1.26	0.85	1.06
17	1.61	0.91	1.34
18	2.42	0.96	1.79

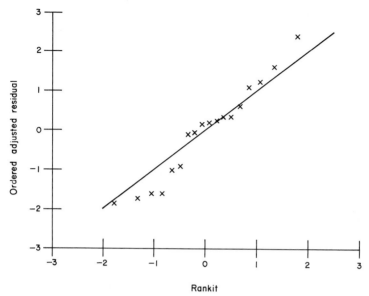

Figure 1.2   Rankit plot of adjusted residuals in Table 1.4.

this pattern does appear present, although positive adjusted residuals do tend to be above the line and negative adjusted residuals tend to be below the line. Some of this behavior may reflect weaknesses in the normal approximation due to small expected values and some may reflect weaknesses in the model not yet identified due to the limited sample size. Exercise 1.1 may help clarify this situation through study of related data.

Given results of the residual analyses and chi-square tests, the log–linear time-trend model appears to be consistent with the data, although some weaknesses in the model have been suggested. In any case, the model provides a much better description of the data than the equiprobability model.

Large-Sample Properties of Estimates

The coefficient β appears large enough to have substantive importance, as can be shown by using the analogy between the model

$$\log m_i = \alpha + \beta i$$

for Table 1.1 and the formally related regression model

$$Y_i = \alpha + \beta i + \varepsilon_i$$

in which the Y_i are hypothetical observations and ε_i are independently distributed $N(0, m_i^{-1})$ random variables.

In the regression model, the estimate

$$b = \frac{\sum_i (i - \theta) Y_i m_i}{\sum_i (i - \theta)^2 m_i}$$

has a normal distribution with mean β and variance

$$\frac{1}{\sum_i (i - \theta)^2 m_i}.$$

In the log–linear time-trend model, the distribution of the corresponding estimate $\hat{\beta}$ of β is approximately a normal distribution with mean β and variance

$$\sigma^2(\hat{\beta}) = \frac{1}{\sum_i (i - \theta)^2 m_i}.$$

The approximation improves as the sample size N increases. Since the approximating normal distribution has mean β and variance $\sigma^2(\hat{\beta})$, the estimate $\hat{\beta}$ is said to have asymptotic mean (AM) β and asymptotic variance (AV) $\sigma^2(\hat{\beta})$. The asymptotic variance $\sigma^2(\hat{\beta})$ may be estimated by substituting the estimate $\hat{m}_i$ and the estimate $\bar{X}$ for the expected value μ. One obtains the estimated asymptotic variance (EAV)

$$s^2(\hat{\beta}) = \frac{1}{\sum_i (i - \bar{X})^2 \hat{m}_i}.$$

The square root $\sigma(\hat{\beta})$ of $\sigma^2(\hat{\beta})$ is called the asymptotic standard deviation (ASD) of $\hat{\beta}$, and the square root $s(\hat{\beta})$ of $s^2(\hat{\beta})$ is called the estimated asymptotic standard deviation (EASD) of $\hat{\beta}$.

Reference to Table 1.3 or 1.4 shows that

$$\sum_i (i - \bar{X})^2 \hat{m}_i = 3543.2.$$

Thus

$$s(\hat{\beta}) = \left(\frac{1}{3543.2} \right)^{1/2} = 0.0168.$$

Since if the time-trend model is valid,

$$(\hat{\beta} - \beta)/s(\hat{\beta})$$

is approximately distributed as standardized normal deviate, an approximate 95% confidence interval for β has lower bound

$$\hat{\beta} - Z_{0.025} s(\hat{\beta}) = -0.0838 - (1.96)(0.0168)$$
$$= -0.117$$

and upper bound

$$\hat{\beta} + Z_{0.025}s(\hat{\beta}) = -0.0838 + (1.96)(0.0168)$$
$$= -0.051.$$

Here for $0 < \alpha < 1$, the probability is α that an $N(0, 1)$ random variable exceeds Z_α. The lower bound corresponds to a ratio p_{i+1}/p_i of

$$e^{-0.117} = 0.890$$

and the upper bound corresponds to a ratio of

$$e^{-0.051} = 0.950.$$

A systematic decrease in reporting probability of 5–11 percent per month is large enough to have serious impact on a study which depends on the ability of subjects to recall events accurately.

The analysis conducted so far cannot prove that the distributional peculiarities in Table 1.1 result from an increasing inability to recall events as the time from occurrence increases, although such a difficulty certainly provides a plausible and simple explanation of the observed behavior. Unexpected causes may be at work, and systematic errors in dating may have some influence. More complete analyses must include consideration of memory behavior of subjects who recall several events and variation in the distribution of months in which events are reported that are related to the type of respondent or type of event. For one example of such an analysis, see Exercise 1.1.

This section has illustrated the general statistical procedures required with log–linear models. Maximum-likelihood equations have been provided, maximum-likelihood estimates have been computed, chi-square tests and residual analyses have been used, and asymptotic variances have been estimated. Much of the computational work has been based on an analogy between the Newton–Raphson algorithm for the log–linear time-trend model and weighted-least-squares estimation for a linear model with a single independent variable. In the following section, a slightly more complex log–linear model is considered. Here analogies will be used between the Newton–Raphson algorithm and weighted-least-squares estimation for two independent variables.

1.2 Self-Classification by Social Class

Log–linear models may be used to describe departures from symmetry rather than from equiprobability. In this section, a symmetry model is introduced and tested by means of chi-square statistics and residual analysis.

A more general log–linear model is then introduced to describe departures from symmetry. The model is tested by chi-square tests, and its parameter estimates are used to assess the importance of departures from symmetry and to assess the predominance of the central classes under study.

Computations of maximum-likelihood estimates and estimation of asymptotic variances are again facilitated by the analogy between the Newton–Raphson algorithm and weighted least squares. Some added complexity arises since the corresponding model for weighted-least-squares estimation involves two independent variables rather than just one; however, the basic procedures of this section are quite similar to those of Section 1.1.

Data for the analyses of this section are from Table 1.6, which summarizes answers to a question on social class membership in the 1975 General Social Survey of the National Opinion Research Center. Subjects for this survey are a representative cross section of persons in the continental United States (Alaska and Hawaii excluded) who are 18 years or older and live in non-institutional settings. The sample is obtained by a complex combination of probability sampling, cluster sampling, and quota sampling. For details, see National Opinion Research Center (1975).

The table has two principal features suggested by a cursory inspection. The table has approximate symmetry. Note that the number of respondents placing themselves in the lower class is comparable to the number placing themselves in the upper class, while the number in the working class category is comparable to the number in the middle class. The bulk of respondents place themselves in either the working class or the middle class.

Table 1.6

Self-Classifications by Social Class
in the 1975 General Social Survey[a,b]

Response	Score	Number responding
Lower class	1	72
Working class	2	714
Middle class	3	655
Upper class	4	41
Total		1482

[a] Subjects are asked the following question: "If you were asked to use one of four names for your social class, which would you say you belong in: the lower class, the working class, the middle class, or the upper class?" The eight subjects who did not provide a classification are excluded from analysis.

[b] National Opinion Research Center (1975, page 41).

The Symmetry Model

Although approximate symmetry may be present, the table is not symmetrical, which can be shown by constructing and testing a simple probability
model. Let X_h be the score of the response of subject h, $1 \leqslant h \leqslant N = 1482$
(see Table 1.6). Thus $X_h = 1$ if subject h has a self-classification of lower
class, and $X_h = 4$ corresponds to an upper class self-classification by subject
h. Let the X_h, $1 \leqslant h \leqslant N$, be independent random variables such that for
each subject h, the probability is $p_i > 0$ that $X_h = i$.
The table is symmetrical if

$$p_1 = p_4$$

and

$$p_2 = p_3.$$

If

$$\alpha = \log N + \tfrac{1}{2}(\log p_1 + \log p_2)$$

and

$$\beta = \tfrac{1}{2}(\log p_1 - \log p_2),$$

then

$$\begin{aligned}
\log m_i &= \alpha + \beta, & i &= 1 \quad \text{or} \quad 4, \\
&= \alpha - \beta, & i &= 2 \quad \text{or} \quad 3.
\end{aligned}$$

Thus the symmetry model is a log–linear model; the logarithms of the means
are linear functions of the parameters α and β.

Maximum-Likelihood Estimates

To find maximum-likelihood estimates, let n_i be the number of subjects h
such that $X_h = i$, and let $m_i = Np_i$ be the expected value of n_i. Thus $n_1 = 72$,
$n_2 = 714$, $n_3 = 655$, and $n_4 = 41$.
As will be shown in Section 1.5 and Exercise 1.3, the maximum-likelihood
estimates $\hat{m}_i$ of m_i and $\hat{p}_i$ of p_i must satisfy the equations

$$\begin{aligned}
\hat{m}_1 &= N\hat{p}_1 = N\hat{p}_4 = \hat{m}_4, \\
\hat{m}_2 &= N\hat{p}_2 = N\hat{p}_3 = \hat{m}_3, \\
\hat{m}_1 &+ \hat{m}_2 + \hat{m}_3 + \hat{m}_4 = N, \\
\hat{m}_1 &- \hat{m}_2 - \hat{m}_3 + \hat{m}_4 = n_1 - n_2 - n_3 + n_4.
\end{aligned}$$

These equations have the solution

$$\hat{m}_1 = \hat{m}_4 = \tfrac{1}{2}(n_1 + n_4),$$
$$\hat{m}_2 = \hat{m}_3 = \tfrac{1}{2}(n_2 + n_3),$$
$$\hat{p}_1 = \hat{p}_4 = \tfrac{1}{2}(n_1 + n_4)/N,$$
$$\hat{p}_2 = \hat{p}_3 = \tfrac{1}{2}(n_2 + n_3)/N.$$

Chi-Square Statistics

The Pearson chi-square statistic is

$$X^2 = \frac{\sum (n_i - \hat{m}_i)^2}{\hat{m}_i}$$

$$= \frac{(n_1 - n_4)^2}{(n_1 + n_4)} + \frac{(n_2 - n_3)^2}{(n_2 - n_3)}.$$

To verify the last expression for X^2, note that

$$\frac{(n_1 - \hat{m}_1)^2}{\hat{m}_1} = \frac{[n_1 - \tfrac{1}{2}(n_1 + n_4)]^2}{[\tfrac{1}{2}(n_1 + n_4)]}$$

$$= \frac{[\tfrac{1}{2}(n_1 - n_4)]^2}{[\tfrac{1}{2}(n_1 + n_4)]} = \frac{\tfrac{1}{2}(n_1 - n_4)^2}{(n_1 + n_4)}.$$

Similarly,

$$\frac{(n_4 - \hat{m}_4)^2}{\hat{m}_4} = \frac{\tfrac{1}{2}(n_1 - n_4)^2}{(n_1 + n_4)}$$

and

$$\frac{(n_2 - \hat{m}_2)^2}{\hat{m}_2} = \frac{(n_3 - \hat{m}_3)^2}{\hat{m}_3} = \frac{\tfrac{1}{2}(n_2 - n_3)^2}{(n_2 + n_3)}.$$

The likelihood-ratio chi-square statistic

$$L^2 = 2\sum n_i \log\left(\frac{n_i}{\hat{m}_i}\right).$$

$$= 2\left[n_1 \log\left(\frac{2n_1}{n_1 + n_4}\right) + n_2 \log\left(\frac{2n_2}{n_2 + n_3}\right) \right.$$

$$\left. + n_3 \log\left(\frac{2n_3}{n_2 + n_3}\right) + n_4 \log\left(\frac{2n_4}{n_1 + n_4}\right) \right].$$

Since there are four possible values of X_h and two estimated parameters, the associated degrees of freedom are $4 - 2 = 2$. If the model holds, then both

X^2 and L^2 are approximately distributed as chi-square random variables with two degrees of freedom.

In this example,

$$X^2 = \frac{(72 - 41)^2}{72 + 41} + \frac{(714 - 655)^2}{714 + 655} = 11.05$$

and

$$L^2 = 2\left[72\log\left(\frac{2 \times 72}{72 + 41}\right) + 714\log\left(\frac{2 \times 714}{714 + 655}\right)\right.$$
$$\left. + 655\log\left(\frac{2 \times 655}{714 + 655}\right) + 41\log\left(\frac{2 \times 41}{72 + 41}\right)\right]$$

$$= 11.16.$$

Since the probability is 0.00409 that a χ_2^2 random variable exceeds 11, these chi-square statistics provide considerable evidence that the symmetry hypothesis does not hold.

Residual analysis

The failure of the symmetry model reflects a tendency for lower rather than higher classifications to be used. The residuals

$$d_i = n_i - \hat{m}_i$$

satisfy the equations

$$d_1 = n_1 - \tfrac{1}{2}(n_1 + n_4) = \tfrac{1}{2}(n_1 - n_4),$$
$$d_2 = n_2 - \tfrac{1}{2}(n_2 + n_3) = \tfrac{1}{2}(n_2 - n_3),$$
$$d_3 = n_3 - \tfrac{1}{2}(n_2 + n_5) = -\tfrac{1}{2}(n_2 - n_3) = -d_2,$$
$$d_4 = n_4 - \tfrac{1}{2}(n_1 + n_4) = -\tfrac{1}{2}(n_1 - n_4) = -d_1,$$

As will be shown in Section 1.5, d_1 has variance

$$c_1 = \tfrac{1}{4}N[p_1 + p_4 - (p_1 - p_4)^2].$$

Under the symmetry model, $p_1 = p_4$ and the variance c_1 is

$$\tfrac{1}{2}Np_1 = \tfrac{1}{2}m_1.$$

The estimated variance of d_1 is then

$$\hat{c}_1 = \tfrac{1}{2}\hat{m}_1 = \tfrac{1}{4}(n_1 + n_4).$$

Thus the adjusted residual

$$r_1 = \tfrac{1}{2}(n_1 - n_4)/[\tfrac{1}{4}(n_1 + n_4)]^{1/2} = (n_1 - n_4)/(n_1 + n_4)^{1/2}.$$

Similarly,

$$r_2 = (n_2 - n_3)/(n_2 + n_3)^{1/2},$$
$$r_3 = -r_2,$$

and

$$r_4 = -r_1.$$

Note that

$$X^2 = r_1{}^2 + r_2{}^2.$$

The adjusted residuals have an obvious pattern:

$$r_1 = (72 - 41)/(72 + 41)^{1/2} = 2.92,$$
$$r_2 = (714 - 655)/(714 + 655)^{1/2} = 1.59,$$
$$r_3 = -r_2 = -1.59,$$
$$r_4 = -r_1 = -2.92.$$

One indication of this tendency toward lower rather than higher scores is the mean

$$\bar{X} = \frac{1}{N} \sum X_h = \frac{1}{N} \sum i n_i.$$

If the symmetry model holds, then $\bar{X}$ has mean 2.5 and variance

$$c = N^{-1}[(1 - 2.5)^2 p_1 + (2 - 2.5)^2 p_2 + (3 - 2.5)^2 p_3 + (4 - 2.5)^2 p_4]$$
$$= N^{-1}(4.5p_1 + 0.5p_2).$$

This variance may be estimated by substitution of $\hat{p}_1$ for p_1 and $\hat{p}_2$ for p_2. Thus the estimated variance is

$$\hat{c} = N^{-1}\left[\frac{4.5(\tfrac{1}{2})(n_1 + n_4)}{N} + \frac{0.5(\tfrac{1}{2})(n_2 + n_3)}{N}\right]$$
$$= N^{-2}[2.25(n_1 + n_4) + 0.25(n_2 + n_3)].$$

The standardized value

$$t = (\bar{X} - 2.5)/\hat{c}^{1/2}$$

has an approximate $N(0, 1)$ distribution under the symmetry model. Since

$$\bar{X} = \frac{1}{1482}(72 \times 1 + 714 \times 2 + 655 \times 3 + 41 \times 4)$$

$$= 2.44872$$

and

$$\hat{c}^{1/2} = \frac{1}{1482}[2.25(72 + 41) + 0.25(714 + 655)]^{1/2}$$

$$= 0.01648,$$

the standardized value $t = -3.11$. The probability that a standard normal deviate has absolute value greater than 3.11 is only 0.00019, so the standardized value t is not consistent with the symmetry model.

A Quadratic Log–Linear Model

Any log–linear model for Table 1.6 must account for the asymmetry of the data and for the preponderance of subjects in the middle two categories. One possibility is to assume that the log frequency $\log m_i$ is a quadratic function of $i - 2.5$, so that for some unknown α', β_1', and β_2',

$$\log m_i = \alpha' + \beta_1'(i - 2.5) + \beta_2'(i - 2.5)^2, \qquad 1 \leqslant i \leqslant 4.$$

To relate this model to the symmetry model, let $\alpha = \alpha' + 1.25\beta_2'$, $\beta_1 = \frac{1}{2}\beta_1'$, and $\beta_2 = \beta_2'$, so that

$$\log m_i = \alpha + \beta_1 x_{i1} + \beta_2 x_{i2},$$

where

$$x_{i1} = 2(i - 2.5)$$

and

$$x_{i2} = (i - 2.5)^2 - 1.25.$$

Readers familiar with orthogonal polynomials should note that x_{i1} and x_{i2} are linear and quadratic orthogonal polynomial scores. More extensive discussion of these scores is given in Volume 2 in Chapter 6. The quadratic model is a generalization of the symmetry model, for if $\beta_1 = 0$, then the symmetry model holds with $\beta = \beta_2$. The coefficient β_1 provides a measure of asymmetry as

$$\log(p_1/p_4) = \log(m_1/m_4) = \log m_1 - \log m_4 = -6\beta_1$$

and

$$\log/p_2/p_3) = \log m_2 - \log m_3 = -2\beta_1.$$

Under the symmetry model, $\log(p_1/p_4) = \log(p_2/p_3) = 0$. The term β_2 measures central concentration, for

$$\frac{1}{2}\log(p_1/p_2) + \frac{1}{2}\log(p_4/p_3) = \frac{1}{2}(\log m_1 - \log m_2 - \log m_3 + \log m_4)$$

$$= 2\beta_2$$

is a measure of the relative sizes of the extreme probabilities p_1 and p_4 and the corresponding central probabilities p_2 and p_3.

Maximum-Likelihood Equations

As shown in Section 1.5, the maximum-likelihood estimates $\hat{m}_i$ of m_i satisfy the equations

$$\sum_i \hat{m}_i = \hat{m}_1 + \hat{m}_2 + \hat{m}_3 + \hat{m}_4$$

$$= n_1 + n_2 + n_3 + n_4 = \sum_i n_i = N, \tag{1.3}$$

$$\sum_i x_{i1}\hat{m}_i = -3\hat{m}_1 - \hat{m}_2 + \hat{m}_3 + 3\hat{m}_4$$

$$= -3n_1 - n_2 + n_3 + 3n_4 = \sum_i x_{i1}n_i, \tag{1.4}$$

and

$$\sum_i x_{i2}\hat{m}_i = \hat{m}_1 - \hat{m}_2 - \hat{m}_3 + \hat{m}_4$$

$$= n_1 - n_2 - n_3 + n_4 = \sum_i x_{i2}n_i. \tag{1.5}$$

These equations imply that the maximum-likelihood estimate

$$\hat{\theta} = \frac{1}{N}\sum_i i\hat{m}_i = \sum_i i\hat{p}_i$$

of the expected value

$$\theta = \sum_i ip_i$$

of X_h is equal to the sample mean $\bar{X}$ and the maximum-likelihood estimate

$$\sigma^2 = \frac{1}{N}\sum_i (i - \hat{\theta})^2\hat{m}_i = \sum_i (i - \hat{\theta})^2\hat{p}_i$$

of the variance

$$\sigma^2 = \sum_i (i - \theta)^2 p_i$$

of X_h is equal to the sample variance

$$s^2 = \frac{1}{N}\sum_h (X_h - \bar{X})^2$$

of the X_h, where s^2 is not corrected for bias. Thus for the quadratic log–linear model, the fitted distribution of the X_h has the same mean and variance as the observed distribution. In contrast, the linear log–linear model of Section 1.1 only leads to the same mean values for the observed and fitted distributions of the X_h; the variances of the two distributions need not coincide.

For further proof and to introduce some notation required in the Newton–Raphson algorithm, let

$$T_{h1} = 2(X_h - 2.5)$$

and

$$T_{h2} = (X_h - 2.5)^2 - 1.25,$$

so that T_{hj}, $1 \leqslant j \leqslant 2$, has expected value

$$\theta_j = \sum_i x_{ij}p_i = N^{-1} \sum_i x_{ij}m_i.$$

For $1 \leqslant j \leqslant 2$, let

$$\bar{T}_j = N^{-1} \sum_h T_{hj} = N^{-1} \sum_i x_{ij}n_i$$

be the sample mean of the T_{hj}, $1 \leqslant h \leqslant N$, and let

$$\hat{\theta}_j = \sum_i x_{ij}\hat{p}_i = N^{-1} \sum_i x_{ij}\hat{m}_i$$

be the maximum-likelihood estimate of the expected value of the T_{hj}. Division of (1.3), (1.4), and (1.5) by N shows that

$$\sum_i \hat{p}_i = 1$$

and

$$\hat{\theta}_j = \bar{T}_j, \qquad 1 \leqslant j \leqslant 2.$$

This equation and the definitions of T_{h1} and T_{h2} imply that the first two observed and fitted moments of the $X_h - 2.5$ coincide; i.e.,

$$\sum_i (i - 2.5)\hat{p}_i = \tfrac{1}{2}\hat{\theta}_1 = \tfrac{1}{2}\,\bar{T}_1 = \frac{1}{N_h}(X_h - 2.5)$$

and

$$\sum_i (i - 2.5)^2\hat{p}_i = \hat{\theta}_2 + 1.25 = \bar{T}_2 + 1.25 = \frac{1}{N_h}\sum (X_h - 2.5)^2.$$

Equivalently, the observed and fitted mean values of the X_h coincide, as do the observed and fitted variances of the X_h.

The Newton–Raphson Algorithm

As in Section 1.1, the Newton–Raphson algorithm may be used to find maximum-likelihood estimates. Again the Newton–Raphson algorithm corresponds to a series of weighted regression analyses. The algorithm is a little more complicated here since there are now two coefficients β_1 and β_2 rather than the single coefficient β of Section 1.1.

As in Section 1.1, computations begin with selection of the initial approximation $m_{i0} = n_i + \frac{1}{2}$ for m_i. Given m_{i0}, the empirical logarithm $y_{i0} = \log m_{i0}$ is computed. Initial estimates β_{10} for $\hat{\beta}_1$ and β_{20} for $\hat{\beta}_2$ are computed as if a regression model

$$y_{i0} = \gamma + \beta_1 x_{i1} + \beta_2 x_{i2} + \varepsilon_i$$

were given in which the ε_i were independent random variables with mean values 0 and variances m_{i0}^{-1}. Thus β_{10} and β_{20} are found by solving the normal equations

$$S_{110}\beta_{10} + S_{120}\beta_{20} = w_1,$$
$$S_{210}\beta_{10} + S_{220}\beta_{20} = w_2.$$

Here S_{jk0}, $1 \leqslant j \leqslant 2$, $1 \leqslant k \leqslant 2$, is the weighted sum of cross-products

$$S_{jk0} = \sum_i (x_{ij} - \theta_{j0})(x_{ik} - \theta_{k0})m_{i0};$$

θ_{j0}, $1 \leqslant j \leqslant 2$, is the weighted average

$$\theta_{j0} = \frac{\sum_i x_{ij}m_{i0}}{\sum_i m_{i0}};$$

and w_{j0}, $1 \leqslant j \leqslant 2$, is the weighted sum of cross-products

$$w_{j0} = \sum_i (x_{ij} - \theta_{j0})y_{i0}m_{i0}.$$

Note that $S_{120} = S_{210}$.

In matrix terminology, one may let

$$S_0 = \begin{bmatrix} S_{110} & S_{120} \\ S_{210} & S_{220} \end{bmatrix},$$

$$\boldsymbol{\beta}_0 = \begin{bmatrix} \beta_{10} \\ \beta_{20} \end{bmatrix},$$

and

$$\mathbf{w}_0 = \begin{bmatrix} w_{10} \\ w_{20} \end{bmatrix}.$$

The normal equations reduce to the equation

$$S_0 \boldsymbol{\beta}_0 = \mathbf{w}_0$$

by the definition of matrix multiplication. If S_0^{-1} is the inverse of S_0, then

$$\boldsymbol{\beta}_0 = S_0^{-1} \mathbf{w}_0.$$

Computation of β_{10} and β_{20} is summarized in Table 1.7 in columns 1–15. Given β_{10} and β_{20}, an approximation α_0 for $\hat{\alpha}$ is obtained by solving the equation

$$\sum \exp(\alpha_0 + v_{i0}) = \exp(\alpha_0) \sum \exp v_{i0} = N,$$

where

$$v_{io} = \beta_{10} x_{i1} + \beta_{20} x_{i2}.$$

If

$$g_0 = \exp(\alpha_0) = N / \sum \exp v_{i0},$$

then

$$m_{il} = g_0 \exp v_{i0}$$

is a new approximation for $\hat{m}_i$. Computations are summarized in columns 16 through 20 of Table 1.7.

The approximations α_0, β_{10}, β_{20}, and m_{i1} appear sufficiently close to the respective estimates $\hat{\alpha}$, $\hat{\beta}_1$, $\hat{\beta}_2$, and $\hat{m}_i$ for all practical purposes. As a proof, consider the new approximations θ_{j1} to $\hat{\theta}_j = \bar{T}_j$, where

$$\theta_{j1} = \sum_i x_{ij} m_{i1} \Big/ \sum_i m_{i1}$$

$$= N^{-1} \sum_i x_{ij} m_{i1}.$$

As shown in columns 21 and 22 of Table 1.7,

$$\theta_{11} - \bar{T}_1 = -0.10212 - (-0.10256) = 0.00044$$

and

$$\theta_{21} - \bar{T}_2 = -0.84587 - (-0.84750) = 0.00163.$$

Normally, in large samples α_{10}, β_{10}, and β_{20} will be quite close to the respective maximum-likelihood estimates; nonetheless, an addition cycle of the algorithm will be used for illustrative purposes.

In this cycle, working logarithms

$$y_{i1} = \log m_{i1} + (n_i - m_{i1})/m_{i1}$$

Table 1.7

Calculation of Maximum-Likelihood Estimates for the Log–Linear Model
$\log m_i = \alpha + \beta_1 x_{i2} + \beta_2 x_{i2}$ for Table 1.6

Score i (1)	Number n_i responding (2)	x_{i1} (3)	x_{i2} (4)	$x_{i1}n_i$ (5)	$x_{i2}n_i$ (6)	m_{i0} (7)	y_{i0} (8)
1	72	-3	1	-216	72	72.5	4.2836
2	714	-1	-1	-714	-714	714.5	6.5716
3	655	1	-1	655	-655	655.5	6.4854
4	41	3	1	123	41	41.5	3.7257
Total	1482	0	0	-152	-1256	1484.0	

$\bar{T}_1 = -152/1482 = -0.10256$; $\bar{T}_2 = -1256/1482 = -0.84750$.

Score i	$x_{i1}m_{i0}$ (9)	$x_{i2}m_{i0}$ (10)	$(x_{i1} - \theta_{10})^2 m_{i0}$ (11)	$(x_{i1} - \theta_{10})(x_{i2} - \theta_{20})m_{i0}$ (12)	$(x_{i2} - \theta_{20})^2 m_{i0}$ (13)
1	-217.5	72.5	608.70	-387.87	247.16
2	-714.5	-714.5	575.63	98.531	16.866
3	655.5	-655.5	796.66	-111.02	15.473
4	124.5	41.5	399.44	237.72	141.48
Total	-152.0	-1256.0	2380.4	-162.65	420.97

$\theta_{10} = -152.0/1484.0 = -0.10243$; $\theta_{20} = -1256.0/1484.0 = -0.84636$.

Score i	$(x_{i1} - \theta_{10})y_{i0}m_{i0}$ (14)	$(x_{12} - \theta_{20})y_{i0}m_{i0}$ (15)	$\beta_{10}x_{i1}$ (16)	$\beta_{20}x_{i2}$ (17)
1	-899.87	573.41	0.19064	-1.2495
2	-4214.5	-721.40	0.063547	1.2495
3	4686.6	-653.15	-0.063547	1.2495
4	479.69	285.48	-0.19064	-1.2495
Total	51.956	-515.66	0.00000	0.0000

$2380.4\beta_{10} - 162.65\beta_{20} = 51.956$; $-162.65\beta_{20} + 420.97\beta_{20} = -515.66$

$$\beta_{10} = \frac{(51.956)(420.97) - (-162.65)(-515.66)}{(2380.4)(420.9) - (-162.65)^2} = -0.063547;$$

$$\beta_{20} = \frac{(2380.4)(-515.66) - (51.956)(-162.65)}{(2380.4)(420.9) - (-162.65)^2} = -1.2495.$$

Table 1.7 (continued)

Score i	v_{i0} (18)	$\exp v_{i0}$ (19)	m_{i1} (20)	$x_{i1}m_{i1}$ (21)	$x_{i2}m_{i2}$ (22)
1	-1.0588	0.34689	67.862	-203.58	67.862
2	1.3130	3.7174	727.30	-727.30	-727.30
3	1.1859	3.2737	640.49	640.49	-640.49
4	-1.4401	0.23690	46.349	139.05	46.349
Total	0.0000	7.5749	1482.0	-151.34	-1253.6

$\theta_{21} = -151.13/1482.0 = -0.10212; \quad \theta_{22} = -1253.6/1482.0 = -0.84587;$
$g_0 = 1482.0/7.5749 = 195.65; \quad a_{11} = -152 - (-151.34) = -0.65915$
$\alpha_0 = \log 195.65 = 5.2763; \quad a_{21} = -1256 - (-1253.6) = -2.4207$

Score i	$(x_{i1} - \theta_{11})^2 m_{i1}$ (23)	$(x_{i1} - \theta_{11})(x_{i2} - \theta_{21})m_{i2}$ (24)	$(x_{i2} - \theta_{21})^2 m_{i2}$ (25)
1	569.88	-363.00	231.22
2	586.34	100.65	17.278
3	777.99	-108.80	15.216
4	446.02	265.40	157.92
Total	2380.2	-105.75	421.63

$2380.2\delta_{11} - 105.75\delta_{21} = -0.65915; \quad -105.75\delta_{11} + 421.63\delta_{21} = -2.4207;$

$$\delta_{11} = \frac{(-0.65915)(421.63) - (-2.4207)(-105.75)}{(2380.2)(421.63) - (-105.75)^2} = -0.00053800;$$

$$\delta_{21} = \frac{(2380.2)(-2.4207) - (-105.75)(-0.65915)}{(2380.2)(421.63) - (-105.75)^2} = -0.0058762;$$

$\beta_{11} = -0.063547 - 0.00053800 = -0.064085; \quad \beta_{21} = -1.2495 - 0.0058762 = 1.2554.$

Score i	$\beta_{11}x_{i1}$ (26)	$\beta_{21}x_{i2}$ (27)	v_{i1} (28)	$\exp v_{i2}$ (29)	m_{i2} (30)
1	0.19225	-1.2554	-1.0631	0.34538	67.234
2	0.064085	1.2554	1.3194	3.7413	728.30
3	-0.064085	1.2554	1.1913	3.2913	640.69
4	-0.19225	-1.2554	-1.4476	0.23513	45.772
Total	0.00000	0.0000	0.0000	7.6131	1482.0

$g_1 = 1482/7.6131 = 194.66; \quad \alpha_1 = \log(194.66) = 5.2713.$

Score i	$x_{i1}m_{i2}$ (31)	$x_{i2}m_{i2}$ (32)	$(x_{i1} - \theta_{12})^2 m_{i2}$ (33)	$(x_{21} - \theta_{12})(x_{i2} - \theta_{22})m_{i2}$ (34)	$(x_{i2} - \theta_{22})^2 m_{i2}$ (35)
1	-201.70	67.234	564.44	-359.90	229.49
2	-728.30	-728.30	586.57	99.678	16.939
3	640.69	-640.69	778.85	-107.73	14.901
4	137.32	45.772	440.60	262.36	156.23
Total	-152.00	-1256.0	2370.5	-105.59	417.56

$\theta_{12} = -152.00/1482.0 = -0.10256; \quad \theta_{22} = -1256.0/1482.0 = -0.84750;$
$s(\hat{\beta}_1) = \{417.56/[2370.5)(417.56) - (-105.59)^2]\}^{1/2} = 0.020656;$
$s(\hat{\beta}_2) = \{2370.5/[2370.5)(417.56) - (-105.59)^2]\}^{1/2} = 0.049215.$

are defined. New approximations β_{11} and β_{21} for $\hat{\beta}_1$ and $\hat{\beta}_2$ are obtained by use of the same computations as would be used to estimate β_1 and β_2 in the regression problem

$$y_{i1} = \alpha + \beta_1 x_{i1} + \beta_2 x_{i2} + \varepsilon_i,$$

where the errors ε_i are independent random variables with respective mean values 0 and variances m_{i1}^{-1}.

Thus β_{11} and β_{21} may be computed by solving the equations

$$S_{111}\beta_{11} + S_{121}\beta_{21} = w_{11},$$
$$S_{211}\beta_{11} + S_{221}\beta_{21} = w_{21},$$

where

$$S_{jk1} = \sum_i (x_{ij} - \theta_{j1})(x_{ik} - \theta_{k1})m_{i1}, \qquad 1 \leqslant j \leqslant 2, \quad 1 \leqslant k \leqslant 2,$$

and

$$w_{j1} = \sum_i (x_{ij} - \theta_{j1})y_{i1}m_{i1}, \qquad 1 \leqslant j \leqslant 2.$$

If

$$S_1 = \begin{bmatrix} S_{111} & S_{121} \\ S_{211} & S_{221} \end{bmatrix},$$

$$\boldsymbol{\beta}_1 = \begin{bmatrix} \beta_{11} \\ \beta_{21} \end{bmatrix}$$

and

$$\mathbf{w}_1 = \begin{bmatrix} w_{11} \\ w_{21} \end{bmatrix},$$

then

$$S_1\boldsymbol{\beta}_1 = \mathbf{w}_1$$

and

$$\boldsymbol{\beta}_1 = S_1^{-1}\mathbf{w}_1.$$

Some algebraic manipulations may be used to show that

$$S_{111}(\beta_{11} - \beta_{10}) + S_{121}(\beta_{21} - \beta_{20}) = a_{11} = N(\bar{T}_1 - \theta_{11}) = \sum x_{i1}n_i - \sum x_{i1}m_{i1}$$

and

$$S_{211}(\beta_{11} - \beta_{10}) + S_{221}(\beta_{21} - \beta_{20}) = a_{21} = N(\bar{T}_2 - \theta_{21}) = \sum x_{i2}n_i - \sum x_{i2}m_{i2}.$$

If $\delta_{j1} = \beta_{j1} - \beta_{j0}$,

$$\boldsymbol{\delta}_1 = \begin{bmatrix} \delta_{11} \\ \delta_{21} \end{bmatrix},$$

and

$$\mathbf{a}_1 = \begin{bmatrix} a_{11} \\ a_{21} \end{bmatrix},$$

then in matrix notation,

$$\boldsymbol{\delta}_1 = S_1^{-1}\mathbf{a}_1$$

and

$$\boldsymbol{\beta}_1 = \boldsymbol{\beta}_0 + \boldsymbol{\delta}_1.$$

Computation of $\delta_{11}, \delta_{21}, \beta_{11}$, and β_{21} is summarized in columns 23 through 25 of Table 1.7.

Given β_{11} and β_{21}, new approximations α_1 for $\hat{\alpha}$ and m_{i2} for $\hat{m}_i$ may be derived. Let

$$v_{i1} = \beta_{11}x_{12} + \beta_{21}x_{i2}$$

and

$$g_1 = N \bigg/ \sum_i \exp v_{i1}.$$

Then

$$\alpha_1 = \log g_1$$

and

$$m_{i2} = g_1 \exp v_{i1}.$$

Computations are summarized in columns 26 through 30 of Table 1.7.

The corrections δ_{11} and δ_{21} are both quite small. The new approximations

$$\theta_{j2} = N^{-1} \sum_i x_{ij}m_{i2}$$

for $\hat{\theta}_j = \bar{T}_j$ are now accurate to five significant figures. All further computations in this section use the approximations α_1 for $\hat{\alpha}$, β_{11} for $\hat{\beta}_1$, β_{21} for $\hat{\beta}_2$, and m_{i2} for $\hat{m}_i$.

Chi-Square Tests

Given the $\hat{m}_i$, chi-square statistics are readily computed. The Pearson chi-square is given by

$$X^2 = (72 - 67.234)^2/67.234 + (714 - 728.30)^2/728.30$$
$$+ (655 - 640.69)^2/640.69 + (41 - 45.772)^2/45.772$$
$$= 1.44$$

and the likelihood-ratio chi-square satisfies the equation

$$L^2 = 2[72\log(72/67.234) + 714\log(714/728.30)$$
$$+ 655\log(655/640.69) + 41\log(41/45.772)]$$
$$= 1.45.$$

Since three parameters α, β_1, and β_2 have been estimated and since each X_h has four possible values, the associated degrees of freedom become $4 - 3 = 1$. The chi-square statistics provide little evidence that the new model does not hold.

Residual Analysis

Residual analysis is not very helpful in examination of the model. The redisuals

$$d_i = n_i - \hat{m}_i$$

are highly constrained by the maximum-likelihood equations. We have

$$d_2 = -3d_1,$$
$$d_3 = 3d_1,$$

and

$$d_4 = -d_1,$$

as can be confirmed by examination of columns 2 and 30 of Table 1.7. Thus

$$X^2 = \sum\left(\frac{d_i^2}{\hat{m}_i}\right) = \frac{d_1^2}{\hat{m}_1} + \frac{(-3d_1)^2}{\hat{m}_2} + \frac{(3d_1)^2}{\hat{m}_3} + \frac{(-d_1)^2}{\hat{m}_1}$$

$$= d_1^2\left(\frac{1}{\hat{m}_1} + \frac{9}{\hat{m}_2} + \frac{9}{\hat{m}_3} + \frac{1}{\hat{m}_4}\right)$$

$$= r_1^2,$$

where

$$r_1 = d_1\left(\frac{1}{\hat{m}_1} + \frac{9}{\hat{m}_2} + \frac{9}{\hat{m}_3} + \frac{1}{\hat{m}_4}\right)^{1/2}$$

has an approximate standard normal distribution. Except for its sign, the adjusted residual r_1 is determined by the Pearson chi-square statistic. The same conclusion applies to other adjusted residuals.

Large-Sample Properties of Estimates

To find the approximate distribution of $\hat{\beta}_1$ and $\hat{\beta}_2$, consider the regression model

$$Y_i = \alpha + \beta_1 x_{i1} + \beta_2 x_{i2} + \varepsilon_i,$$

where the ε_i are independent $N(0, m_i^{-1})$ variables and the Y_i are (hypothetical) observed responses. Let

$$S_{jk} = \sum_i (x_{ij} - \theta_j)(x_{ik} - \theta_k)m_i$$

and

$$w_j = \sum_i (x_{ij} - \theta_j)Y_i m_i.$$

Then the usual weighted regression estimates b_1 and b_2 are found by solving the equations

$$S_{11}b_1 + S_{12}b_2 = w_1,$$
$$S_{21}b_1 + S_{22}b_2 = w_2.$$

Let S be the matrix

$$\begin{bmatrix} S_{11} & S_{12} \\ S_{21} & S_{22} \end{bmatrix};$$

let

$$\mathbf{w} = \begin{bmatrix} w_1 \\ w_2 \end{bmatrix}$$

and

$$\mathbf{b} = \begin{bmatrix} b_1 \\ b_2 \end{bmatrix}.$$

Then one can write

$$S\mathbf{b} = \mathbf{w}.$$

If

$$S^{-1} = \begin{bmatrix} S^{11} & S^{12} \\ S^{21} & S^{22} \end{bmatrix} = \frac{1}{S_{11}S_{22} - S_{12}^2}\begin{bmatrix} S_{22} & -S_{21} \\ -S_{12} & S_{11} \end{bmatrix}$$

is the inverse of S, then

$$\mathbf{b} = S^{-1}\mathbf{w}.$$

The estimate $\mathbf{b}$ has a bivariate normal distribution $N(\boldsymbol{\beta}, S^{-1})$ with mean

$$\boldsymbol{\beta} = \begin{bmatrix} \beta_1 \\ \beta_2 \end{bmatrix}$$

and covariance matrix S^{-1}. Thus b_1 has an $N(\beta_1, S^{11})$ distribution and b_2 has an $N(\beta_2, S^{22})$ distribution. If c_1 and c_2 are constants, then

$$c_1 b_1 + c_2 b_2$$

has an $N(c_1\beta_1 + c_2\beta_2, c_1{}^2 S^{11} + 2c_1 c_2 S^{12} + c_2{}^2 S^{22})$ distribution.

A simple relationship exists between the distribution of the weighted least-squares estimate $\mathbf{b}$ and the maximum-likelihood estimate $\hat{\boldsymbol{\beta}}$. While $\mathbf{b}$ has an exact $N(\boldsymbol{\beta}, S^{-1})$ distribution, $\hat{\boldsymbol{\beta}}$ has an approximate $N(\boldsymbol{\beta}, S^{-1})$ distribution. The vector $\boldsymbol{\beta}$ is called the asymptotic mean of $\hat{\boldsymbol{\beta}}$, and the matrix S^{-1} is called the asymptotic covariance matrix of $\hat{\boldsymbol{\beta}}$.

To estimate S^{-1}, let $\hat{S}_{jk}$ be the maximum-likelihood estimate of S, let

$$\hat{S} = \begin{bmatrix} \hat{S}_{11} & \hat{S}_{12} \\ \hat{S}_{21} & \hat{S}_{22} \end{bmatrix}$$

and

$$\hat{S}^{-1} = \begin{bmatrix} \hat{S}^{11} & \hat{S}^{12} \\ \hat{S}^{21} & \hat{S}^{22} \end{bmatrix}$$

be the inverse of $\hat{S}$. Then $\hat{S}^{-1}$ is the estimated asymptotic covariance matrix of $\hat{\boldsymbol{\beta}}$.

Computation of $\hat{S}_{jk}$ is accomplished through the formula

$$\hat{S}_{jk} = \sum_i (x_{ij} - \hat{\theta}_j)(x_{ik} - \hat{\theta}_k)\hat{m}_i$$

$$= \sum_i (x_{ij} - \bar{T}_j)(x_{ik} - \bar{T}_k)\hat{m}_i.$$

Given the $\hat{S}_{jk}$, $\hat{S}^{-1}$ is found by the formulas

$$\hat{S}^{11} = \hat{S}_{22}/|\hat{S}|,$$
$$\hat{S}^{12} = \hat{S}^{21} = -\hat{S}_{12}/|\hat{S}|,$$
$$\hat{S}^{22} = \hat{S}_{11}/|\hat{S}|,$$

where $|\hat{S}|$, the determinant of $\hat{S}$, is

$$|\hat{S}| = \hat{S}_{11}\hat{S}_{22} - \hat{S}_{12}^2.$$

The estimated asymptotic standard deviation $s(\hat{b}_1)$ is $(\hat{S}^{11})^{1/2}$, and $s(\hat{b}_2)$ is $(\hat{S}^{22})^{1/2}$. Calculations are summarized in the last three columns of Table 1.7.

To interpret the estimates $\hat{\beta}_1$ and $s(\hat{\beta}_1)$, recall that the model under study implies that

$$\log(p_1/p_4) = \log p_1 - \log p_4 = -6\beta_1$$

and

$$\log(p_2/p_3) = \log p_2 - \log p_3 = -2\beta_1,$$

so that β_1 provides an asymmetry measure. The estimate $\hat{\beta}_1 = -0.0641$ corresponds to estimated odds ratios

$$\frac{\hat{p}_1}{\hat{p}_4} = \exp(-6\hat{\beta}_1) = \exp[-6(-0.064085)] = 1.47$$

and

$$\frac{\hat{p}_2}{\hat{p}_3} = \exp(-2\hat{\beta}_1) = 1.14.$$

The first ratio corresponds to a rate of classification into the lower class 1.47 times as great as the rate of classfication in the upper class. The second ratio corresponds to a rate of classification into the working class 1.14 times as great as the rate of classification into the middle class.

An approximate 95 percent confidence interval for β_1 has lower bound

$$\hat{\beta}_1 - 1.96s(\hat{\beta}_1) = -0.064085 - 1.96(0.020656) = -0.105$$

and upper bound

$$\hat{\beta}_1 + 1.96s(\hat{\beta}_1) = -0.064085 + 1.96(0.020656) = -0.0236.$$

These bounds correspond to those of

$$\exp[-6(-0.0236)] = 1.15$$

and

$$\exp[-6(-0.105)] = 1.87$$

for p_1/p_4, and to

$$\exp[-2(-0.0236)] = 1.05$$

and

$$\exp[-2(-0.105)] = 1.23$$

for p_2/p_3. The asymmetry thus appears rather modest in size.

As previously noted, the coefficient β_2 provides a measure of the tendency for respondents to use the intermediate classifications working class and middle class rather than the extreme classifications lower class and upper class as $2\beta_2$ is the average of the contrasts $\log(p_2/p_1)$ and $\log(p_3/p_4)$. The geometric mean

$$\left(\frac{p_2}{p_1}\frac{p_3}{p_4}\right)^{1/2}$$

of p_2/p_1 and p_3/p_4 is then $\exp(-2\beta_2)$, which has a maximum-likelihood estimate

$$\exp(-2\hat{\beta}_2) = \exp[-2(-1.2554)] = 12.3.$$

This geometric mean is quite large and is fairly reliably determined. An approximate 95 percent confidence interval for β_2 has lower bound

$$\hat{\beta}_2 - 1.96s(\hat{\beta}_2) = -1.2554 - 1.96(0.049215) = -1.35$$

and upper bound

$$\hat{\beta}_2 + 1.96s(\hat{\beta}_2) = -1.2554 + 1.96(0.049215) = -1.16.$$

The corresponding bounds for $\exp(-2\beta_2)$ are then

$$\exp[-2(-1.16)] = 10.2$$

and

$$\exp[-2(-1.35)] = 14.9.$$

Thus the most striking feature of Table 1.6 is the concentration of responses in the central categories lower class (2) and middle class (3) rather than in the extreme categories working class (1) and upper class (4). This phenomenon is suggested by the large geometric mean of $\hat{p}_2/\hat{p}_1$ and $\hat{p}_3/\hat{p}_4$. In contrast, asymmetry, which is reflected by the estimates $\hat{p}_1/\hat{p}_4$ and $\hat{p}_2/\hat{p}_3$, appears to be modest in magnitude.

In this section, maximum-likelihood estimation, chi-square tests, residual analysis, and approximate confidence intervals have been considered for a log–linear model for departures from symmetry. This model is slightly more complex than the log–linear time-trend model of Section 1.1 since three parameters—α, β_1, and β_2—must be estimated rather than the two parameters α and β of Section 1.1. The change in complexity of analysis corresponds to the change in regression analysis from study of one dependent and one independent variable to study of one dependent and two independent variables. This comparison is especially appropriate in comparing the Newton–Raphson algorithms of Sections 1.1 and 1.2 and in comparing estimation of asymptotic variances in the two sections. Matrix inversion and solution of simultaneous linear equations arise in this section, as they do in ordinary regression analyses with two independent variables. As in simple linear regression, no matrix inversions or simultaneous linear equations have had to be considered in Section 1.1.

Both Sections 1.1 and 1.2 involve multinomial data and ordered categories. Data from one-way tables need not be multinomial and categories need not be ordered. In Sections 1.3 and 1.4, Poisson data are considered in which ordering is much more difficult to exploit.

1.3 Suicide and the Poisson Distribution

The Poisson distribution is a classical probability distribution for frequency counts. Feller (1968, pages 153–164) provides examples of its use, historical background, and references. Log–linear models are also used with Poisson data. In this section, some simple log–linear models are used to examine variations in the means of independent Poisson observations. These models resemble the equiprobability model of Section 1.1 and the symmetry model of Section 1.2. As in the models of earlier sections, explicit expressions for maximum-likelihood equations are available, and chi-square statistics and adjusted residuals are readily computed.

Illustrations of analysis of Poisson data in this section and in the next section use data on reported suicides in the United States. The analysis in this section is based on Durkheim's (1951[1897], pages 106–122) observation of large seasonal variations in suicide rates. To examine whether this observation, which is based on nineteenth-century data, applies to modern American society, consider the data in Table 1.8. The monthly variations observed in this table in suicides per day are modest in size; however, the largest rate is 16 percent greater than the smallest rate, so that the variations obtained are not negligible. However, it is necessary to consider the extent to which the observed fluctuations in daily rates can be ascribed to chance.

To examine this problem, it is helpful to begin by numbering the months from 1 to 12 and letting n_i be the number of suicides in month i, so that $n_1 = 1867$ is the number of suicides in January. The simplest probability model which can be constructed for the data assumes that the n_i are independent Poisson observations with respective means m_i. In this model, for each month i, the probability is

$$(m_i)^k \exp(-m_i)/k!$$

that the number of suicides n_i in the month is equal to $k \geqslant 0$. The symbol $k!$ is the product of all integers less than or equal to k. Thus $0! = 1, 1! = 1, 2! = 2, 3! = 6$, etc.

Table 1.8

Distribution of Suicides in the United States in 1970
by Months[a]

Month	Number	Rate per day
January	1867	60.2
February	1789	63.9
March	1944	62.7
April	2094	69.8
May	2097	67.6
June	1981	66.0
July	1887	60.9
August	2024	65.3
September	1928	64.3
October	2032	65.6
November	1978	65.9
December	1859	60.0
Total	23,480	64.3

[a] From National Center for Health Statistics (1970,
1–174 and 1–175).

To consider the plausibility of this model, let the total United States
population at the start of 1970 be numbered from 1 to h. Let X_{gi} be 1 if the
individual g, $1 \leqslant g \leqslant h$, is reported to have committed suicide in month i,
and let X_{gi} be 0 otherwise. Let p_{gi} be the probability that X_{gi} is 1, and let
X_{fi} and X_{gj} be statistically independent whenever f and g differ. (X_{gi} and
X_{gj} are *not* independent if p_{gi} and p_{gj} are positive, for individual g can presumably commit at most one successful suicide.) Note that

$$n_i = \sum_g X_{gi},$$

assuming that suicides are impossible among infants less than a year old.
The n_i are approximately distributed as independent Poisson observations
with means

$$m_i = \sum_g p_{gi}.$$

The approximation is increasingly accurate as the size of the p_{gi} decreases.
For the purposes of analyses in this section, the requirement is that for each i,

$$\sum_g p_{gi}^2 \Big/ \sum_q p_{gi}$$

be close to 0. Given the fact that the annual suicide rate is on the order of 1 in 10,000 this assumption is probably reasonable. As will become evident in the analyses, the independence assumption may require more critical examination.

The Model of Constant Rates

The model for no seasonal effects on the suicide rate assumes that the expected rate $f_i = m_i/z_i$ of suicides per day in month i is an unknown constant k for all months. Here z_i is the number of days in month i. In the logarithmic scale,

$$\log f_i = \alpha,$$

where α is the unknown logarithm of the expected number k of suicides per day. Thus the model for no seasonal effects is a very simple log–linear model. The form is similar to that of the equiprobability model of Section 1.1, where

$$\log m_i = \alpha$$

for each i.

Maximum-Likelihood Equations

Given the assumption that the n_i are independent Poisson observations, it follows that from general results of Section 1.5 that the maximum-likelihood estimates $\hat{m}_i$ of m_i, $\hat{k}$ of k, and $\hat{\alpha}$ of α satisfy the equations

$$\sum \hat{m}_i = N = \sum n_i$$

and

$$\hat{m}_i = \hat{k} z_i = z_i \exp \hat{\alpha}.$$

Thus

$$\hat{k} \sum z_i = N.$$

Since there are 365 days in 1970, the maximum-likelihood estimate $\hat{f}_i$ of the suicide rate per day f_i of month i is

$$\hat{f}_i = \hat{k} = 23{,}480/365 = 64.3.$$

One finds that

$$\hat{\alpha} = \log(\hat{k}) = \log(64.3) = 4.16.$$

The estimated expected values

$$\hat{m}_i = N z_i / 365$$

are given in Table 1.9.

Chi-Square Tests

To test the adequacy of this model, note that

$$X^2 = \sum (n_i - \hat{m}_i)^2 / \hat{m}_i = 47.4$$

and

$$L^2 = 2 \sum n_i \log(n_i / \hat{m}_i) = 47.4.$$

Since there are 12 months and only one parameter α need be estimated, the degrees of freedom number 11. An inspection of a table of the chi-square distribution shows that X^2 and L^2 are much too large to be consistent with the model of no seasonal variation.

Adjusted Residuals

Adjusted residuals can be examined to determine whether these large chi-square statistics reflect a failure in the hypothesis that the n_i are Poisson or a failure in the hypothesis that the suicide rate is constant over all months

Table 1.9

Observed Monthly Suicides, Estimated Expected Monthly Suicides, and Adjusted Residuals for the Model of Constant Suicide Rates as Applied to Table 1.8

Month	Observed Number n_i	Days z_i in month	Estimated expected number $\hat{m}_i$	Residual d_i	Estimated standard deviation $\hat{c}_i^{1/2}$	Adjusted residual r_i
January	1867	31	1994.2	−127.19	42.718	−2.98
February	1789	28	1801.2	−12.205	40.780	−0.30
March	1944	31	1994.2	−50.192	42.718	−1.17
April	2094	30	1929.9	164.14	42.086	3.90
May	2097	31	1994.2	102.81	42.178	2.41
June	1981	30	1929.9	51.137	42.086	1.22
July	1887	31	1994.2	−107.19	42.718	−2.51
August	2024	31	1994.2	29.808	42.718	0.70
September	1928	30	1929.9	−1.8630	42.086	−0.04
October	2032	31	1994.2	37.808	42.718	0.89
November	1978	30	1929.9	48.137	42.086	1.14
December	1859	31	1994.2	−135.19	42.718	−3.16
Total	23,480	365	23,480	0.0000		

of the year. The adjusted residuals r_i are computed from the formula

$$r_i = (n_i - \hat{m}_i)/\hat{c}_i^{1/2},$$

where

$$\hat{c}_i = \hat{m}_i\left(1 - \frac{\hat{m}_i}{N}\right) = \hat{m}_i\left(1 - \frac{z_i}{365}\right).$$

To obtain this formula, note that the residual $d_i = n_i - \hat{m}_i$ can be written as

$$d_i = n_i - \hat{m}_i = n_i - \frac{z_i}{365}\sum_j n_j = \sum_j \left(\delta_{ij} - \frac{z_i}{365}\right)n_j,$$

where j has ranges over the integers 1 to 12 and

$$\delta_{ij} = 1, \quad i = j,$$
$$= 0, \quad i \neq j.$$

For example,

$$d_1 = \sum_j\left(\delta_{ij} - \frac{31}{365}\right)n_j = \left(1 - \frac{31}{365}\right)n_1 - \frac{31}{365}n_2 - \frac{31}{365}n_5 - \cdots - \frac{31}{365}n_{12}.$$

The variance of a Poisson observation n_i with mean m_i is m_i. Thus d_i has variance

$$\sum_j\left(\delta_{ij} - \frac{z_i}{365}\right)^2 m_j = \sum_j \delta_{ij}^2 m_j - 2\left(\frac{z_i}{365}\right)\sum \delta_{ij}m_j + \left(\frac{z_i}{365}\right)^2 \sum m_j$$

$$= m_i - 2\left(\frac{z_i}{365}\right)m_i + \left(\frac{z_i}{365}\right)^2 \sum m_j.$$

Under the model of constant suicide rates,

$$m_i = z_i k$$

and

$$\sum m_j = \left(\sum z_j\right)k = 365k.$$

Thus

$$\frac{z_i}{365}\sum m_j = \frac{z_i}{365}(365k) = m_i$$

and c_i, the variance of d_i, is

$$m_i - 2\left(\frac{z_i}{365}\right)m_i + \frac{z_i}{365}m_i = m_i\left(1 - \frac{z_i}{365}\right).$$

Substitution of $\hat{m}_i$ for m_i yields the estimated variance $\hat{c}_i$. If the model holds, then each adjusted residual r_i has an approximate $N(0, 1)$ distribution. Since the observed n_i are very large, the approximation is quite accurate in this example.

The adjusted residuals are displayed in Table 1.9. Three adjusted residuals are very large, about 3 or more in magnitude. They correspond to January, April, and December. Residuals for May and July are also large. The difficulty at this point is to determine the extent to which systematic seasonal effects cause model failure and the extent to which other factors may be involved.

The Model of Constant Rates within Seasons

A simple check on seasonal variation can be achieved by dividing the months into four equal portions according to the four seasons. Some arbitrariness in division is necessarily involved; however, it is reasonable to let winter include December, January, and February. Then spring includes March, April, and May, summer June, July, and August, and autumn September, October, and November.

To consider the extent to which the large chi-square statistics reflect seasonal effects, the model

$$
\begin{aligned}
f_i &= k_1, & i &= 1, \quad 2, \quad \text{or} \quad 12, \\
&= k_2, & 3 &\leqslant i \leqslant 5, \\
&= k_3, & 6 &\leqslant i \leqslant 8, \\
&= k_4, & 9 &\leqslant i \leqslant 11,
\end{aligned}
$$

may be tried. Thus the suicide rate is assumed constant during each season. One may write the model in log–linear form as

$$
\begin{aligned}
\log f_i &= \alpha + \beta_1, & i &= 1, \quad 2, \quad \text{or} \quad 12, \\
&= \alpha + \beta_2, & 3 &\leqslant i \leqslant 5, \\
&= \alpha + \beta_3, & 6 &\leqslant i \leqslant 8, \\
&= \alpha, & 9 &\leqslant i \leqslant 11.
\end{aligned}
$$

Thus $\alpha = \log k_4$, $\beta_1 = \log k_1 - \log k_4$, $\beta_2 = \log k_2 - \log k_4$, and $\beta_3 = \log k_3 - \log k_4$.

Maximum-Likelihood Equations

Under this model, the maximum-likelihood equations are

$$
\sum \hat{m}_i = N,
$$
$$
\hat{m}_3 + \hat{m}_4 + \hat{m}_5 = n_3 + n_4 + n_5,
$$
$$
\hat{m}_6 + \hat{m}_7 + \hat{m}_8 = n_6 + n_7 + n_8,
$$
$$
\hat{m}_9 + \hat{m}_{10} + \hat{m}_{11} = n_9 + n_{10} + n_{11}.
$$

Subtraction of the last three equations from the first equation gives us

$$\hat{m}_{12} + \hat{m}_1 + \hat{m}_2 = n_{12} + n_1 + n_2.$$

Since

$$\hat{m}_i = z_i \hat{k}_1, i = 1, \quad 2, \quad \text{or} \quad 12,$$
$$\hat{m}_{12} + \hat{m}_1 + \hat{m}_2 = (z_{12} + z_1 + z_2)\hat{k}_1 = n_{12} + n_1 + n_2.$$

Thus

$$\hat{k}_1 = \frac{n_{12} + n_1 + n_2}{z_{12} + z_1 + z_2} = \frac{1859 + 1867 + 1789}{31 + 31 + 28} = 61.278.$$

Similarly,

$$\hat{k}_2 = \frac{n_3 + n_4 + n_5}{z_3 + z_4 + z_5} = \frac{1944 + 2094 + 2097}{31 + 30 + 31} = 66.685,$$

$$\hat{k}_3 = \frac{n_6 + n_7 + n_8}{z_6 + z_7 + z_8} = \frac{1981 + 1887 + 2024}{30 + 31 + 31} = 64.043,$$

$$\hat{k}_4 = \frac{n_9 + n_{10} + n_{11}}{z_9 + z_{10} + z_{11}} = \frac{1928 + 2032 + 1978}{30 + 31 + 30} = 65.253.$$

Note that $\hat{k}_1$ is the total reported number of suicides in the winter months January, February, and December divided by the total number of days in these months, so that $\hat{k}_1$ is the average rate of suicides per day in winter. Similar observations apply to $\hat{k}_2$, $\hat{k}_3$, and $\hat{k}_4$.

Given $\hat{k}_1$, $\hat{k}_2$, $\hat{k}_3$, and $\hat{k}_4$, the maximum-likelihood estimates $\hat{m}_i$ are found from the formula

$$\begin{aligned}
\hat{m}_i &= z_i \hat{k}_1, & i = 12, \quad 1, \quad \text{or} \quad 2, \\
&= z_i \hat{k}_2, & 3 \leqslant i \leqslant 5, \\
&= z_i \hat{k}_3, & 6 \leqslant i \leqslant 8, \\
&= z_i \hat{k}_4, & 9 \leqslant i \leqslant 11.
\end{aligned}$$

Thus

$$\begin{aligned}
\hat{m}_1 &= \hat{m}_{12} = 31\hat{k}_1 = 1899.6 \\
\hat{m}_2 &= \phantom{\hat{m}_{12} =} 28\hat{k}_1 = 1715.8 \\
\hat{m}_3 &= \hat{m}_5 = 31\hat{k}_2 = 2067.2 \\
\hat{m}_4 &= \phantom{\hat{m}_5 =} 30\hat{k}_2 = 2000.5 \\
\hat{m}_6 &= \phantom{\hat{m}_5 =} 30\hat{k}_3 = 1921.3 \\
\hat{m}_7 &= \hat{m}_8 = 31\hat{k}_3 = 1985.3 \\
\hat{m}_9 &= \hat{m}_{11} = 31\hat{k}_4 = 2022.8
\end{aligned}$$

The maximum-likelihood estimates $\hat{\alpha}$, $\hat{\beta}_1$, $\hat{\beta}_2$, and $\hat{\beta}_3$ satisfy

$$\hat{\alpha} = \log \hat{k}_4 = 4.1783,$$
$$\hat{\beta}_1 = \log \hat{k}_1 - \log \hat{k}_4 = -0.06285,$$
$$\hat{\beta}_2 = \log \hat{k}_2 - \log \hat{k}_4 = 0.02171,$$
$$\hat{\beta}_3 = \log \hat{k}_3 - \log \hat{k}_4 = 0.01871.$$

Chi-Square Statistics

Since four parameters α, β_1, β_2, and β_4 are estimated, and there are 12 months, the chi-square statistics X^2 and L^2 have $12 - 4 = 8$ degrees of freedom. Since the observed values of X^2 and L^2 are 24.9 and 25.0, respectively, the model of constant suicide rates is untenable. The significance level is about 0.002. Thus suicide rates vary within seasons or the Poisson distribution does not describe the monthly numbers of suicides.

Although a model which fits the data has not been achieved, progress has been made in reducing the value of the chi-square statistics. These statistics are now about 25, whereas they were about 47 in the model of constant suicide rates. The R^2 statistic comparing the new model to the model of constant suicide rates is

$$R^2 = (47.4 - 25.0)/47.4 = 0.48.$$

The F statistic for the model of constant suicide rates is

$$\tfrac{1}{11} 47.4 = 4.3,$$

while the F statistic for the model of seasonal variation is

$$\tfrac{1}{8} 25.0 = 3.1.$$

Thus some progress has been made, although the F statistic remains far from 1.

Further progress in reduction of chi-square statistics is difficult to achieve without some appearance of arbitrariness. The failure of the seasonal model can be attributed to the relatively low suicide rate in March and to the relatively low rate in July. If the months are divided into unequal portions, so that December, January, February, and March are one portion, April, May and June are a second portion, July by itself is a third portion, and August, September, October, and November are the fourth portion, and if suicide rates are assumed constant for all months in each of the four portions, then X^2 is 9.15 and L^2 is 9.14. Since the degrees of freedom number 8, the fits are very satisfactory.

This division of the year into four unequal parts is so arbitrary that little confidence can be placed on it. To see the problem more clearly, consider the suicide data of 1968 and 1969 (Table 1.10). Using the same model success-

Table 1.10

Distribution of Suicides in the United States in 1968 and 1969 by Months[a]

Month	Number of suicides		Rate per day	
	1968	1969	1968	1969
January	1720	1831	55.5	59.1
February	1712	1609	59.0	57.5
March	1924	1973	62.1	63.6
April	1882	1944	62.7	64.8
May	1870	2003	60.3	64.6
June	1680	1774	56.0	59.1
July	1868	1811	60.3	58.4
August	1801	1873	58.1	60.4
September	1756	1862	58.5	62.1
October	1760	1897	56.8	61.2
November	1666	1866	55.5	62.2
December	1733	1921	55.9	62.0
Total	21,372	22,364	58.4	62.0

[a] From National Center for Health Statistics (1968, pages 1–131; 1969, pages 1–132).

fully applied to the 1970 data, one now obtains Pearson chi-squares of 29.7 for 1968 data and 22.4 for 1969 data. The corresponding likelihood ratio chi-squares are 29.7 and 22.5. Thus the apparently successful fit has vanished when applied to different data. Nonetheless, one area of seasonal stability can be observed in each year. Within a given year, the suicide rates in August, September, October, and November do not show much variation. The stability can be checked by means of the chi-square statistics

$$X^2 = (n_8 - 31N_4/122)^2/(31N_4/122)$$
$$+ (n_9 - 30N_4/122)^2/(30N_4/122)$$
$$+ (n_{10} - 31N_4/122)^2/(31N_4/122)$$
$$+ (n_{11} - 30N_4/122)^2/(30N_4/122)$$

and

$$L^2 = 2\{n_8 \log[n_8/(31N_4/122)]$$
$$+ n_9 \log[n_9/(30N_4/122)]$$
$$+ n_{10} \log[n_{10}/(31N_4/122)]$$
$$+ n_{11} \log[n_{11}/(30N_4/122)]\},$$

where

$$N_4 = n_8 + n_9 + n_{10} + n_{11}.$$

Each chi-square statistic has 3 degrees of freedom. The values of X^2 for 1968, 1969, and 1970 are, respectively, 2.9, 1.0, and 0.7. The values of L^2 and X^2 in no case differ by as much as 0.01.

At this point, other stabilities from year to year can be noted. For example, in each year April is the month with the highest suicide rate. The results for April and the results from the months August through November suggest several conclusions. The chi-square statistics for homogeneity of rates in the August–November interval suggest that the distribution of the number of suicides in a month is adequately represented by the Poisson distribution. The expected rate of suicides per day is not constant within a given year. Variations in this rate are partially accounted for by seasonal effects that apply in each year and partially by local disturbances which are not identified by the data used in this analysis. An analysis of the relative contributions of these effects is deferred until Chapter 2.

Despite remaining ambiguities concerning the nature of monthly variations in suicide rates, it is very clear that current American variations are far less than those observed by Durkheim. The largest ratio in monthly suicide rates which is observed in a single year is between April and December of 1970. The ratio is 1.16, and its natural logarithm is 0.15. The estimated standard error of this natural logarithm is

$$[1^2/2094 + (-1)^2/1859]^{1/2} = 0.03.$$

Durkheim (1951 [1897], page 112) examines the monthly distribution of suicides in France (1866–1870), Italy (1883–1888), and Prussia (1876–1878, 1880–1882, 1885–1889). The ratios between rates in June and rates in December are 1.75 in France, 1.72 in Italy, and 1.72 in Prussia. The corresponding natural logarithms are 0.56, 0.54, and 0.54. Thus observed variations in monthly suicide rates are quite small relative to those Durkheim observes.

This section illustrates use of log–linear models of constant rates to assess variability of means of Poisson counts and to assess sources of variability. As in Sections 1.1 and 1.2, maximum-likelihood equations set linear combinations of observed frequencies equal to the corresponding linear combinations of estimated expected frequencies. In contrast to Sections 1.1 and 1.2, analysis in this section has relied on maximum-likelihood estimates with explicit expressions, so that iterative computations have been unnecessary. Nonetheless, chi-square statistics, adjusted residuals, and approximate confidence intervals have remained valuable tools. Thus the statistical tools used in this section with Poisson data have been similar to those used in the preceding two sections with multinomial data.

1.4 Distribution of Reported Suicides by Region

In Section 1.3, variations in rates associated with Poisson observations were largely assessed through examination of chi-square statistics for various simple log–linear models. Similar methods are also found in this section; a new approach, simultaneous confidence intervals, is also introduced, which allows a specific identification of those variations in rates that cannot be explained by chance fluctuations. The data used in this section are described in Table 1.11. This table provides a classification of reported suicides in the United States in 1970 by division of occurrence. The suicide rates appear to vary somewhat from region to region, with a low of 8.5 per 100,000 in the Middle Atlantic States and a high of 17.5 per 100,000 in the Pacific states.

Table 1.11

Distribution of Suicides in the United States in 1970 by Division of Occurrence[a]

Division[b]	Division number	Number of suicides reported	Enumerated population in 1970 census	Rate per 100,000
New England[c]	1	1119	11,841,663	9.4
Middle Atlantic[d]	2	3165	37,199,040	8.5
E. N. Central[e]	3	4308	40,252,476	10.7
W. N. Central[f]	4	1790	16,319,187	11.0
S. Atlantic[g]	5	3729	30,671,337	12.2
E. S. Central[h]	6	1335	12,803,470	10.4
W. S. Central[i]	7	2075	19,320,560	10.7
Mountain[j]	8	1324	8,281,562	16.0
Pacific[k]	9	4635	26,522,631	17.5
Total		23,480	203,211,926	11.6

[a] From National Center for Health Statistics (1970, pages 1–49, 1–65, and 6–18).

[b] Geographical divisions in this table and in other tables in this book consist of the following states:

[c] New England: Connecticut, Maine, Massachusetts, New Hampshire, Rhode Island, and Vermont.

[d] Middle Atlantic: New Jersey, New York, and Pennsylvania.

[e] East North Central: Illinois, Indiana, Michigan, Ohio, and Wisconsin.

[f] West North Central: Iowa, Kansas, Minnesota, Missouri, Nebraska, North Dakota, and South Dakota.

[g] South Atlantic: Delaware, District of Columbia, Florida, Georgia, Maryland, North Carolina, South Carolina, Virginia, and West Virginia.

[h] East South Central: Alabama, Kentucky, Mississippi, and Tennessee.

[i] West South Central: Arkansas, Louisiana, Oklahoma, and Texas.

[j] Mountain: Arizona, Colorado, Idaho, Montana, Nevada, New Mexico, Utah, and Wyoming.

[k] Pacific: Alaska, California, Hawaii, Oregon, and Washington.

It is desirable to determine which differences between regions in suicide rates are not due to chance fluctuations.

To examine these suicides rates, let n_i be the number of reported suicides in division i. Assume that the n_i are independent Poisson random variables with respective means m_i. Let z_i be the enumerated population of region i, divided by 100,000, and let

$$f_i = m_i/z_i$$

be the expected suicide rate per 100,000 persons in region i.

The Model of Constant Rates

If the expected suicide rates f_i are all equal to some constant k, then one obtains the log–linear model

$$\log f_i = \alpha,$$

where $\alpha = \log k$. This model may be tested in the same manner as the model of constant monthly suicide rates developed in Section 1.3. The maximum-likelihood estimate $\hat{k}$ of k is

$$\sum n_i / \sum z_i = 23{,}480/2032.1 = 11.554$$

and

$$\hat{\alpha} = \log \hat{k} = 2.4471.$$

The maximum-likelihood estimate of

$$m_i = z_i f_i$$

is

$$\hat{m}_i = z_i \hat{k}.$$

These estimates are listed in Table 1.12.

The Pearson chi-square statistic X^2 is 1354.8 and the likelihood ratio chi-square statistic L^2 is 1263.9. Since there are nine divisions and one parameter to be estimated, the degrees of freedom are only $9 - 1 = 8$. Thus the model of equal suicide rates is quite untenable.

The West-versus-East Model

Much of the observed variation in suicide rates involves a contrast between the Mountain and Pacific states, which are the most westerly states, and the

Table 1.12

Maximum-Likelihood Estimates of Expected Numbers of Suicides in the United States
in 1920 by Division of Occurrence

Division	Observed number	Estimated expected number: constant rate model	Estimated expected number: two-rate model
New England	1119	1368.2	1232.0
Middle Atlantic	3165	4298.1	3870.2
E. N. Central	4308	4650.9	4187.8
W. N. Central	1790	1885.6	1697.8
S. Atlantic	3729	3543.9	3191.0
E. S. Central	1335	1479.4	1332.1
W. S. Central	2075	2232.4	2010.1
Mountain	1324	956.89	1417.9
Pacific	4635	13064.5	4541.1
Total	23,480	23,480.0	23,480.0

remaining states. Consider the log–linear model

$$\log f_i = \alpha + \beta, \quad 1 \leqslant i \leqslant 7$$
$$= \alpha, \quad 8 \leqslant i \leqslant 9,$$

in which f_i is a constant k_1 for $1 \leqslant i \leqslant 7$ and f_i is a constant k_2 for $8 \leqslant i \leqslant 9$. Note that

$$\alpha = \log k_2$$

and

$$\beta = \log k_1 - \log k_2.$$

The maximum-likelihood estimate of k_1 is

$$\hat{k}_1 = \frac{n_1 + n_2 + n_3 + n_4 + n_5 + n_6 + n_7}{z_1 + z_2 + z_3 + z_4 + z_5 + z_6 + z_7}$$

$$= \frac{1119 + 3165 + 4308 + 1790 + 3729 + 1335 + 2075}{118.42 + 371.99 + 402.52 + 163.19 + 306.71 + 128.03 + 193.21}$$

$$= 10.404$$

and that of k_2 is

$$\hat{k}_2 = \frac{n_8 + n_9}{z_8 + z_9} = \frac{1324 + 4635}{82.816 + 265.226} = 17.122.$$

The remaining maximum-likelihood estimates are

$$\hat{\alpha} = \log \hat{k}_2 = \log 17.122 = 2.8403,$$
$$\hat{\beta} = \log \hat{k}_1 - \log \hat{k}_2 = \log 10.343 - \log 17.122 = -0.49815$$

and

$$\hat{m}_i = z_i \hat{k}_1, \quad 1 \leqslant i \leqslant 7,$$
$$= z_i \hat{k}_2, \quad 8 \leqslant i \leqslant 9.$$

The $\hat{m}_i$ are listed in Table 1.12.

The Pearson chi-square statistic X^2 is now 248.31, L^2 is 252.5, and the degrees of freedom $9 - 2 = 7$. The revised model still fails to fit the data; however, the F statistic for the model of constant suicide rate is

$$\tfrac{1}{8} \times 1263.9 = 158.0,$$

and the F statistic for the new model is

$$\tfrac{1}{7} \times 252.5 = 36.1.$$

The R^2 statistic comparing the new model to the model of constant suicide rates becomes

$$1 - \frac{252.5}{1263.9} = 0.80.$$

Thus a large portion, although not all, of the factors of the model of constant rates for all regions is accounted for by the contrast between the Mountain and Pacific divisions on the one hand and the rest of the country on the other hand.

The Model of Constant Rates for North Central Regions

The divisions East North Central, West North Central, East South Central, and West South Central have similar suicide rates. The hypothesis that the expected suicide rates are the same for these regions is readily tested. Under this hypothesis,

$$f_i = k, \quad i = 3, 4, 6, \quad \text{or} \quad 7;$$

equivalently

$$\log f_i = \alpha, \quad i = 3, 4, 6, \quad \text{or} \quad 7.$$

The estimated joint suicide rate $\hat{k}$ for these four divisions is

$$\hat{k} = \frac{n_3 + n_4 + n_6 + n_7}{z_3 + z_4 + z_6 + z_7} = \frac{4308 + 1790 + 1335 + 2075}{402.52 + 163.19 + 128.03 + 193.21} = 10.720.$$

The Pearson chi-square statistic for testing this hypothesis is

$$X^2 = \frac{(n_3 - \hat{k}z_3)^2}{\hat{k}z_3} + \frac{(n_4 - \hat{k}z_4)^2}{\hat{k}z_4} + \frac{(n_6 - \hat{k}z_6)^2}{\hat{k}z_6} + \frac{(n_7 - \hat{k}z_4)^2}{\hat{k}z_7}$$

$$= 1.99$$

The likelihood-ration chi-square is

$$L^2 = 2\left[n_3 \log\left(\frac{n_3}{\hat{k}z_3}\right) + n_4 \log\left(\frac{n_6}{\hat{k}z_4}\right) + n_6 \log\left(\frac{n_6}{\hat{k}z_6}\right) + n_7 \log\left(\frac{n_7}{\hat{k}z_7}\right) \right]$$

$$= 1.99.$$

Since there are four frequencies and one parameter α, the degrees of freedom are $4 - 1 = 3$. The two chi-square statistics thus provide no indication that suicide rates vary for the different central divisions.

Confidence Intervals

To ascertain which differences exist between suicide rates in different divisions, first note that the log rate

$$\log \hat{f}_i = \log(n_i / z_i)$$

has an approximate $N(\log f_i, m_i^{-1})$ distribution (see Bishop, Fienberg, and Holland (1975, pages 486–500)). The asymptotic variance m_i^{-1} can be estimated by n_i^{-1}. Thus a difference

$$\log \hat{f}_i - \log \hat{f}_j = \log(\hat{f}_i / \hat{f}_j)$$

has an approximate $N(\log(f_i/f_j), m_i^{-1} + m_j^{-1})$ distribution. The asymptotic variance $m_i^{-1} + m_j^{-1}$ may be estimated by $n_i^{-1} + n_j^{-1}$.

If two specific divisions i and j, say New England and Middle Atlantic, are of interest, a 95 percent confidence interval for $\log(f_i/f_j)$ has lower bound

$$\log(\hat{f}_i/\hat{f}_j) - 1.96(n_i^{-1} + n_j^{-1})^{1/2}$$

and upper bound

$$\log(\hat{f}_i/\hat{f}_j) + 1.96(n_i^{-1} + n_j^{-1})^{1/2}.$$

In the case of New England and the Middle Atlantic division, a 95 percent confidence interval for $\log(f_1/f_2)$ has lower bound

$$\log\left(\frac{1119}{118.42}\frac{371.99}{3165}\right) - 1.96\left(\frac{1}{1119} + \frac{1}{3165}\right)^{1/2} = 0.037$$

and upper bound

$$\log\left(\frac{1119}{118.42}\frac{371.99}{3165}\right) + 1.96\left(\frac{1}{1119} + \frac{1}{3165}\right)^{1/2} = 0.173.$$

Corresponding bounds for the ratio f_1/f_2 are

$$e^{0.037} = 1.037$$

and

$$e^{0.173} = 1.189.$$

Simultaneous Confidence Intervals

A different sort of confidence intervals may be of interest if they are being used to scan the list of suicide rates to determine differences in expected rates. In this situation, ordinary confidence intervals are not very satisfactory. To illustrate, assume that each rate f_i is the same and approximate confidence intervals for $\log(f_i/f_j)$ are found with respective lower bounds

$$\log(\hat{f}_i/\hat{f}_j) - 1.96(n_i^{-1} + n_j^{-1})^{1/2}$$

and upper bounds

$$\log(\hat{f}_i/\hat{f}_j) + 1.96(n_i^{-1} + n_j^{-1})^{1/2}.$$

If 0 is not within the bounds for $\log(f_i/f_j)$, then the confidence interval suggests that $\log(f_i/f_j) \neq 0$, so that $f_i \neq f_j$.

If confidence intervals are computed for $\log(f_i/f_j)$, $1 \leqslant j < i \leqslant 9$, then $36 \times 0.05 = 1.8$ is the approximate expected number of pairs i and j with 0 not within the bounds for $\log(f_i/f_j)$. Thus one is very likely to conclude that some f_i and f_j differ, even if no such differences exist.

To resolve this problem of spurious differences, 95 percent Bonferroni simultaneous confidence intervals may be used (see Goodman (1964a)). Instead of the multiplier $Z_{0.05/2} = 1.96$, the multiplier $Z_{0.05/(2 \times 36)} = 3.20$ is used. If all expected rates f_i are equal, then the probability does not exceed 0.05 that 0 is not within the confidence bounds

$$\log(\hat{f}_i/\hat{f}_j) - 3.20(n_i^{-1} + n_j^{-1})^{1/2}$$

and

$$\log(\hat{f}_i/\hat{f}_j) + 3.20(n_i^{-1} + n_j^{-1})^{1/2}.$$

Thus the probability does not exceed 0.05 that one would erroneously conclude the two expected rates f_i and f_j were different.

Bonferroni 95 percent simultaneous confidence intervals are listed in Table 1.13. Note that if the bounds for $\log(f_i/f_j)$ are L_{ij} and U_{ij}, then the bounds for f_i/f_j are $\exp L_{ij}$ and $\exp U_{ij}$. These bounds confirm the earlier observation that the suicide rates in the Mountain and Pacific regions are

Table 1.13

Bonferroni 95 Percent Simultaneous Confidence Intervals for Relative Suicide Rates
for Different Divisions

				$\log(f_i/f_j)$		f_i/f_j	
Division 1	i	Division 2	j	Lower bound	Upper bound	Lower bound	Upper bound
Middle Atlantic	2	New England	1	−0.216	0.064	0.806	1.006
E. N. Central	3	New England	1	0.017	0.232	1.017	1.261
		Middle Atlantic	2	0.155	0.304	1.167	1.357
W. N. Central	4	New England	1	0.027	0.271	1.027	1.311
		Middle Atlantic	2	0.159	0.349	1.173	1.417
		E. N. Central	3	−0.065	0.115	0.937	1.121
S. Atlantic	5	New England	1	0.143	0.361	1.154	1.434
		Middle Atlantic	2	0.280	0.434	1.323	1.544
		E. N. Central	3	0.056	0.199	1.058	1.220
		W. N. Central	4	0.011	0.195	1.011	1.215
E. S. Central	6	New England	1	−0.031	0.228	0.969	1.256
		Middle Atlantic	2	0.099	0.308	1.104	1.360
		E. N. Central	3	−0.126	0.074	0.881	1.077
		W. N. Central	4	−0.166	0.065	0.847	1.067
		S. Atlantic	5	−0.256	−0.052	0.774	0.950
W. S. Central	7	New England	1	0.009	0.247	1.009	1.280
		Middle Atlantic	2	0.143	0.323	1.153	1.382
		E. N. Central	3	−0.082	0.089	0.921	1.093
		W. N. Central	4	−0.124	0.082	0.883	1.086
		S. Atlantic	5	−0.212	−0.036	0.809	0.964
		E. S. Central	6	−0.083	0.142	0.921	1.152
Mountain	8	New England	1	0.396	0.656	1.486	1.927
		Middle Atlantic	2	0.526	0.735	1.692	2.087
		E. N. Central	3	0.301	0.502	1.351	1.652
		W. N. Central	4	0.261	0.493	1.298	1.637
		S. Atlantic	5	0.171	0.376	1.187	1.457
		E. S. Central	6	0.303	0.552	1.354	1.736
		W. S. Central	7	0.285	0.510	1.330	1.666
Pacific	9	New England	1	0.508	0.721	1.662	2.057
		Middle Atlantic	2	0.646	0.794	1.908	2.211
		E. N. Central	3	0.423	0.558	1.526	1.747
		W. N. Central	4	0.377	0.555	1.457	1.742
		S. Atlantic	5	0.292	0.433	1.340	1.542
		E. S. Central	6	0.417	0.616	1.517	1.851
		W. S. Central	7	0.402	0.571	1.495	1.771
		Mountain	8	−0.011	0.189	0.989	1.208

somewhat higher than those in other areas. The smallest bound for f_i/f_j, $8 \leqslant i \leqslant 9$, $1 \leqslant j \leqslant 7$, is 1.187, a bound associated with the South Atlantic division and the Mountain division. The bounds also confirm that the extreme variations in suicide rates are large. Note the bounds of 1.908 and 2.211 for f_9/f_2. Here the Pacific and Middle Atlantic regions are compared.

The divisions can be divided into four groups. The suicide rates are lowest for New England and the Middle Atlantic region, and the simultaneous confidence interval for $\log(f_2/f_1)$ does include 0. The next lowest suicide rates are in the central divisions. The rates for these four divisions are all similar, as has already been noted.

The confidence intervals for relative suicide rates indicate that in each central division, the expected suicide rate is at least 10 percent greater than the expected rate in the Middle Atlantic division. Contrasts between New England and the central divisions are less decisive. The South Atlantic division falls into a class by itself. Zero is not contained in any confidence interval for $\log(f_i/f_j)$ that involves this division. The suicide rate in the South Atlantic division is larger than in the central divisions or in the New England or Middle Atlantic states but somewhat smaller than in the Pacific or Mountain divisions. The last group consists of the Pacific and Mountain divisions, both of which have rather high suicide rates. Except for the South Atlantic region, the overall pattern is an increase in expected suicide rates as one proceeds westward.

This section resembles the preceding section in its use of a Poisson model and in use of log–linear models with maximum-likelihood estimates expressed in closed form; however, some change in approach occurs since simultaneous confidence intervals have been constructed for all differences between regions for log suicide rates.

These simultaneous intervals provide a means of searching for identification of variations in rates which takes into account problems of selection. This problem of development of methods of inference that provide protection against selection of particular features from tables with many cells is a common one in analysis of frequency data; consequently, it will be faced throughout this book.

1.5   General Techniques for Log–Linear Models

The models considered in this chapter are all examples of log–linear models. In this section, some general properties of log–linear models are summarized and references are provided.

In all examples in this chapter, $s \geq 1$ frequencies n_i, $1 \leq i \leq s$, are observed. These frequencies have respective expected values $m_i > 0$. Associated with each i are $q \geq 0$ known independent variables x_{ij}, $1 \leq j \leq q$, and a scale factor $z_i > 0$. For some unknown parameters α and β_j, $1 \leq j \leq q$,

$$\log(m_i/z_i) = \alpha + \sum_{j=1}^{q} \beta_j x_{ij}$$

$$= \alpha + \beta_1 x_{i2} + \cdots + \beta_q x_{iq}. \tag{1.6}$$

(If $q = 0$, then $\log(m_i/z_i) = \alpha$.)

If each z_i is 1, then (1.6) defines a linear model in the log means $\log m_i$. Such a choice of z_i, which occurs in Sections 1.1 and 1.2, is usually made when multinomial data are studied. On the other hand, Poisson data such as those of Sections 1.3 and 1.4 typically involve counts of uncommon events from a reference population with size proportional to z_i. The quotient $f_i = m_i/z_i$ is then an expected rate of an event for units of the population, and (1.6) defines a model linear in the log rates $\log f_i$.

To illustrate application of (1.6), consider the log–linear models considered in Sections 1.1 to 1.4. In Section 1.1, two cases are considered. In the equiprobability model, $q = 0$ and each $z_i = 1$, $1 \leq i \leq s = 18$. Thus

$$\log m_i = \alpha.$$

In the log–linear time-trend model, $q = 1$, each $z_i = 1$, and $x_{i1} = i$. One has

$$\log m_i = \alpha + \beta_1 x_{i1} = \alpha + \beta i$$

In Section 1.2, a model is considered with $q = 2$, $s = 4$, $z_i = 1$ for each i, $x_{11} = -3, x_{21} = -1, x_{31} = 1, x_{41} = 3, x_{12} = x_{42} = 1$, and $x_{22} = x_{32} = -1$. One has $\log m_i = \alpha + \beta_1 x_{i1} + \beta_2 x_{i2}$.

In Sections 1.3 and 1.4, a model is considered in which

$$\log f_i = \log(m_i/z_i) = \alpha.$$

Multinomial Models

Two types of probability models have been considered. In Sections 1.1 and 1.2, multinomial models are used. Here independent identically distributed random variables X_h, $1 \leq h \leq N$, are given. The probability is $p_i > 0$ that $X_h = i$, and n_i is the number of h such that $X_h = i$. The table $\mathbf{n} = \{n_i : 1 \leq i \leq s\}$ is then said to have a multinomial distribution with sample size N and probabilities $\mathbf{p} = \{p_i : 1 \leq i \leq s\}$.

Although detailed knowledge of the multinomial distribution is not required in this book, a few elementary properties can be used to derive some results concerning residuals in simple log–linear models and concerning some maximum-likelihood estimates $\hat{m}_i$. Each count n_i has a binomial distribution with expected value $m_i = Np_i$ and variance $Np_i(1 - p_i)$. If c_i, $1 \leqslant i \leqslant s$, are constants, then $\sum c_i n_i$ has expected value $N\sum c_i p_i$ and variance

$$N[\sum c_i^2 p_i - (\sum c_i p_i)^2].$$

For example, $n_1 - n_2 = (1)n_1 + (-1)n_2$ has expected value

$$N[(1)p_1 + (-1)p_2] = N(p_1 - p_2)$$

and variance

$$N\{(1)^2 p_1 + (-1)^2 p_2 - [(1)p_1 - (-1)p_2]^2\}$$
$$= N[(p_1 + p_2) - (p_1 - p_2)^2].$$

The general formula is not difficult to verify. Let $T_h = c_i$ if $X_h = i$. Then T_h has expected value

$$\sum c_i p_i$$

and variance

$$\sum c_i^2 p_i - (\sum c_i p_i)^2.$$

The sum $\sum c_i n_i$ is equal to $\sum T_h$. Thus $\sum c_i n_i$ has expected value equal to N times the expected value of T_h and a variance equal to N times the variance of T_h.

Poisson Models

In Sections 1.3 and 1.4, Poisson models are used. Here each n_i has an independent Poisson distribution with mean m_i. In this case, each n_i has variance m_i and $\sum c_i n_i$ has variance $\sum c_i^2 m_i$. To verify these results requires more sophisticated arguments than those used in the multinomial case.

Maximum-Likelihood Equations

The means m_i and the parameters α and β_j, $1 \leqslant j \leqslant q$, have been estimated in this chapter by the method of maximum likelihood. This method is a general estimation technique discussed in detail in advanced statistical texts such as Rao (1973, pages 353–374); however, knowledge of the general properties of maximum-likelihood estimates is not really needed for their application to log–linear models.

Under either of the sampling models discussed in this section, the maximum-likelihood estimates $\hat{m}_i$ of m_i, $\hat{\alpha}$ of α, and $\hat{\beta}_j$ of β_j satisfy the equations

$$\log(\hat{m}_i/z_i) = \hat{\alpha} + \sum_{j=1}^{q} \hat{\beta}_j x_{ij}, \tag{1.7}$$

$$\sum \hat{m}_i = \sum n_i = N, \tag{1.8}$$

$$\sum_i x_{ij}\hat{m}_i = \sum_i x_{ij}n_i, \qquad 1 \leqslant j \leqslant q. \tag{1.9}$$

For derivations, see Birch (1963) and Haberman (1973a; 1974a, page 374).

In the case of multinomial sampling, interpretation of these equations may be aided by the following argument. Let X_h, $1 \leqslant h \leqslant N$, be independent and identically distributed random variables such that $p_i > 0$ is the probability that $X_h = i$. Let n_i be the number of h such that $X_h = i$. Let T_{hj} be x_{ij}, $1 \leqslant j \leqslant q$, if $X_h = i$. Let $\hat{p}_i = \hat{m}_i/N$ be the maximum-likelihood estimate of p_i. Then the second equation reduces to the requirement that the sum $\sum \hat{p}_i$ of the estimated probabilities must be 1. The third equation reduces to the requirement that the estimated mean

$$\hat{\theta}_j = \sum_i x_{ij}\hat{p}_i \tag{1.10}$$

of the score T_{hj} is the sample mean

$$\bar{T}_j = \frac{1}{N}\sum_h T_{hj} = \frac{1}{N}\sum_i x_{ij}n_i \tag{1.11}$$

of these scores. Interpretations of this kind have been used in Sections 1.1 and 1.2. For example, consider the model

$$\log m_i = \alpha + \beta i$$

of Section 1.1. Here $q = 1$ and $x_{i1} = i$. The score T_{h1} is X_h. The maximum-likelihood equations include requirements that

$$\sum \hat{p}_i = 1 \tag{1.12}$$

and

$$\hat{\theta}_1 = \sum i\hat{p}_i = \bar{T}_1 = \bar{X} = \frac{1}{N}\sum in_i. \tag{1.13}$$

As another example, consider the model of Section 1.2 in which $\log m_i = \alpha + \beta_1 x_{i1} + \beta_2 x_{i2}$, $x_{11} = -3$, $x_{21} = -1$, $x_{31} = 1$, $x_{41} = 3$, $x_{12} = x_{42} = 1$, and $x_{22} = x_{32} = -1$. We now have the equations

$$\sum \hat{p}_i = 1,$$

$$\hat{\theta}_1 = \sum_i x_{i1}\hat{p}_i = -3\hat{p}_1 - \hat{p}_2 + \hat{p}_3 + 3\hat{p}_4$$

$$= \bar{T}_1 = \frac{1}{N}(-3n_1 - n_2 + n_3 + 3n_4),$$

and

$$\hat{\theta}_2 = \sum_i x_{i2}\hat{p}_i = \hat{p}_1 - \hat{p}_2 - \hat{p}_3 + \hat{p}_4$$

$$= \bar{T}_2 = \frac{1}{N}(n_1 - n_2 - n_3 + n_4).$$

In the case of the Poisson sampling model, a similar argument is possible. A Poisson sampling model can be described in terms of N independent and identically distributed random variables X_h, $1 \leq h \leq N$, such that

$$p_i = m_i/M > 0$$

is the probability that $X_h = i$,

$$M = \sum m_i,$$

and N has a Poisson distribution with mean M. This result is well known but rarely formally derived. For example, see Lehmann (1959, page 57). Given the random variables X_h, one may let T_{hj} be x_{ij}, $1 \leq j \leq q$, if $X_h = i$. The estimated probabilities $\hat{p}_i = \hat{m}_i/\hat{M} = \hat{m}_i/N$, have sum 1, and the estimated expected values

$$\hat{\theta}_j = \sum_i x_{ij}\hat{p}_i = N^{-1}\sum_i x_{ij}\hat{m}_i \qquad (1.14)$$

of the T_{hj} are still equal to the sample averages

$$\bar{T}_j = \frac{1}{N}\sum_h T_{hj} = \frac{1}{N}\sum_i x_{ij}n_i.$$

Solution of Maximum-Likelihood Equations

Maximum-likelihood equations sometimes have explicit solutions. In other cases, iterative algorithms such as the Newton–Raphson algorithm can be used to obtain solutions.

Explicit solutions to maximum-likelihood equations have been used in Sections 1.1 to 1.4. The simplest case arises if $q = 0$. Then

$$\log(\hat{m}_i/z_i) = \hat{\alpha}$$

and

$$\sum \hat{m}_i = \sum n_i = N.$$

Thus

$$\hat{m}_i = z_i \exp \hat{\alpha}$$

and

$$\sum \hat{m}_i = \left(\sum z_i\right) \exp \hat{\alpha}.$$

Consequently,

$$\hat{k} = \exp \hat{\alpha} = \frac{N}{\sum z_i}, \qquad \hat{\alpha} = \log \hat{k},$$

and

$$\hat{m}_i = \hat{k} z_i.$$

For example, in the equiprobability model of Section 1.1, $z_i = 1$ for $1 \leqslant i \leqslant 18$. Thus

$$\hat{k} = \frac{N}{18}, \qquad \hat{\alpha} = \log\left(\frac{N}{18}\right),$$

and

$$\hat{m}_i = \left(\frac{1}{18}\right) N.$$

The Newton–Raphson algorithm may be used for iterative calculations. This algorithm is a very old procedure for the solution of nonlinear equations. For general discussion of the algorithm, see Ostrowski (1966, pages 183–194). Its use with log–linear models goes back in special cases to Fisher and Yates (1963[1938], page 16) and Garwood (1941). For more general applications, see Bock (1970, 1975, pages 525–526) and Haberman (1974a, pages 47–64; 1974b). For computer programs, see the Appendix or Bock and Yates (1973).

The Newton–Raphson algorithm developed in Sections 1.1 and 1.2 reduces to a series of weighted regressions. Assume $q \geqslant 1$ and let $m_{i0} > 0$ be an initial approximation to $\hat{m}_i$. For example, m_{i0} might be n_i or $n_i + \frac{1}{2}$. Let

$$y_{i0} = \log(m_{i0}/z_i), \tag{1.15}$$

so that if the $z_i = 1$, then $y_i = \log m_{i0}$. To obtain initial approximations β_{j0} to $\hat{\beta}_j$, $1 \leqslant j \leqslant q$, consider a regression problem

$$y_{i0} = \alpha + \sum_{j=1}^{q} \beta_j x_{ij} + \varepsilon_i,$$

where the ε_i are independent random variables with respective means 0 and variances m_{i0}^{-1}.

The approximations β_{j0} are computed as if this regression problem were under study. Thus

$$\sum_{k=1}^{q} S_{jk0}\beta_{j0} = w_{j0}, \qquad 1 \leqslant j \leqslant q, \tag{1.16}$$

where

$$S_{jk} = \sum_{i} (x_{ij} - \theta_{j0})(x_{ik} - \theta_{k0})m_{i0}, \tag{1.17}$$

$$\theta_{j0} = \sum_{i} x_{ij}m_{i0} \Big/ \sum_{i} m_{i0}, \tag{1.18}$$

and

$$w_{j0} = \sum_{i} (x_{ij} - \theta_{j0})y_{i0}m_{i0}. \tag{1.19}$$

If $\boldsymbol{\beta}_0$ is the vector with coordinates β_{j0}, $1 \leqslant j \leqslant q$, if $\mathbf{w}_0$ has coordinates w_{j0}, $1 \leqslant j \leqslant q$, and if S_0 is the matrix with elements S_{jk0}, $1 \leqslant j \leqslant q$, $1 \leqslant k \leqslant q$, then

$$S_0\boldsymbol{\beta}_0 = \mathbf{w}_0 \tag{1.20}$$

and

$$\boldsymbol{\beta}_0 = S_0^{-1}\mathbf{w}_0. \tag{1.21}$$

These calculations were performed in Section 1.1 with $q = 1$ and $x_{i1} = i$. In that section, $\boldsymbol{\beta}_0$ reduces to the scalar β_0, $\mathbf{w}_0$ reduces to the scalar w_0, etc. In Section 1.2, these calculations were performed with $q = 2$, $x_{11} = -3$, $x_{21} = -1$, $x_{31} = 1$, $x_{41} = 3$, $x_{12} = x_{42} = 1$, and $x_{22} = x_{32} = -1$.

Given $\boldsymbol{\beta}_0$, an initial approximation α_0 for $\hat{\alpha}$ and a new approximation m_i for $\hat{m}_i$ can be found. Let

$$v_{i0} = \sum_{j=1}^{q} \beta_{j0}x_{ij} \tag{1.22}$$

and

$$g_0 = N \Big/ \sum_{i} z_i \exp v_{i0}, \tag{1.23}$$

so that if each $z_i = 1$, then

$$g_0 = N \Big/ \sum \exp v_{i0}. \tag{1.24}$$

Then

$$\alpha_0 = \log g_0$$

and

$$m_{i1} = z_i g_0 \exp v_{i0}. \tag{1.25}$$

If each $z_i = 1$,

$$m_{i1} = g_0 \exp v_{i0}. \tag{1.26}$$

Given approximations m_{it} for some $t \geq 1$, new approximations α_t, β_{jt}, $1 \leq j \leq q$, and $m_{i(t+1)}$ are computed in the following manner. Let

$$y_{it} = \log(m_{it}/z_i) + (n_i - m_{it})/m_{it}, \tag{1.27}$$

so that if the z_i are 1,

$$y_{it} = \log m_{it} + (n_i - m_{it})/m_{it}. \tag{1.28}$$

Compute β_{jt}, $1 \leq j \leq q$, as in a regression problem

$$y_i = \alpha + \sum_j \beta_j x_{ij} + \varepsilon_i,$$

where the ε_i are independent random variables with respective means 0 and variance m_{it}^{-1}. Thus

$$\sum_{k=1}^q S_{jkt} \beta_{jt} = w_{jt}, \qquad 1 \leq j \leq q, \tag{1.29}$$

where

$$S_{jkt} = \sum_i (x_{ij} - \theta_{jt})(x_{ik} - \theta_{kt}) m_{it}, \tag{1.30}$$

$$\theta_{jt} = \frac{\sum_i x_{ij} m_{it}}{\sum_i m_{it}}, \tag{1.31}$$

and

$$w_{jt} = \sum_i (x_{ij} - \theta_{jt}) y_{it} m_{it}. \tag{1.32}$$

In matrix terms,

$$S_t \boldsymbol{\beta}_t = \mathbf{w}_t \tag{1.33}$$

and

$$\boldsymbol{\beta}_t = S_t^{-1} \mathbf{w}_t. \tag{1.34}$$

In practice, computations are simplified if one lets

$$a_{jt} = N(\bar{T}_j - \theta_{jt}) = \sum_i x_{ij} n_i - \sum_i x_{ij} m_{it} \tag{1.35}$$

and

$$\delta_{jt} = \beta_{jt} - \beta_{j(t-1)}. \tag{1.36}$$

Then

$$\sum_{k=1}^{q} S_{jkt}\, \delta_{kt} = a_{jt}, \qquad 1 \leqslant j \leqslant q, \tag{1.37}$$

so that

$$S_t \boldsymbol{\delta}_t = \mathbf{a}_t \tag{1.38}$$

and

$$\boldsymbol{\beta}_{t+1} = \boldsymbol{\beta}_t + S_t^{-1}\mathbf{a}_t. \tag{1.39}$$

Given the β_{jt}, $1 \leqslant j \leqslant q$, approximations α_t for $\hat{\alpha}$ and $m_{i(t+1)}$ for $\hat{m}_i$ are found through the formulas

$$v_{it} = \sum_j \beta_{jt} x_{ij}, \tag{1.40}$$

$$g_t = N \Big/ \sum_i z_i \exp v_{it}, \tag{1.41}$$

$$\alpha_t = \log g_t, \tag{1.42}$$

and

$$m_{i(t+1)} = z_i g_t \exp v_{it}. \tag{1.43}$$

As is apparent from Sections 1.1 and 1.2, convergence of the approximations β_{jt} to $\hat{\beta}_j$, α_t to $\hat{\alpha}$, and $\hat{m}_{it}$ to m_i is generally quite rapid. Even approximations such as β_{j0} or m_{i0} are often sufficiently accurate for all practical purposes. This situation is especially likely if the expected values m_i are all large, say greater than 50.

A few technical dangers should be noted in concluding this description of the Newton–Raphson algorithm. The most important reservation is that it has been assumed that each matrix S_t has an inverse. This condition, which holds in all examples in this chapter can be assured through appropriate definition of the variables x_{ij} so that coefficients α and β_j, $1 \leqslant j \leqslant q$, are uniquely determined by the $\log(m_i/z_i)$, $1 \leqslant i \leqslant s$.

As an example, consider a model in which

$$\begin{aligned} m_i &= k_1, & i &= 1, \\ &= k_2, & i &> 1. \end{aligned}$$

This model can be written as

$$\log m_i = \alpha + \beta_1 x_{i1} + \beta_2 x_{i2},$$

where

$$x_{i1} = 1, \quad i = 1,$$
$$\quad = 0, \quad i > 1,$$
$$x_{i2} = 0, \quad i = 1,$$
$$\quad = 1, \quad i > 1.$$

A possible choice of α, β_1, and β_2 is

$$\alpha = \log k_2,$$
$$\beta_1 = \log k_1 - \log k_2,$$
$$\beta_2 = 0.$$

Unfortunately, other choices of α, β_1, and β_2 are also possible. For example, let

$$\alpha = 0,$$
$$\beta_1 = \log k_1,$$

and

$$\beta_2 = \log k_2.$$

Since the α, β_1, and β_2 are not unique, it is not possible to use the version of the Newton–Raphson algorithm presented in this section.

The difficulty can be avoided by writing the model as

$$\log m_i = \alpha + \beta_1 x_{i1},$$

where x_{i1} is defined as before. Then α must be $\log k_2$ and β_1 must be $\log k_1 - \log k_2$. Thus the Newton–Raphson algorithm can be used.

A second technical problem is that it is possible for maximum-likelihood estimates not to exist or for the algorithm to fail to converge. Neither of these problems is likely to cause much difficulty. For technical discussions, see Haberman (1973a; 1974a, pages 37–38, 47–64, 395–407).

Large-Sample Properties of Maximum-Likelihood Estimates

In the log–linear model

$$\log(m_i/z_i) = \alpha + \sum_{j=1}^{q} \beta_j x_{ij},$$

the distribution of the vector $\hat{\boldsymbol{\beta}}$ with coordinates $\hat{\beta}_j$, $1 \leqslant j \leqslant q$, is approximately the same as the distribution of the weighted-least-squares estimate

b for the regression model

$$Y_i = \alpha + \sum_{j=1}^{q} \beta_j x_{ij} + \varepsilon_i,$$

where the Y_i are hypothetical observed dependent variables and the ε_i are independently distributed $N(0, m_i^{-1})$ variables. Here

$$\sum_k S_{jk} b_k = w_j, \tag{1.44}$$

where

$$S_{jk} = \sum_i (x_{ij} - \theta_j)(x_{ik} - \theta_k) m_i, \tag{1.45}$$

$$w_j = \sum_i (x_{ij} - \theta_j) Y_i m_i, \tag{1.46}$$

and

$$\theta_j = \sum_i x_{ij} m_i \Big/ \sum_i m_i. \tag{1.47}$$

As is well known from regression analysis, the vector **b** with coordinates $b_j, 1 \leqslant j \leqslant q$, has a multivariate normal distribution with mean $\boldsymbol{\beta}$ and covariance matrix S^{-1}. Here S^{-1}, the $q \times q$ matrix with elements $S^{jk}, 1 \leqslant j \leqslant q, 1 \leqslant k \leqslant q$, is the inverse of S, the $q \times q$ matrix with elements $S_{jk}, 1 \leqslant j \leqslant q, 1 \leqslant k \leqslant q$, so that

$$\sum_k S_{jk} S^{kl} = 1, \qquad j = l, \tag{1.48}$$

$$= 0, \qquad j \neq l.$$

The maximum-likelihood estimate $\hat{\boldsymbol{\beta}}$ thus has an approximate $N(\boldsymbol{\beta}, S^{-1})$ distribution. The vector $\boldsymbol{\beta}$ is called the asymptotic mean of $\hat{\boldsymbol{\beta}}$, and S^{-1} is called the asymptotic covariance matrix of $\hat{\boldsymbol{\beta}}$. The coordinate $\hat{\beta}_j, 1 \leqslant j \leqslant q$, has an approximate $N(\beta_j, S^{jj})$ distribution, so that $\hat{\beta}_j$ has asymptotic mean β_j and asymptotic variance

$$\sigma^2(\hat{\beta}_j) = S^{jj}. \tag{1.49}$$

If $c_j, 1 \leqslant j \leqslant q$, are constants and

$$\tau = \sum_j c_j \beta_j,$$

then

$$\hat{\tau} = \sum_j c_j \hat{\beta}_j$$

has asymptotic mean τ and asymptotic variance

$$\sigma^2(\hat{\tau}) = \sum_j \sum_k c_j c_k S^{jk}. \tag{1.50}$$

The maximum-likelihood estimate $\hat{S}_{jk}$ satisfies the equation

$$\hat{S}_{jk} = \sum_i (x_{ij} - \hat{\theta}_j)(x_{ik} - \hat{\theta}_k)\hat{m}_i$$

$$= \sum_i (x_{ij} - \bar{T}_j)(x_{ik} - \bar{T}_k)\hat{m}_i. \tag{1.51}$$

This estimate, divided by N, is the estimated covariance of T_{hj} and T_{hk} for each h, $1 \leqslant h \leqslant N$. Given $\hat{S}_{jk}$, one can find the $q \times q$ matrix $\hat{S}$ with coordinates $\hat{S}_{jk}$, $1 \leqslant j \leqslant q$, $1 \leqslant k \leqslant q$, and the inverse matrix $\hat{S}^{-1}$ with coordinates $\hat{S}^{jk}$, $1 \leqslant j \leqslant q$, $1 \leqslant k \leqslant q$. The estimated asymptotic variance of $\hat{\beta}_j$ is then $s^2(\hat{\beta}_j) = \hat{S}^{jj}$. Examples of computations of estimated asymptotic variances have been provided in Sections 1.1 and 1.2 in the cases $q = 1$ and $q = 2$. The program in the Appendix in Volume 2 computes estimated asymptotic standard deviations $s(\hat{\beta}_j)$ as a by-product of calculations of maximum-likelihood estimates. Such computations are facilitated by the fact that in the Newton–Raphson algorithm, S_{jkt} converges to $\hat{S}_{jk}$ as the m_{it} converges to $\hat{m}_i$. The Bock and Yates (1973) program MULTIQUAL also computes estimated asymptotic standard deviations.

Given an estimate

$$\hat{\tau} = \sum c_j \hat{\beta}_j$$

of a parameter

$$\tau = \sum c_j \beta_j$$

and the estimated asymptotic standard deviation

$$s(\hat{\tau}) = \left[\sum_j \sum_k c_j c_k \hat{S}^{jk} \right]^{1/2}, \tag{1.52}$$

an approximate level $(1 - \alpha)$ confidence interval for τ has lower bound

$$\hat{\tau} - Z_{\alpha/2} s(\hat{\tau})$$

and upper bound

$$\hat{\tau} + Z_{\alpha/2} s(\hat{\tau}).$$

In the case of β_j, the bounds are

$$\hat{\beta}_j - Z_{\alpha/2} s(\hat{\beta}_j) = \hat{\beta}_j - Z_{\alpha/2}(\hat{S}^{jj})^{1/2}$$

and

$$\hat{\beta}_j + Z_{\alpha/2} s(\hat{\beta}_j) = \hat{\beta}_j + Z_{\alpha/2}(\hat{S}^{jj})^{1/2}.$$

The formulas used in this section also apply if β_{j0} is substituted for $\hat{\beta}_j$ for each j and if S_{jk0} is substituted for $\hat{S}_{jk}$ for each j and k. Here β_{j0} and S_{jk0} are defined as in the Newton–Raphson algorithm and m_{i0} is n_i or $n_i + \frac{1}{2}$. The differences $|\hat{\beta}_j - \beta_{j0}|$ tend to become very small in large samples, so that β_{j0} is often a sufficiently close approximation for $\hat{\beta}_j$ that further iterations are unnecessary.

The Saturated Model

In one special case, the saturated model, use of the regression analogy of this section lead to simple results in which matrix inversions can be avoided. In this model, no restrictions are placed on m_i/z_i, so that

$$\log(m_i/z_i) = \alpha + \beta_i, \qquad 1 \leqslant i \leqslant s - 1,$$
$$= \alpha, \qquad\qquad i = s.$$

Thus

$$\log(m_i/z_i) = \alpha + \sum_{j=1}^{q-1} \beta_j x_{ij},$$

where

$$x_{ij} = 1, \qquad i = j,$$
$$= 0, \qquad i \neq j.$$

It is easily seen that

$$\sum \hat{m}_i = \sum n_i = N$$

and

$$\sum_i x_{ij} \hat{m}_i = \hat{m}_j = \hat{n}_j = \sum_i x_{ij} n_i, \qquad 1 \leqslant j \leqslant s - 1.$$

Thus

$$\hat{m}_i = n_i, \qquad 1 \leqslant i \leqslant s,$$
$$\hat{\alpha} = \log(n_s/z_s),$$

and

$$\hat{\beta}_j = \log(n_j/z_j) - \log(n_s/z_s).$$

To find asymptotic variances, consider the regression model

$$Y_i = \alpha + \beta_i + \varepsilon_i, \quad 1 \leqslant i \leqslant s - 1,$$
$$= \alpha + \varepsilon_i, \qquad i = s,$$

where the Y_i are hypothetical dependent variables and the ε_i are independent $N(0, m_i^{-1})$ random variables. The regression estimates b_j of β_j and a of α are

$$b_j = Y_j - Y_s, \quad 1 \leqslant j \leqslant s - 1,$$
$$a = Y_s.$$

Note that if

$$\hat{Y}_i = a + b_i, \quad 1 \leqslant i \leqslant s - 1,$$
$$= a, \quad i = s,$$

then the residual sum of squares

$$\sum m_i (Y_i - \hat{Y}_i)^2 = 0.$$

The variance of b_j is $m_j^{-1} + m_s^{-1}$, so $\sigma^2(\hat{\beta}_j) = m_j^{-1} + m_s^{-1}$. Since the numbering of categories is arbitrary, it follows more generally that $\log(n_i/z_i) - \log(n_j/z_j)$ has asymptotic variance $m_i^{-1} + m_j^{-1}$ if $i \neq j$. This formula is difficult to discover using matrix multiplication. It has been used in Sections 1.3 and 1.4 to compare a log suicide rate

$$\log \hat{f}_i = \log(n_i/z_i)$$

to another log rate

$$\log \hat{f}_j = \log(n_j/z_j).$$

The asymptotic variance of the difference $\log \hat{f}_i - \log \hat{f}_j$ has an estimate $n_i^{-1} + n_j^{-1}$. Thus a level $(1 - \alpha)$ confidence interval for $\log f_i - \log f_j = \log(m_i/z_i) - \log(m_j/z_j)$ has bounds

$$\log \hat{f}_i - \log \hat{f}_j - Z_{\alpha/2}(n_i^{-1} + n_j^{-1})^{1/2}$$

and

$$\log \hat{f}_i - \log \hat{f}_j + Z_{\alpha/2}(n_i^{-1} + n_j^{-1})^{1/2}.$$

More generally, if $\sum c_i = 0$, the maximum-likelihood estimate

$$\hat{\tau} = \sum c_i \log(\hat{m}_i/z_i) = \sum c_i \log(n_i/z_i)$$

of

$$\tau = \sum c_i \log(\hat{m}_i/z_i)$$

has asymptotic variance

$$\sigma^2(\hat{\tau}) = \sum c_i^2/m_i.$$

This formula applies to $\sigma^2(\hat{\beta}_j)$, for $\hat{\beta}_j = \sum c_i \log(\hat{m}_i/z_i)$ with $c_j = 1$, $c_s = -1$, and $c_i = 0$ for $i \neq j$, $i \neq s$. Thus $\sum c_i^2/m_i = 1/m_j + 1/m_s$.

To verify the general formula, note that

$$\hat{\tau} = \left(\sum_{i=1}^{s} c_i \right) \hat{\alpha} + \sum_{i=1}^{s-1} c_i \hat{\beta}_j = \sum_{i=1}^{s-1} c_i \hat{\beta}_i$$

and

$$\tau = \sum_{i=1}^{s-1} c_i \beta_i.$$

Since

$$c_s = - \sum_{i=1}^{s-1} c_i,$$

the regression analog $t = \sum c_i Y_i$ satisfies the equation

$$t = \sum_{i=1}^{s-1} c_i b_i = \sum_{i=1}^{s-1} c_i (Y_i - Y_s) = \sum_{i=1}^{s-1} c_i Y_i - \left(\sum_{i=1}^{s-1} c_i \right) Y_s = \sum_{i=1}^{s} c_i Y_i.$$

Thus t has variance

$$\sum c_i^2 / m_i.$$

Consequently, $\hat{\tau}$ has asymptotic variance $\sum c_i^2 / m_i$.

Conditions for Application of Results

All formulas used here require that S and $\hat{S}$ have inverses. These conditions hold whenever the Newton–Raphson algorithm can be used. Thus no special restrictions apply.

If the model is valid, the approximations in this section for the distributions of maximum-likelihood estimates become increasingly accurate for a given model as the expected values m_i become large. Relatively little is really known about how large is large. Haberman (1974a, page 145) suggests that it should be quite sufficient for most purposes to have each m_i at least 25. At the other extreme, Haberman (1974a, page 145) also suggests that if

$$m_i < 25,$$

then approximations are questionable. If the number s of categories is large, $\sum m_i$ is large, but each m_i is relatively small, then it may be the case that the normal approximation for $\hat{\beta}$ is quite accurate but β_0 is not a sufficiently accurate approximation to $\hat{\beta}$ so that further iterations can be avoided. Haberman (1974a, page 94) provides an example of this problem.

Formulas can also be provided for the large-sample distribution of $\hat{\alpha}$; however, they are rarely needed and depend on which sampling model is used. Consequently, they are omitted.

For detailed but mathematically difficult verifications of results of this section, see Haberman (1974a, pages 75–97). Special cases of results presented here go back to Fisher and Yates (1963 [1938], page 16). Birch (1963), Bock (1970; 1975, pages 524–525), and Haberman (1974b) provide statements of results analogous to those presented here. Plackett (1962) is helpful in the case of the saturated model.

Chi-Square Tests

The standard chi-square test statistics for a log–linear model are

$$X^2 = \sum (n_i - \hat{m}_i)^2 / \hat{m}_i$$

and

$$L^2 = 2\sum n_i \log(n_i/\hat{m}_i).$$

Both statistics have approximate chi-square distributions with $s - q - 1$ degrees of freedom if the model holds. As usual, it is assumed that α and β_j, $1 \leq j \leq q$, are determined by the logarithms $\log(m_i/z_i)$, $1 \leq i \leq s$, which is always the case in this chapter.

The approximations become increasingly accurate as all means m_i become large, and the difference $|X^2 - L^2|$ tends to become increasingly small as the m_i increase. A common recommendation is that approximations are adequate if all m_i exceed 5. An alternate recommendation due to Cochran (1954) is that all m_i exceed 1 and at least 80% of the m_i exceed 5. Note that chi-square statistics can have satisfactory approximations for their distributions for smaller m_i than can maximum-likelihood estimates.

These recommendations are actually based on limited evidence. Yarnold (1970) has conducted a detailed investigation of

$$X^2 = \sum (n_i - Nq_i)^2 / (Nq_i)$$

for the log–linear model

$$\log(m_i/q_i) = \alpha.$$

If $s \geq 3$ and if r products Nq_i are less than 5, then Yarnold recommends that all Nq_i should be at least $5r/s$. In the case $q_i = 1/s$, Good, Gover, and Mitchell (1970) obtain results that suggest that for $s \geq 5$, approximations for the distribution of X^2 and L^2 are generally adequate for expected values $N/s \geq 1$.

If the model holds, the statistic

$$F = \frac{1}{s - q - 1} L^2$$

has an approximate expected value of 1. If the model does not hold, then the expected value of F exceeds 1. Consequently, Goodman (1971a) and

Haberman (1974b) have proposed F as an index of goodness of fit for comparing different models.

An alternate and more formal method for comparison of models is based on decomposition of the likelihood-ratio chi-square. Let $\hat{m}_i$ be the maximum-likelihood estimate of m_i under the usual model

$$\log(m_i/z_i) = \alpha + \sum_{j=1}^{q} \beta_j x_{ij}$$

and let

$$L^2 = 2\sum n_i \log(n_i/\hat{m}_i)$$

be the corresponding likelihood-ratio chi-square. Let $\hat{m}_i{}'$ be the maximum-likelihood estimate of m_i under the model

$$\log(m_i/z_i) = \alpha + \sum_{j=1}^{q'} \beta_j x_{ij},$$

where $q' < q$, and let

$$L'^2 = 2\sum n_i \log(n_i/\hat{m}_i')$$

be the likelihood-ratio chi-square for this model. If the second model holds, then $L'^2 - L^2$ and L^2 are approximately independent and $L'^2 - L^2$ has an approximate chi-square distribution with $q - q'$ degrees of freedom. Use of $L'^2 - L^2$ rather than L'^2 leads to a more sensitive test of the second model if the first model holds, at least approximately.

An example of a decomposition of the likelihood-ratio chi-square L'^2 into components $L'^2 - L^2$ and L^2 is found in Section 1.1. In that section, L'^2 was associated with the equiprobability model

$$\log p_i = \alpha,$$

so that $q' = 0$, and L^2 was associated with the log–linear time trend model

$$\log p_i = \alpha + \beta_i,$$

so that $q = 1$. Thus $L'^2 - L^2$ had $1 - 0 = 1$ degree of freedom.

The ratio $R^2 = (L'^2 - L^2)/L'^2$ is proposed by Goodman (1971a) as an index of the relative quality of the fits to the data provided by the two models. The statistic is bounded by 0 and 1, just as the R^2 statistic of regression analysis. An R^2 close to 1 suggests that the model

$$\log(m_i/z_i) = \alpha + \sum_{j=1}^{q} \beta_j x_{ij}$$

is much better than the model

$$\log(m_i/z_i) = \alpha + \sum_{j=1}^{q'} \beta_j x_{ij}.$$

If R^2 is close to 0, then the two models provide approximately equal fits. In this chapter, R^2 has been used only in the case $q' = 0$; however, other choices will arise in later chapters.

The test statistics X^2 and L^2 have a long history. Pearson (1900) originated X^2 and Fisher (1924), Neyman and Pearson (1928), and Wilks (1935, 1938) contributed to early development of L^2. For an extensive bibliography on chi-square tests, see Lancaster (1969). Cochran's (1952, 1954) review articles on chi-square tests remain valuable. For detailed derivations of results concerning application of chi-square tests to log–linear models, see Haberman (1974a, pages 97–122).

Residual Analysis

Residual analysis for log–linear models has been discussed in Haberman (1973b; 1974a, pages 138–144; 1976). General background can be found in Cox and Snell (1968). For alternate approaches to those used here, see Bishop, Fienberg, and Holland (1975, pages 136–141). For general background on use of residuals in regression analysis, see Draper and Smith (1966, pages 86–103).

In the regression model

$$Y_i = \alpha + \sum_{j=1}^{q} \beta_j x_{ij} + \varepsilon_i$$

in which the ε_i are independent $N(0, m_i^{-1})$ random variables, the residuals

$$R_i = Y_i - \bar{Y} - \sum_{j=1}^{q} b_j(x_{ij} - \theta_j).$$

Here

$$\bar{Y} = \frac{1}{N} \sum Y_i$$

and

$$\theta_j = \sum_i x_{ij} m_i \bigg/ \sum_i m_i.$$

The estimates b_j are defined as in the discussion of large-sample properties of maximum-likelihood estimates. If the model holds, $m_i^{1/2}R_i$ has a normal distribution with mean 0 and variance

$$\frac{c_i}{m_i} = 1 - \frac{m_i}{N} - \sum_j \sum_k (x_{ij} - \theta_j)(x_{ik} - \theta_k)S^{jk}.$$

In log–linear models, the standardized residual $(n_i - \hat{m}_i)/\hat{m}_i^{1/2}$ has approximately the same distribution as $m_i^{1/2}R_i$. If $c_i > 0$ and

$$\hat{c}_i = \hat{m}_i\left[1 - \frac{\hat{m}_i}{N} - \sum_j \sum_k (x_{ij} - \bar{T}_j)(x_{ik} - \bar{T}_k)\hat{S}^{jk}\right],$$

then the adjusted residual

$$r_i = (n_i - \hat{m}_i)/\hat{c}_i^{1/2}$$

has an approximate standard normal distribution. The adjusted residuals in Section 1.1 illustrate application of these formulas if $q = 1$ and $x_{i1} = i$.

The program in the Appendix in Volume 2 provides adjusted residuals. The MULTIQUAL program of Bock and Yates (1973) provides standardized residuals. If s is much larger than q, the standardized and adjusted residuals are rather similar. The average variance of the standardized residuals $(n_i - \hat{m}_i)/\hat{m}_i^{1/2}$ is $1 - (q + 1)/s$.

A more general form of residual analysis has received some attention in Armitage (1955), Cochran (1954, 1955), Fisher (1970[1925], pages 101–111), and Haberman (1976). Here constants e_i, $1 \leqslant i \leqslant s$, are assigned to the s categories and

$$t = \frac{\sum e_i(n_i - \hat{m}_i)}{\hat{c}^{1/2}}$$

is found, where $\hat{c}$, an estimate of the variance of $\sum e_i(n_i - \hat{m}_i)$, satisfies

$$\hat{c} = \sum_i e_i^2 \hat{m}_i - N^{-1}\left(\sum_i e_i \hat{m}_i\right)^2 - \sum_j \sum_k \hat{f}_j \hat{f}_k \hat{S}^{jk}$$

and where

$$\hat{f}_j = \sum_i e_i(x_{ij} - \hat{\theta}_j) = \sum_i e_i(x_{ij} - \bar{T}_j).$$

If the model holds, then t has an approximate $N(0, 1)$ distribution.

This kind of residual is used in Section 1.1 with the equiprobability model. Here $q = 0$, $e_i = i/N$, $\hat{m}_i = N/18$ and $s = 18$.

As shown in Section 1.1,

$$\sum e_i(n_i - \hat{m}_i) = \bar{X} - 9.5$$

and

$$\hat{c} = \sum_i e_i^2 \hat{m}_i - N^{-1} \left(\sum_i e_i \hat{m}_i \right)^2$$

$$= \frac{1}{18N^2} \sum i^2 - \frac{1}{324N} \left(\sum i \right)^2$$

$$= \frac{(18+1)(18-1)}{12N}$$

$$= \frac{323}{12N}.$$

Large-sample approximations for the adjusted residuals r_i are probably adequate if each m_i exceeds 25. If the m_i are smaller, then the residuals still provide information concerning patterns of deviations from the model and they do provide a rough indication of the size of errors. The more general residuals t can be used with smaller samples if several e_i are not zero. Precise recommendations are difficult to formulate given current knowledge.

EXERCISES

1.1 In the study described in Section 1.1, most subjects mentioned more than one stressful event. It is of some interest to see whether the log–linear time-trend model used with Table 1.1 can be applied to all subjects remembering more than one event. Unfortunately, new complications arise if more than one event is mentioned by a subject since these events may be related. Thus the independence assumption used in this chapter may be unjustified. A crude way to rectify the problem is to take one event at random for each subject mentioning an event and to record the corresponding number of months before interview. This process has been used to produce Table 1.14.

Compute the likelihood-ratio and Pearson chi-square statistics for the equiprobability model for Table 1.14. Compute the maximum-likelihood estimates $\hat{\beta}$ and $\hat{m}_i$ for the log–linear time-trend model

$$\log p_i = \log(m_i/N) = \alpha + \beta i$$

for Table 1.14. Find the EASD $s(\hat{\beta})$.

Based on these estimates, compute the Pearson and likelihood-ratio chi-square statistics and the adjusted residuals r_i. Considering on the chi-square statistics for the two models, is there evidence that either the equiprobability model or the log–linear time-trend model fails to fit the data? Do the adjusted residual r_i indicate any specific failings of the log–linear time-trend model?

Table 1.14

Distribution by Months Prior to Interview
of Stressful Events Reported by Subjects:
One Event Randomly Selected for Each
Subject Reporting Events from 1 to 18 Months
Prior to Interview[a]

Months i before interview	Number of Subjects
1	49
2	55
3	42
4	43
5	35
6	35
7	42
8	31
9	37
10	21
11	35
12	40
13	29
14	22
15	29
16	12
17	15
18	15
Total	587

[a] Compiled from data file used in Uhlenhuth, Balter, Lipman, and Haberman (1977) and Uhlenhuth, Lipman, Balter, and Stern (1974).

On the assumption that the log–linear time-trend model fits the data, construct an approximate 95 percent confidence interval for the common ratio $p_{i+1}/p_i = e^{\beta}$, $1 \leq i \leq 17$.

Solution

The Pearson chi-square for the equiprobability model is 75.6, and the likelihood ratio chi-square is 81.1. Since the degrees of freedom are 17, the equiprobability model is quite untenable, just as in Section 1.1. The X^2 and L^2 statistics for the log–linear time-trend model are respectively 23.8 and 23.7. Since there are 16 degrees of freedom, the significance level is about 10 percent. Thus there is weak evidence against the log–linear time-trend model.

The maximum-likelihood estimate $\hat{\beta} = -0.0612$ and $s(\hat{\beta}) = 0.0082$. Approximate 95 percent confidence intervals for β have respective lower and upper bounds

$$\hat{\beta} - 1.96s(\hat{\beta}) = -0.0773$$

and

$$\hat{\beta} + 1.96s(\hat{\beta}) = -0.0451.$$

The corresponding bounds for e^{β} are $\exp(-0.0773) = 0.926$ and $\exp(-0.0451) = 0.956$.

The estimates $\hat{m}_i$ and the adjusted residuals r_i are shown in Table 1.15. The adjusted residual r_{12} of 2.70 is rather suspicious since it suggests that respondents may have a tendency to think events occurred one year ago. The significance level of the statistic $|r_{12}|$ is about 0.01. If one has *a priori* suspicions about events reported 12 months ago, this significance level is low enough to be troublesome. If no suspicions exist, then the residual is somewhat less bothersome. The expected number of $|r_i|$ greater than 2.58 is about $18(0.01) = 0.18$ if the model holds. Thus one such residual is not unusual.

Table 1.15

Estimated Means and Adjusted Residuals for the Log–Linear Time-Trend Model for Table 1.14

Month i before interview	Expected count m_i	Adjusted residual r_i
1	52.2	−0.51
2	49.1	0.94
3	46.2	−0.67
4	43.4	−0.07
5	40.9	−0.96
6	38.4	−0.58
7	36.2	1.00
8	34.0	−0.53
9	32.0	0.91
10	30.1	−1.71
11	28.3	1.30
12	26.6	2.70
13	25.0	0.83
14	23.6	−0.34
15	22.2	1.54
16	20.8	−2.07
17	19.6	−1.12
18	18.4	−0.87

Table 1.16

*Monthly Distributions of Homicides
in the United States in 1970[a]*

Month	Number of homicides
January	1318
February	1229
March	1327
April	1257
May	1424
June	1399
July	1475
August	1559
September	1417
October	1507
November	1400
December	1534
Total	16,848

[a] From National Center for Health Statistics
(1970, pages 1–174 and 1–175).

1.2   The monthly distribution of homicides in the United States in 1970, as reported in National Center for Health Statistics (1970) is listed in Table 1.16. Are these data consistent with a log–linear model of the form

$$\log(m_i/z_i) = \alpha + \beta i,$$

where the monthly counts n_i are independent Poisson random variables and z_i is the number of days in month i? Use chi-square tests and residual analyses as a basis for your answer. Whatever your conclusions, construct a 95 percent confidence interval for β. What does this confidence interval imply about the monthly homicide rate?

Solution

The degrees of freedom are $12 - 2 = 10$, $L^2 = 18.16$, and $X^2 = 18.16$. The significance level of both chi-square statistics is about 5 percent, so evidence against the model does exist. As an aid in checking calculations, the estimates $\hat{m}_i$ are shown in Table 1.17, along with adjusted residuals r_i. Three of 12 adjusted residuals exceed 2 in absolute value, a fact which suggests deficiencies in the model. The positive values of adjusted residuals for May through August suggest an increase in homicide in summer; the negative

Table 1.17

Maximum-Likelihood Estimates and Adjusted Residuals
for Table 1.16

Month	Estimated expected mean	Adjusted residual
January	1323.2	−0.17
February	1211.9	0.55
March	1360.6	1.00
April	1335.2	−2.29
May	1399.0	0.70
June	1372.9	0.74
July	1438.6	1.00
August	1458.8	2.76
September	1431.5	−0.41
October	1500.0	0.20
November	1472.0	−2.13
December	1542.4	−0.26

signs for November, December, and January suggest a corresponding decrease in winter.

The coefficient β has maximum-likelihood estimate $\hat{\beta} = 0.01394$ and $s(\hat{\beta}) = 0.00224$. Thus an approximate 95 percent confidence interval for β has lower bound

$$0.01394 - 1.96(0.00224) = 0.00955$$

and upper bound

$$0.01394 + 1.96(0.00224) = 0.01833.$$

If the proposed log–linear model were to hold, then $100(e^{\beta} - 1)$ would be the constant monthly percentage increase

$$\frac{(m_{i+1}/z_{i+1}) - (m_i/z_i)}{m_i/z_i}$$

in the rate per day of homicides for each month i, $1 \leqslant i \leqslant 11$. Note that

$$100(e^{0.00955} - 1) = 0.96$$

and

$$100(e^{0.01833} - 1) = 1.85.$$

Thus the rate of increase is around 1 to 2 percent a month.

1.3 In the symmetry model of Section 1.2, show that

$$\hat{m}_1 = \hat{m}_4 = \tfrac{1}{2}(n_1 + n_4)$$

and

$$\hat{m}_2 = \hat{m}_3 = \tfrac{1}{2}(n_2 + n_3).$$

Also show that

$$d_1 = \tfrac{1}{2}(n_1 - n_4)$$

has variance $\tfrac{1}{2}Np_1$ and $d_2 = \tfrac{1}{2}(n_2 - n_3)$ has variance $\tfrac{1}{2}Np_2$ under the symmetry model.

Solution

Note that under the symmetry model,

$$\log m_i = \alpha + \beta_2 x_{i2}.$$

Thus

$$\hat{m}_1 = \hat{m}_4, \qquad \hat{m}_2 = \hat{m}_3,$$

$$\hat{m}_1 + \hat{m}_2 + \hat{m}_3 + \hat{m}_4 = n_1 + n_2 + n_3 + n_4,$$

and

$$\hat{m}_1 - \hat{m}_2 - \hat{m}_3 + \hat{m}_4 = n_1 - n_2 - n_3 + n_4.$$

These simultaneous equations have the unique solution

$$\hat{m}_1 = \hat{m}_4 = \tfrac{1}{2}(n_1 + n_4),$$
$$\hat{m}_2 = \hat{m}_3 = \tfrac{1}{2}(n_2 + n_3).$$

Since $p_1 = p_4$ and

$$d_1 = \sum c_i n_i,$$

where $c_1 = \tfrac{1}{2}, c_4 = -\tfrac{1}{2}$, and $c_2 = c_3 = 0$, the variance of d_1 is

$$N\{\sum c_i^2 p_i - (\sum c_i p_i)^2\} = N[\tfrac{1}{4}p_1 + \tfrac{1}{4}p_4 - (\tfrac{1}{2}p_1 - \tfrac{1}{2}p_4)^2]$$
$$= \tfrac{1}{2}Np_1.$$

A similar argument applies to d_2.

1.4 Subjects in the 1975 General Social Survey were asked to state their political views. Results are summarized in Table 1.18. Test the symmetry hypothesis that the population probabilities p_i, $1 \le i \le 7$, satisfy the constraints $p_1 = p_7, p_2 = p_6$, and $p_3 = p_5$. This model assumes that subjects are as likely to be extremely liberal as extremely conservative, as likely to

Table 1.18

Political Views of Subjects in the 1975 General Social Survey[a]

Response[b]	Code	Number responding
Extremely liberal	1	46
Liberal	2	179
Slightly liberal	3	196
Moderate, middle of the road	4	559
Slightly conservative	5	232
Conservative	6	150
Extremely conservative	7	35
Don't know	8	81
No answer	9	12
Total		1490

[a] National Opinion Research Center (1975, page 36).

[b] The question was as follows: "We hear a lot of talk these days about liberals and conservatives. I'm going to show you a seven-point scale on which the *political* views that people might hold are arranged from extremely liberal—point 1—to extremely conservative—point 7. Where would you place yourself on this scale?"

be liberal as conservative, and as likely to be slightly liberal as slightly conservative. Use adjusted residuals to check for specific failings of the hypothesis.

Solution

The maximum-likelihood equations are

$$\hat{m}_1 = \hat{m}_7 = \tfrac{1}{2}(n_1 + n_7),$$
$$\hat{m}_2 = \hat{m}_6 = \tfrac{1}{2}(n_2 + n_6),$$
$$\hat{m}_3 = \hat{m}_5 = \tfrac{1}{2}(n_3 + n_5),$$
$$\hat{m}_i = n_i, \qquad i = 4, 8, \quad \text{or} \quad 9.$$

Thus

$$X^2 = \sum (n_i - \hat{m}_i)^2/\hat{m}_i = r_1{}^2 + r_2{}^2 + r_3{}^3,$$

where

$$r_1 = -r_4 = (n_1 - n_7)/(n_1 + n_7)^{1/2},$$
$$r_2 = -r_6 = (n_2 - n_6)/(n_2 + n_6)^{1/2},$$
$$r_3 = -r_5 = (n_3 - n_5)/(n_3 + n_5)^{1/2}.$$

The likelihood-ratio chi-square

$$L^2 = 2\left[n_i \log\left(\frac{2n_1}{n_1 + n_7}\right) + n_2 \log\left(\frac{2n_2}{n_2 + n_6}\right) + n_3 \log\left(\frac{2n_3}{n_3 + n_5}\right) \right.$$
$$\left. + n_5 \log\left(\frac{2n_5}{n_3 + n_5}\right) + n_6 \log\left(\frac{2n_6}{n_2 + n_6}\right) + n_7 \log\left(\frac{2n_1}{n_1 + n_7}\right) \right].$$

There are three degrees of freedom. Note the similarities to formulas in Section 1.2 and note that n_4, n_8, and n_9 are irrelevant to the analysis.

In this example,

$$r_1 = 1.22,$$
$$r_2 = 1.60,$$
$$r_3 = -1.74,$$
$$X^2 = 7.08,$$

and

$$L^2 = 7.09.$$

The significance level is between 5 and 10 percent, so there is a modest amount of evidence that the table is not symmetrical. The residuals do not indicate any specific problems with the model. It is not possible on the basis of this table to say that the American people of the time of the survey tended to place themselves on the liberal side of the scale or on the conservative side of the scale.

1.5 Let n_i, $1 \leqslant i \leqslant s$, be independent Poisson observations with respective expected values m_i, $1 \leqslant i \leqslant s$. Let

$$\bar{n} = s^{-1}\sum n_i.$$

Show that $\bar{n}$ is the maximum-likelihood estimate of k under the model

$$m_i = k, \quad 1 \leqslant i \leqslant s,$$

and

$$X^2 = \bar{n}^{-1}\sum (n_i - \bar{n})^2$$

is the Pearson chi-square statistic for this model. Show that the degrees of freedom is $s - 1$. (The statistic X^2 is presented in Fisher, Thornton, and Mackenzie (1922). The term variance test is often used to describe the chi-square test based on X^2 since the n_i have sample variance

$$(s - 1)^{-1}\sum (n_i - \bar{n})^2.)$$

Solution

One has a special case of the model

$$\log(m_i/z_i) = \alpha$$

in which $z_i = 1$ and $\alpha = \log k$. Thus

$$\hat{m}_i = k = N \bigg/ \sum z_i = \frac{1}{s} \sum n_i = \bar{n}$$

and

$$X^2 = \sum (n_i - \hat{m}_i)^2/\hat{m}_i = \frac{1}{\bar{n}} \sum (n_i - \bar{n})^2.$$

Since there are s frequencies and one parameter α to be estimated, there are $s - 1$ degrees of freedom.

1.6 Durkheim (1951[1897], page 118) reports on a classification by Guerry of suicides in France by day of the week. The distribution used is shown in Table 1.19. Based on these data, Durkheim concludes that suicide diminishes at the end of the week, beginning on Friday. He also notes that the suicide rate is not lower on Sunday than on Saturday. Durkheim's conclusions are not based on any formal statistical analysis. To see how well his claims hold up, first use chi-square statistics to test the model that the expected number m_i of suicides on day i is the same for each day of the week. Next consider

Table 1.19
Distribution of Suicides by Day of the Week[a,b]

Day	Number of Suicides	Percentage
Monday	1001	15.20
Tuesday	1035	15.71
Wednesday	982	14.91
Thursday	1033	15.68
Friday	905	13.74
Saturday	737	11.19
Sunday	894	13.57
Total	6587	100.00

[a] Computed from Table XV of Durkheim (1951[1897], page 118).
[b] The percentage 14.91 in the table is a slight correction of the percentage 14.90 given by Durkheim. A very small arithmetic error appears to have occurred.

the models

$$m_i = k_1, \quad 1 \leqslant i \leqslant 4,$$
$$ = k_2, \quad 5 \leqslant i \leqslant 7,$$

and

$$m_i = k_1, \quad 1 \leqslant i \leqslant 4,$$
$$ = k_{i-3}, \quad 5 \leqslant i \leqslant 7.$$

Compute F statistics for these three models. Use the R^2 statistic to compare the model of constant m_i to each of the other two models. Based on the available chi-square statistics, which of these models is tenable? Using the most restrictive tenable model, estimate m_i/m_6, the ratio of the expected suicide rates for Monday and Saturday. Find an approximate 95 percent confidence interval for m_1/m_6.

Solution

As noted in Exercise 1.5, X^2 for the model of constant rates is

$$\frac{1}{\bar{n}} \sum (n_i - \bar{n})^2 = 71.95.$$

The L^2 statistic is

$$2 \sum n_i \log(n_i/\bar{n}) = 74.92.$$

The degrees of freedom is $7 - 1 = 6$, so variations in rates clearly exist for different days.

For the second model,

$$\hat{m}_i = \hat{k}_1 = \frac{1001 + 1035 + 982 + 1033}{4} = 1012.75, \quad 1 \leqslant i \leqslant 4,$$

$$\phantom{\hat{m}_i} = \hat{k}_2 = \frac{905 + 737 + 894}{3} = 845.33, \quad 5 \leqslant i \leqslant 7.$$

The chi-square statistics are $X^2 = 22.86$ and $L^2 = 23.35$. The degrees of freedom are $7 - 2 = 5$. Thus the second model is not tenable.

For the third model,

$$\hat{m}_i = \hat{k}_1 = \frac{1001 + 1035 + 982 + 1033}{4} = 1012.75, \quad 1 \leqslant i \leqslant 4,$$

$$\hat{m}_i = n_i = \hat{k}_{i-3}, \quad 5 \leqslant i \leqslant 7.$$

Thus

$$X^2 = \hat{k}_1^{-1} \sum_{i=1}^{4} (n_i - \hat{k}_1)^2 = 1.96$$

and

$$L^2 = 2 \sum_{i=1}^{4} n_i \log(n_i/\hat{k}_1) = 1.97.$$

The degrees of freedom are $7 - 4 = 3$, so the third model fits the data.

The F statistic for the three models are 12.5, 4.7, and 0.7. The R^2 statistic for the first versus second model is 0.69. The R^2 statistic for the first versus third model is 0.97. Thus much of the problem in the model of constant rate involves a constrast between the first four days and the last three days; however, some of the error results from variation in suicide rates among the last three days.

Under the third model,

$$\hat{m}_1/\hat{m}_6 = \hat{k}_1/n_6 = 1.37.$$

To find a confidence interval, consider the regression model

$$Y_i = \alpha + \varepsilon_i, \qquad\qquad 1 \leqslant i \leqslant 4,$$
$$= \alpha - \beta_{i-4} + \varepsilon_i, \qquad 5 \leqslant i \leqslant 7,$$

where the ε_i are independent $N(0, m_i^{-1})$ variables. The coefficient β_2 has a weighted-least-squares estimate

$$Y_6 - \tfrac{1}{4}(Y_1 + Y_2 + Y_3 + Y_4)$$

if $m_i = k_1$, $1 \leqslant i \leqslant 4$. Thus the variance of β_2 is

$$m_6^{-1} + \tfrac{1}{4}k_1^{-1}.$$

The third model is the log–linear model,

$$\log m_i = \alpha, \qquad\qquad 1 \leqslant i \leqslant 4,$$
$$= \alpha - \beta_{i-4}, \qquad 5 \leqslant i \leqslant 7.$$

Since

$$\log(m_1/m_6) = \log m_1 - \log m_6 = \alpha - (\alpha - \beta_2) = \beta_2,$$

the estimated asymptotic standard deviation of $\log(\hat{m}_1/\hat{m}_6) = \hat{\beta}_2$ is

$$\hat{m}_6^{-1} + \tfrac{1}{4}\hat{k}_1^{-1} = n_6^{-1} + \tfrac{1}{4}\hat{k}_1^{-1} = 0.0400.$$

Since

$$\log(\hat{m}_1/\hat{m}_6) = 0.318,$$

$\log(m_1/m_6)$ has an approximate 95 percent confidence interval with lower bound

$$0.318 - 1.96(0.04001) = 0.239$$

and upper bound

$$0.318 + 1.96(0.04001) = 0.396.$$

The corresponding bounds for m_1/m_6 are

$$e^{0.239} = 1.27$$

and

$$e^{0.396} = 1.49.$$

Note that Durkheim has captured the basic features of the table rather successfully.

2 Complete Two-Way Tables

A basic problem in the analysis of qualitative data is an examination of two polytomous variables to see if they are dependent. As shown in Section 2.6, the model of independence of two polytomous variables is equivalent to a log–linear model. Maximum-likelihood estimates for this log–linear model have explicit expressions, so that the chi-square tests and adjusted residuals described in Chapter 1 are easily applied to the independence model. As in Chapter 1, use of adjusted residuals provides a more subtle analysis of sources of departure from independence than is available from Pearson's (1904) familiar chi-square test of independence.

Quantitative assessment of the size of departures from independence is at least as important as determination that departures from independence do exist. In this book, such assessment is normally made through examination of the linear combinations of log cell means known as log cross-product ratios. Their estimation and interpretation is discussed throughout this chapter. Traditional measures of association for polytomous variables are not examined here, although a number of references are provided.

In Sections 2.1 to 2.3, we examine increasingly complex two-way tables. In Section 2.1, two dichotomous variables are cross-classified. Since chi-square tests here have only a single degree of freedom, testing independence is simplified; because only one log cross-product ratio need be considered, common measures of association are equivalent. In Section 2.2 independence is tested in a table with one dichotomous and one polytomous variable. Some simplifications in the Pearson chi-square statistic and in adjusted residuals are available; however, chi-square tests have multiple degrees of freedom.

In Section 2.3, two polytomous variables are cross-classified. Here many log cross-product ratios and many adjusted residuals must be considered. Consequently, problems of simultaneous inference arise, just as in Section 1.4.

In Section 2.4, we consider testing the homogeneity of several independent samples. The hypothesis tested is whether the distribution of the other variable depends on the sample from which it is drawn; despite this change in hypothesis, the same methods of Sections 2.1 to 2.3 apply equally to this problem.

In Section 2.5, the Rasch (1960) model for Poisson data is introduced. This log–linear model has the same form as the one for independence of polytomous variables. However, here the model is applied to Poisson counts, and since it involves explicit use of scale factors of the type described in Sections 1.3 to 1.5, maximum-likelihood estimates cannot be expressed in closed form. Consequently, iterative calculations are needed. The Newton–Raphson algorithm remains available, just as in Chapter 1. In addition, an alternate algorithm, iterative proportional fitting, is available. This new algorithm is easily implemented but does not converge as rapidly as does the Newton–Raphson algorithm and does not provide help in computation of adjusted residuals or estimated asymptotic standard deviations.

In Section 2.6, some general theory is introduced for two-way tables. Special emphasis is placed on the additive log–linear models used in the preceding five sections.

2.1 A Two-by-Two Example

Two-by-two contingency tables are cross-classifications of observations from two dichotomous variables. Their behavior can be characterized by reference to conditional and marginal probabilities or, equivalently, to conditional and marginal odds ratios and log odds ratios. This section is partly concerned with estimation of these quantities. Dependence of the two variables may be assessed by differences in conditional probabilities or differences in conditional log odds ratios. This section also considers estimates of these differences. The difference in log odds ratios leads to the log cross-product ratio. As noted in this section, the log cross-product ratio is equivalent to Yule's (1900, 1912) classical measures of association of dichotomous variables. Since the log cross-product ratio is a linear function of the logarithms of the cell means of a two-by-two contingency table and since independence of the two dichotomous variables is equivalent to a log cross-product ratio of zero, the hypothesis of independence can be shown to be equivalent to an additive log–linear model. Maximum-likelihood estimates and chi-square statistics are easily computed. Special simplifications for the two-by-two case make the Pearson chi-square unusually attractive. Continuity corrections are available to improve assessment of significance levels, and a procedure exists for computation of exact significance levels. The following example may be used to illustrate analysis of two-by-two tables.

Traits Desired in Fathers

Lynd and Lynd (1956[1929], page 524) report ratings by high school students of traits most desirable in parents. The data were obtained through questionnaires given in 1924 to white students in Middletown in sophomore, junior, and senior English classes. Students were asked to indicate which two of ten listed traits they considered most desirable in their fathers and mothers. Ratings for fathers are discussed here, while those for mothers are considered in Exercise 2.1. One trait considered for fathers was being a college graduate. Results summarized in Table 2.1 are classified by sex of respondent and by mention of this trait. It is of some interest to see if sex of respondent is related to response and to see how large such a relationship may be if it does exist.

Table 2.1

*High School Student Ratings of Traits Most Desirable
in a Father*[a]

	Sex of respondent		
Being a college graduate	Male	Female	Total
Mentioned	86	55	141
Not mentioned	283	360	643
Total	369	415	784

[a] From Lynd and Lynd (1956 [1929], page 524).

To describe the relationship between sex and response, let A_h be 1 if subject h mentions being a college graduate and let A_h be 2 otherwise. Let B_h be 1 if respondent h is male and 2 if respondent h is female. Assume that the $N = 784$ pairs (A_h, B_h), $1 \le h \le N$, are independently and identically distributed, and let $p_{ij} > 0$ be the probability that $A_h = i$ and $B_h = j$. Let n_{ij} be the number of subjects h for whom $A_h = i$ and $B_h = j$, so that $n_{11} = 86$, $n_{21} = 283$, $n_{12} = 55$, and $n_{22} = 360$. Note that n_{ij} has mean $m_{ij} = N p_{ij}$.

Marginal Probabilities

A number of probabilities and ratios of probabilities are helpful in the description of Table 4.1. Let the index i, $1 \leq i \leq 2$, correspond to a value of A_h and let j, $1 \leq j \leq 2$, correspond to a value of B_h. The marginal probability

$$p_i^A = p_{i1} + p_{i2} = \sum_j p_{ij}$$

is the probability that $A_h = i$, so that p_1^A is the probability that respondent h mentions being a college graduate. To compare the two marginal probabilities p_1^A and p_2^A, the odds ratios

$$q_{12}^A = \frac{p_1^A}{p_2^A}$$

or

$$q_{21}^A = \frac{p_2^A}{p_1^A} = \frac{1}{q_{12}^A}$$

may be used. Since

$$p_1^A + p_2^A = 1,$$

knowledge of q_{12}^A implies knowledge of p_1^A and p_2^A, for

$$p_1^A = \frac{q_{12}^A}{(q_{12}^A + 1)}$$

and

$$p_2^A = \frac{1}{(q_{12}^A + 1)}.$$

Similarly,

$$p_j^B = p_{1j} + p_{2j} = \sum_i p_{ij}$$

is the probability that respondent h has sex $B_h = j$. The odds ratio

$$q_{12}^B = \frac{p_1^B}{p_2^B}$$

or

$$q_{21}^B = \frac{p_2^B}{p_1^B} = \frac{1}{q_{12}^B}$$

compares the two marginal probabilities $p_1{}^B$ and $p_2{}^B$. The ratio q_{12}^B determines the probabilities $p_1{}^B$ and $p_2{}^B$, for

$$p_1{}^B = q_{12}^B(q_{12}^B + 1)$$

and

$$p_2{}^B = \frac{1}{(q_{12}^B + 1)}.$$

Logarithms of odds, or logits, are often used to improve statistical properties of estimates and to obtain measures not restricted to positive values. Logits will be extensively used in Chapter 5. Haldane (1955), Woolf (1955), Anscombe (1956), and Gart and Zweifel (1967) are helpful references for properties of logits required in this section. Let

$$\tau_{12}^A = \log q_{12}^A,$$
$$\tau_{21}^A = \log q_{21}^A = -\tau_{12}^A,$$
$$\tau_{12}^B = \log q_{12}^B,$$

and

$$\tau_{21}^B = \log q_{21}^B = -\tau_{12}^B.$$

Then τ_{12}^A is the log odds that $A_h = 1$ rather than 2, while τ_{12}^B is the log odds that $B_h = 1$ rather than 2.

Estimation of Marginal Parameters

To estimate marginal probabilities, marginal odds ratios, or marginal logits, n_{ij}/N may be substituted for p_{ij} in all formulas. Estimated asymptotic standard deviations may then be derived by a statistical procedure called the δ method which is described by Bishop, Fienberg, and Holland (1975, 486–502). Since the mathematical arguments used are more advanced than those used in this book, only results are presented here; they have been available for some time, as is evident from Yule (1900) and Haldane (1955). The marginal probability $p_i{}^A$ may be estimated by $\hat{p}_i{}^A = n_i{}^A/N$, where

$$n_i{}^A = \sum_j n_{ij} = n_{i1} + n_{i2}$$

is the number of respondents h such that $A_h = i$. Thus $n_i{}^A/N$ is the fraction of subjects h with $A_h = i$. In this example, $n_1{}^A/N = 141/784 = 0.180$ is the fraction of subjects mentioning being a college graduate. The standard deviation of n_i^A/N is $\sigma(\hat{p}_i^A) = [p_i^A(1 - p_i^A)/N]^{1/2}$, which in turn has an estimate

$$s(\hat{p}_i^A) = \left[\frac{(n_i^A/N)\left(\dfrac{1 - \{n_i^A/N\}}{N}\right)}{N}\right] = \left(\frac{n_1^A n_2^A}{N^3}\right)^{1/2}.$$

In this example, n_1^A/N has an estimated standard deviation 0.014. The odds ratio q_{12}^A has an estimate $\hat{q}_{12}^A = n_1^A/n_2^A$ with an estimated asymptotic standard deviation (EASD) of

$$s(\hat{q}_{12}^A) = \frac{n_1^A}{n_2^A}\left(\frac{1}{n_1^A} + \frac{1}{n_2^A}\right)^{1/2},$$

while the log odds ratio τ_{12}^A has an estimate $\tau_{12}^A = \log(n_1^A/n_2^A)$ with an EASD of

$$s(\hat{\tau}_{12}^A) = \left(\frac{1}{n_1^A} + \frac{1}{n_2^A}\right)^{1/2}.$$

In this example, q_{12}^A has the estimate

$$\hat{q}_{12}^A = \frac{141}{643} = 0.219$$

and the EASD

$$s(\hat{q}_{12}^A) = \frac{141}{643}\left(\frac{1}{141} + \frac{1}{643}\right)^{1/2} = 0.020.$$

Thus only about 0.2 students mention being a high school graduate for every student who does not mention this trait.

The log odds τ_{12}^A has an estimate

$$\hat{\tau}_{12}^A = \log 0.219 = -1.517$$

and an EASD of

$$s(\hat{\tau}_{12}^A) = \left(\frac{1}{141} + \frac{1}{643}\right)^{1/2} = 0.093.$$

The negative sign of $\hat{\tau}_{12}^A$ indicates the predominance of subjects who do not mention being a high school graduate.

In smaller samples, it may be advisable to replace n_1^A by $n_1^A + \frac{1}{2}$, n_2^A by $n_2^A + \frac{1}{2}$, and N by $N + 1$. This procedure reflects recommendations of

Haldane (1955) and Gart and Zweifel (1967). In this example, the difference is negligible; for instance, note that

$$\log\left(\frac{141.5}{643.5}\right) = -1.515$$

and

$$\left(\frac{1}{141.5} + \frac{1}{643.5}\right)^{1/2} = 0.093.$$

Similarly, the marginal probabilities $p_j{}^B$ may be estimated by $\hat{p}_j{}^B = n_j{}^B/N$, where

$$n_j{}^B = \sum_i n_{ij} = n_{1j} + n_{2j}$$

is the number of respondents h of sex $B_h = j$. In this example, $n_1{}^B/N = 369/784 = 0.471$ is the fraction of male respondents. The estimated standard deviation of $n_1{}^B/N$ is

$$s(\hat{p}_j{}^B) = \left(\frac{n_1{}^B n_2{}^B}{N^3}\right)^{1/2},$$

which in this example is

$$\left(\frac{369 \times 415}{784^3}\right)^{1/2} = 0.018.$$

The respondents are fairly evenly divided by sex. An approximate 95 percent confidence interval for $p_1{}^B$ has bounds

$$0.471 - 1.96(0.018) = 0.436$$

and

$$0.471 + 1.96(0.018) = 0.506.$$

If desired, $\hat{q}_{12}^B$ may be estimated by $\hat{q}_{12}^B = n_1{}^B/n_2{}^B$, which has an EASD of

$$s(\hat{q}_{12}^B) = \frac{n_1{}^B}{n_2{}^B}\left(\frac{1}{n_1{}^B} + \frac{1}{n_2{}^B}\right)^{1/2}.$$

The corresponding estimate for τ_{12}^B is $\hat{\tau}_{12}^B = \log(n_1{}^B/n_2{}^B)$, which has EASD

$$s(\hat{\tau}_{12}^B) = \left(\frac{1}{n_1{}^B} + \frac{1}{n_2{}^B}\right)^{1/2}.$$

In Table 4.1,

$$\hat{q}^B_{12} = \frac{369}{415} = 0.889,$$

$$s(\hat{q}^B_{12}) = \frac{369}{415}\left(\frac{1}{369} + \frac{1}{415}\right)^{1/2} = 0.064,$$

$$\hat{\tau}^B_{12} = \log 0.889 = -0.117,$$

$$s(\hat{\tau}^B_{12}) = \left(\frac{1}{369} + \frac{1}{415}\right)^{1/2} = 0.072.$$

The relatively even division of respondents by sex is reflected in the estimate of q^B_{12} near 1 and the estimate of τ^B_{12} near 0.

Conditional Probabilities

Since the effect of sex on response is of interest, conditional probabilities must be considered. The conditional probability of response $A_h = i$ given sex $B_h = j$ is

$$p^{A\cdot B}_{i\cdot j} = \frac{p_{ij}}{p^B_j}$$

The corresponding odds ratio comparing $p^{A\cdot B}_{1\cdot j}$ to $p^{A\cdot B}_{2\cdot j}$ is

$$q^{A\cdot B}_{12\cdot j} = \frac{p^{AB}_{1\cdot j}}{p^{AB}_{2\cdot j}} = \frac{p_{1j}/p^B_j}{p_{2j}/p^B_j} = \frac{p_{1j}}{p_{2j}} = \frac{Np_{1j}}{Np_{2j}} = \frac{m_{1j}}{m_{2j}},$$

where for $1 \leqslant i \leqslant 2$, $1 \leqslant j \leqslant 2$, $m_{ij} = Np_{ij}$ is the expected value of n_{ij}. This odds ratio determines $p^{A\cdot B}_{1\cdot j}$ and $p^{A\cdot B}_{2\cdot j}$, for

$$p^{A\cdot B}_{1\cdot j} = \frac{q^{A\cdot B}_{12\cdot j}}{(q^{A\cdot B}_{12\cdot j} + 1)}$$

and

$$p^{A\cdot B}_{2\cdot j} = \frac{1}{q^{A\cdot B}_{12\cdot j} + 1}.$$

The logarithm of this odds ratio is

$$\tau^{A\cdot B}_{12\cdot j} = \log q^{A\cdot B}_{12\cdot j} = \log m_{1j} - \log m_{2j}.$$

Similarly,

$$q^{A\cdot B}_{21\cdot j} = \frac{p^{A\cdot B}_{2\cdot j}}{p_{1\cdot j}} = \frac{m_{2j}}{m_{1j}} = \frac{1}{q^{A\cdot B}_{12\cdot j}}$$

and

$$\tau_{21\cdot j}^{A\cdot B} = \log q_{21\cdot j}^{A\cdot B} = \log m_{2j} - \log m_{1j} = -\tau_{12\cdot j}^{A\cdot B}.$$

One may estimate $p_{i\cdot j}^{A\cdot B}$ by

$$\hat{p}_{i\cdot j}^{A\cdot B} = \frac{n_{ij}}{n_j^B},$$

which has an EASD

$$s(\hat{p}_{i\cdot j}^{A\cdot B}) = \left[\frac{1}{n_j^B}\left(\frac{n_{ij}}{n_j^B}\right)\left(1 - \frac{n_{ij}}{n_j^B}\right)\right]^{1/2} = \left(\frac{n_{1j}n_{2j}}{(n_j^B)^3}\right)^{1/2}.$$

In Table 4.1, the estimated conditional probability of a male student mentioning being a college graduate is

$$\hat{p}_{1\cdot 1}^{A\cdot B} = \frac{86}{369} = 0.233.$$

The corresponding EASD is

$$s(\hat{p}_{1\cdot 1}^{A\cdot B}) = \left(\frac{86 \times 283}{369^3}\right)^{1/2} = 0.022.$$

The equivalent conditional probability estimate for females is

$$\hat{p}_{1\cdot 2}^{A\cdot B} = \frac{55}{415} = 0.133$$

and the EASD $s(\hat{p}_{1\cdot 2}^{A\cdot B})$ is

$$\left(\frac{55 \times 360}{415^3}\right)^{1/2} = 0.017.$$

Thus male students appear somewhat more likely than female students to mention the desirability of a father being a college graduate.

To estimate $q_{12\cdot j}^{A\cdot B}$, let

$$\hat{q}_{12\cdot j}^{A\cdot B} = \frac{n_{ij}}{n_{2j}},$$

so that

$$s(\hat{q}_{12\cdot j}^{A\cdot B}) = \left(\frac{n_{1j}}{n_{2j}}\right)\left(\frac{1}{n_{1j}} + \frac{1}{n_{2j}}\right)^{1/2}.$$

Similarly,

$$\hat{\tau}_{12\cdot j}^{A\cdot B} = \log\left(\frac{n_{1j}}{n_{2j}}\right)$$

and

$$s(\hat{\tau}_{12 \cdot j}^{A \cdot B}) = \left(\frac{1}{n_{1j}} + \frac{1}{n_{2j}}\right)^{1/2}.$$

For example,

$$\hat{q}_{12 \cdot 1}^{A \cdot B} = \frac{86}{283} = 0.304,$$

$$s(\hat{q}_{12 \cdot 1}^{A \cdot B}) = \frac{86}{283}\left(\frac{1}{86} + \frac{1}{283}\right)^{1/2} = 0.037,$$

$$\hat{\tau}_{12 \cdot 1}^{A \cdot B} = \log 0.304 = -1.191,$$

and

$$s(\hat{\tau}_{12 \cdot 1}^{A \cdot B}) = \left(\frac{1}{86} + \frac{1}{283}\right)^{1/2} = 0.123.$$

Primary interest in this problem is on influence of sex on response; however, for the sake of completeness, it should be noted that

$$p_{j \cdot i}^{B \cdot A} = \frac{p_{ij}}{p_i^A}$$

is the conditional probability that a respondent is of sex $B_h = j$ given a response $A_h = i$. Odds ratios and log odds are defined by the equations

$$q_{12 \cdot j}^{B \cdot A} = \frac{1}{q_{21 \cdot j}^{B \cdot A}} = \frac{p_{1 \cdot i}^{B \cdot A}}{p_{2 \cdot i}^{B \cdot A}} = \frac{m_{i1}}{m_{i2}}$$

and

$$\tau_{12 \cdot i}^{B \cdot A} = -\tau_{21 \cdot i}^{B \cdot A} = \log q_{12 \cdot i}^{B \cdot A} = \log m_{i1} - \log m_{i2}.$$

The estimates of these quantities and the corresponding estimates of asymptotic standard deviations are

$$\hat{p}_{j \cdot i}^{B \cdot A} = \frac{n_{ij}}{n_i^A},$$

$$s(\hat{p}_{j \cdot i}^{B \cdot A}) = \left[\frac{n_{i1} n_{i2}}{(n_i^A)^3}\right]^{1/2},$$

$$\hat{q}_{12 \cdot i}^{B \cdot A} = \frac{n_{i1}}{n_{i2}},$$

$$s(\hat{q}_{12 \cdot i}^{B \cdot A}) = \frac{n_{i1}}{n_{i2}}\left(\frac{1}{n_{i1}} + \frac{1}{n_{i2}}\right)^{1/2},$$

$$\hat{\tau}_{12 \cdot i}^{B \cdot A} = \log\left(\frac{n_{i1}}{n_{i2}}\right),$$

and

$$s(\hat{\tau}^{B \cdot A}_{12 \cdot i}) = \left(\frac{1}{n_{i1}} + \frac{1}{n_{i2}} \right)^{1/2}.$$

Independence and Additivity

The conditional probability estimates $\hat{p}^{A \cdot B}_{1 \cdot 1}$ and $\hat{p}^{A \cdot B}_{1 \cdot 2}$ are somewhat different, thus raising the possibility that respondent's sex and response are not independent. Various ways exist to investigate the question. Here the model of independence will be shown to be a log–linear model. Chi-square statistics will be constructed, and deviations from independence will be measured.

First note that if sex of respondent and response are independent, then

$$p_{ij} = p_i^A p_j^B, \qquad 1 \leqslant i \leqslant 2, \quad 1 \leqslant j \leqslant 2.$$

Equivalently, the conditional probability $p^{A \cdot B}_{i \cdot 1}$ of response $A_h = i$ given that respondent h is male ($B_h = 1$) is the same as the conditional probability $p^{A \cdot B}_{i \cdot 2}$ of response $A_h = i$ given that respondent h is female ($B_h = 2$). Thus independence of sex and response holds if and only if

$$\tau^{A \cdot B}_{12 \cdot 1} = \tau^{A \cdot B}_{12 \cdot 2} = \tau^A_{12}.$$

Let the log cross-product ratio $\tau^{AB}_{(12)(12)}$ be defined as the difference

$$\begin{aligned}
\tau^{A \cdot B}_{(12) \cdot 1} - \tau^{A \cdot B}_{(12) \cdot 2} &= \tau^{B \cdot A}_{(12) \cdot 1} - \tau^{B \cdot A}_{(12) \cdot 2} \\
&= \log m_{11} - \log m_{21} - \log m_{12} + \log m_{22}.
\end{aligned}$$

Then the independence model is equivalent to the log–linear model in which the $\log m_{ij}$ are restrained by the linear equation

$$\tau^{AB}_{(12)(12)} = \log m_{11} - \log m_{21} - \log m_{12} + \log m_{22} = 0.$$

The restraint $\tau^{AB}_{(12)(12)} = 0$ is equivalent to the additivity condition that for some unique parameters λ, λ_1^A, λ_2^A, λ_1^B, and λ_2^B, the log means $\log m_{ij}$ satisfy the equation

$$\log m_{ij} = \lambda + \lambda_i^A + \lambda_j^B$$

and the parameters λ_1^A, λ_2^A, λ_1^B, and λ_2^B satisfy the equations

$$\sum \lambda_i^A = \lambda_1^A + \lambda_2^A = 0$$

and

$$\sum \lambda_j^B = \lambda_1^B + \lambda_2^B = 0.$$

An analogous model appears in the analysis of variance for a two-by-two factorial design. Here for some unknown parameters λ, λ_i^A, $1 \leqslant i \leqslant 2$, and λ_j^B, $1 \leqslant j \leqslant 2$,

$$\sum \lambda_i^A = \sum \lambda_j^B = 0$$

and the observations Y_{ij}, $1 \leqslant i \leqslant 2$, $1 \leqslant j \leqslant 2$, have expected values $\lambda + \lambda_i{}^A + \lambda_j{}^B$.

To verify that the condition $\tau_{(12)(12)}^{AB} = 0$ corresponds to an additive log–linear model, note that for some unique λ, $\lambda_i{}^A$, $1 \leqslant i \leqslant 2$, $\lambda_j{}^B$, $1 \leqslant j \leqslant 2$, and λ_{ij}^{AB}, $1 \leqslant i \leqslant 2$, $1 \leqslant j \leqslant 2$,

$$\log m_{ij} = \lambda + \lambda_i{}^A + \lambda_j{}^B + \lambda_{ij}^{AB},$$

$$\sum \lambda_i{}^A = \lambda_1{}^A + \lambda_2{}^A = 0,$$

$$\sum \lambda_j{}^B = \lambda_1{}^B + \lambda_2{}^B = 0,$$

$$\sum_j \lambda_{ij}^{AB} = \lambda_{i1}^{AB} + \lambda_{i1}^{AB} = 0, \; 1 \leqslant i \leqslant 2,$$

and

$$\sum_i \lambda_{ij}^{AB} = \lambda_{1j}^{AB} + \lambda_{2j}^{AB} = 0, \; 1 \leqslant j \leqslant 2.$$

As is well known,

$$\lambda = \tfrac{1}{4}(\log m_{11} + \log m_{21} + \log m_{12} + \log m_{22}),$$

$$\lambda_1{}^A = -\lambda_2{}^A = \tfrac{1}{4}(\log m_{11} - \log m_{21} + \log m_{12} - \log m_{22}),$$

$$\lambda_1{}^B = -\lambda_2{}^B = \tfrac{1}{4}(\log m_{11} + \log m_{21} - \log m_{12} - \log m_{22}),$$

and

$$\lambda_{11}^{AB} = -\lambda_{21}^{AB} = -\lambda_{12}^{AB} = \lambda_{22}^{AB} = \tfrac{1}{4}(\log m_{11} - \log m_{21} - \log m_{12} + \log m_{22})$$
$$= \tfrac{1}{4}\tau_{(12)(12)}^{AB}.$$

Thus $\tau_{(12)(12)}^{AB} = 0$ if and only if the additive model

$$\log m_{ij} = \lambda + \lambda_i{}^A + \lambda_j{}^B$$

is satisfied for some λ, $\lambda_i{}^A$, and $\lambda_j{}^B$ such that

$$\sum \lambda_i{}^A = \sum \lambda_j{}^B = 0.$$

For further details, the reader may wish to refer to Bishop and Fienberg (1969), Bishop, Fienberg, and Holland (1975, pages 11–31), and Goodman (1970).

The $\lambda_i{}^A$ parameters are called main effects of variable A (response of subject). For $i \neq i'$, the difference $\lambda_i{}^A - \lambda_{i'}{}^A = 2\lambda_i{}^A$ is

$$\tfrac{1}{2}(\tau_{ii'\cdot 1}^{A \cdot B} + \tau_{ii'\cdot 2}^{A \cdot B}),$$

the average logarithm $\tau_{ii'\cdot j}^{A \cdot B}$ of the conditional odds ratio $q_{ii'\cdot j}^{A \cdot B}$. Under the the additive (or independence) model,

$$\lambda_i{}^A - \lambda_{i'}{}^A = \tau_{ii'}^A.$$

Thus the $\lambda_i{}^A$ provide a measure of the relative prevalence of the two responses.

Similarly, the λ_j^B are main effects of variable B (sex of respondent). For $j \neq j'$, $\lambda_j^B - \lambda_{j'}^B = 2\lambda_j^B$ is

$$\tfrac{1}{2}(\tau_{jj' \cdot 1}^{B \cdot A} + \tau_{jj' \cdot 2}^{B \cdot A}),$$

the average of the conditional log odds $\tau_{jj' \cdot i}^{B \cdot A}$, $1 \leq i \leq 2$. Under independence,

$$\lambda_j^B - \lambda_{j'}^B = \tau_{jj'}^B.$$

The λ_{ij}^{AB} are the interactions between variables A and B. For $i \neq i'$, $j \neq j'$,

$$\lambda_{ij}^{AB} - \lambda_{i'j}^{AB} - \lambda_{ij'}^{AB} + \lambda_{i'j'}^{AB} = 4\lambda_{ij}^{AB}$$

is

$$\tau_{(ii')(jj')}^{AB} = \log m_{ij} - \log m_{i'j} - \log m_{ij'} + \log m_{i'j'}.$$

Under independence, the λ_{ij}^{AB} are zero.

Maximum-Likelihood Estimation under Independence

Under the independence model, it is well known that the marginal probability p_i^A has maximum-likelihood estimate $\hat{p}_{iA} = n_i^A/N$, so that $\hat{p}_i^A$ is the observed fraction of subjects with response i. Similarly,

$$\hat{p}_j^B = n_j^B/N$$

is the observed fraction of subjects with sex j. The maximum-likelihood estimate $\hat{p}_{ij}$ of p_{ij} is then $\hat{p}_i^A \hat{p}_j^B = (n_i^A/N)(n_j^B/N)$, and the maximum-likelihood estimate $\hat{m}_{ij}$ of m_{ij} is

$$\hat{m}_{ij} = N\hat{p}_{ij} = \frac{n_i^A n_j^B}{N}.$$

These observations go back to Neyman and Pearson (1928) within the context of maximum-likelihood estimation. The estimates $\hat{p}_i^A$, $\hat{p}_j^B$, $\hat{p}_{ij}$, and $\hat{m}_{ij}$ were used earlier without regard to their being maximum-likelihood estimates. For verification of those equations for maximum-likelihood estimates, see Section 2.6.

Tests of Independence

To test the independence model, the Pearson (1904) chi-square statistic

$$X^2 = \sum_i \sum_j \left(n_{ij} - \frac{n_i^A n_j^B}{N} \right)^2 \bigg/ \left(\frac{n_i^A n_j^B}{N} \right)$$

$$= \frac{N(n_{11}n_{22} - n_{12}n_{21})^2}{n_1^A n_2^A n_1^B n_2^B}$$

or the likelihood-ratio chi-square statistic

$$L^2 = 2 \sum_i \sum_j n_{ij} \log\left(\frac{n_{ij}N}{n_i{}^A n_j{}^B}\right)$$

$$= 2\left[n_{11} \log\left(\frac{n_{11}N}{n_1{}^A n_1{}^B}\right) + n_{12} \log\left(\frac{n_{12}N}{n_1{}^A n_2{}^B}\right) \right.$$

$$\left. + n_{21} \log\left(\frac{n_{21}N}{n_2{}^A n_1{}^B}\right) + n_{22} \log\left(\frac{n_{22}N}{n_2{}^A n_2{}^N}\right) \right]$$

of Wilks (1935) may be used, just as in Chapter 1.

Under independence, both statistics have approximate chi-square distributions with one degree of freedom, as first noted by Fisher (1922) in the case of X^2; however, ease of computation so much favors use of X^2 that the likelihood-ratio chi-square is not often used in this case.

In Table 2.1,

$$X^2 = \frac{784(86 \times 360 - 55 \times 283)^2}{141 \times 643 \times 369 \times 415} = 13.4.$$

Thus there is very strong evidence that sex of respondent and response are related. The approximate significance level is less than 0.0003.

The Pearson chi-square statistic may be used to indicate the general direction of the observed relationship. The adjusted residual

$$r_{11} = n_{11} - \left(\frac{n_1{}^A n_1{}^B}{N}\right) \bigg/ \left(\frac{n_1{}^A n_2{}^A n_1{}^N n_2{}^B}{N}\right)^{1/2}$$

to be described in Section 2.6 satisfies the equation

$$X^2 = r_{11}^2.$$

As in Chapter 1, the adjusted residual is approximately distributed as a standard normal deviate under independence. In this example, $r_{11} = 3.66$. The positive sign indicates that the desirability of more male respondents desire fathers who are college graduates than would be the case if sex were unrelated to response.

Measuring Association in a 2 × 2 Table

The chi-square statistic X^2 has shown that sex and response are related. To assess the size of the relationship, the cross-product ratio

$$q_{(12)(12)}^{AB} = \frac{m_{11}m_{22}}{m_{21}m_{12}} = \frac{p_{11}p_{22}}{p_{21}p_{12}} = \frac{q_{(12)\cdot 1}^{A\cdot B}}{q_{(12)\cdot 2}^{A\cdot B}}$$

of Yule (1900) or its logarithm

$$\tau^{AB}_{(12)(12)} = \log m_{11} - \log m_{21} - \log m_{12} + \log m_{22}$$
$$= \tau^{A \cdot B}_{(12) \cdot 1} - \tau^{A \cdot B}_{(12) \cdot 2}$$

may be used.

The logarithm will be emphasized here since it has an unrestricted range and possesses the simple property that

$$\tau^{AB}_{(12)(12)} = -\tau^{AB}_{(12)(21)} = -\tau^{AB}_{(21)(12)} = \tau^{AB}_{(21)(21)}.$$

Thus relabeling of categories only affects the sign of the log cross-product ratio.

Note that under independence, $q^{AB}_{(12)(12)} = 1$ and $\tau^{AB}_{(12)(12)} = 0$. If males are more likely than females to mention being a college graduate, then $q^{AB}_{(12)(12)} > 1$ and $\tau^{AB}_{(12)(12)} > 0$. Similarly, if males are less likely than females to mention this trait, then $q^{AB}_{(12)(12)} < 1$ and $\tau^{AB}_{(12)(12)} < 0$. Also note that $(q^{AB}_{(12)(12)} - 1)/(q^{AB}_{(12)(12)} + 1)$ is the coefficient of association Q of Yule (1900) and $[(q^{AB}_{(12)(12)})^{1/2} - 1]/[(q^{AB}_{(12)(12)})^{1/2} + 1]$ is the coefficient of colligation Y of Yule (1912). Thus several common measures of association are simple functions of $\tau^{AB}_{(12)(12)}$. This book will emphasize $q^{AB}_{(12)(12)}$ and $\tau^{AB}_{(12)(12)}$ since interpretation of these measures is straightforward.

To estimate $\tau^{AB}_{(12)(12)}$, one may use the estimate

$$\hat{\tau}^{AB}_{(12)(12)} = \log n_{11} - \log n_{12} - \log n_{21} + \log n_{22}.$$

As noted in Woolf (1955), the asymptotic variance

$$\sigma^2(\hat{\tau}^{AB}_{(12)(12)}) = \frac{1}{m_{11}} + \frac{1}{m_{12}} + \frac{1}{m_{21}} + \frac{1}{m_{22}}$$

may be estimated by

$$s^2(\hat{\tau}^{AB}_{(12)(12)}) = \frac{1}{n_{11}} + \frac{1}{n_{12}} + \frac{1}{n_{21}} + \frac{1}{n_{22}}.$$

Haldane (1955) and Gart and Zweifel (1967) provide results that suggest addition of $\frac{1}{2}$ to each cell frequency n_{ij} prior to computation of $\hat{\tau}^{AB}_{(12)(12)}$ and $s^2(\hat{\tau}^{AB}_{(12)(12)})$.

In the example under study, $\tau^{AB}_{(12)(12)}$ is estimated to be

$$\hat{\tau}^{AB}_{(12)(12)} = \log 86.5 - \log 55.5 - \log 283.5 + \log 360.5$$

$$= \log \frac{(86.5)(360.5)}{(55.5)(283.5)} = 0.684$$

and

$$s(\hat{\tau}^{AB}_{(12)(12)}) = \left(\frac{1}{86.5} + \frac{1}{55.5} + \frac{1}{283.5} + \frac{1}{360.5} \right) = 0.189.$$

An approximate 95 percent confidence interval for $\tau^{AB}_{(12)(12)}$ has lower bound

$$0.684 - 1.96(0.189) = 0.313$$

and upper bound

$$0.684 + 1.96(0.189) = 1.06.$$

The corresponding interval for $q^{AB}_{(12)(12)}$ has bounds

$$e^{0.313} = 1.37$$

and

$$e^{1.06} = 2.87.$$

Thus boys are somewhat more likely than girls to mention being a college graduate as a desirable trait in a father. The size of the difference is not accurately determined.

Comparison of Conditional Probabilities

Association can also be explored by estimation of differences $p^{A \cdot B}_{1 \cdot 1} - p^{A \cdot B}_{1 \cdot 2}$ in the conditional probabilities of boys and girls desiring fathers who are college graduates. The estimate $\hat{p}^{A \cdot B}_{1 \cdot 1} - \hat{p}^{A \cdot B}_{1 \cdot 2}$ has EASD

$$s(\hat{p}^{A \cdot B}_{1 \cdot 1} - \hat{p}^{A \cdot B}_{1 \cdot 2}) = \left[s^2(\hat{p}^{A \cdot B}_{1 \cdot 1}) + s^2(\hat{p}^{A \cdot B}_{1 \cdot 2}) \right]^{1/2}$$

$$= \left[\frac{n_{11} n_{21}}{(n_1{}^B)^3} + \frac{n_{12} n_{22}}{(n_2{}^B)^3} \right]^{1/2}$$

In this case, $\hat{p}^{A \cdot B}_{1 \cdot 1} - \hat{p}^{A \cdot B}_{1 \cdot 2} = 0.101$ and $s(\hat{p}^{A \cdot B}_{1 \cdot 1} - \hat{p}^{A \cdot B}_{1 \cdot 2}) = 0.028$. An approximate 95 percent confidence interval for $\hat{p}^{A \cdot B}_{1 \cdot 1} - \hat{p}^{A \cdot B}_{1 \cdot 2}$ has bounds

$$0.101 - 1.96(0.028) = 0.046$$

and

$$0.101 + 1.96(0.028) = 0.155,$$

so that boys are estimated to be from 4.6 to 15.5 percent more likely than girls to mention being a college graduate as a desirable trait in a father.

Differences in conditional probabilities do not provide the same information as does the cross-product ratio since $p_{1 \cdot 1}^{A \cdot B} - p_{1 \cdot 2}^{A \cdot B}$ does not take account of the sizes of the probabilities being compared. The difference of 0.101 can be achieved if $p_{1 \cdot 1}^{A \cdot B} = 0.233$ and $p_{1 \cdot 2}^{A \cdot B} = 0.133$, in which case $\tau_{(12)(12)}^{AB} = 0.684$, or if $p_{1 \cdot 1}^{A \cdot B} = 0.102$ and $p_{1 \cdot 2}^{A \cdot B} = 0.001$, in which case $\tau_{(12)(12)}^{AB} = 4.732$. The very large value of $\tau_{(12)(12)}^{AB}$ in the latter case reflects the fact that mentioning being a college graduate is almost entirely confined to male respondents.

Since Yule (1900, 1912), the failure of $p_{1 \cdot 1}^{A \cdot B} - p_{1 \cdot 2}^{A \cdot B}$ to reflect the sizes of the probabilities being compared has led to use of measures equivalent to the log cross-product ratio $\tau_{(12)(12)}^{AB}$. Nonetheless, it should be noted that differences in probabilities may still prove informative.

The Continuity Correction

If the significance level of X^2 is found by use of the chi-square distribution, then the reported significance level tends to be smaller than the significance level for X^2 obtained by the Fisher (1970[1925], pages 96–97) exact test of independence. Since the significance level of the Fisher test is relatively difficult to find when suitable tables are unavailable, the continuity correction of Yates (1934) has been proposed to improve the chi-square approximation to the significance level of X^2. Both the Fisher test and the Yates continuity correction are described in this subsection.

In the Fisher test, inferences are conditional on the observed row marginal totals $n_1{}^A$ and $n_2{}^A$ and on the observed column marginal totals $n_1{}^B$ and $n_2{}^B$. The conditional significance level is the sum of

$$p(N, n_1{}^A, n_1{}^B, x) = \frac{n_1{}^B!(N - n_1{}^B)!n_1{}^A!(N - n_1{}^A)!}{N!x!(n_1{}^B - x)!(n_1{}^A - x)!(N - n_1{}^A - n_1{}^B + x)!}$$

over all integers $x \geqslant 0$ such that

$$\left| x - \frac{n_1{}^A n_1{}^B}{N} \right| \geqslant \left| n_{11} - \frac{n_1{}^A n_1{}^B}{N} \right|,$$

$$x \leqslant n_1{}^A,$$

$$x \leqslant n_1{}^B,$$

and

$$x \geqslant n_1{}^A + n_1{}^B - N.$$

The significance level is independent of the probabilities $p_{ij} = p_i{}^A p_j{}^B$ if the independence model holds.

In practice, the exact test is not often used. Tables of $p(N, n_1{}^A, n_1{}^B, x)$ are required such as those of Lieberman and Owen (1960). Their tables, which cover 694 pages, are primarily devoted to the case $N \leqslant 100$. Computations without tables are tedious if N is large. However, a modification of X^2 due to Yates (1934) may be used to improve the correspondence in significance levels. Yates recommends that significance levels be found by use of

$$X_c^2 = \frac{N(|n_{11}n_{22} - n_{12}n_{21}| - \frac{1}{2}N)^2}{n_1{}^A n_2{}^A n_1{}^N n_2{}^B}.$$

Some theoretical basis for this correction can be obtained from Nicholson (1956).

In Table 2.1,

$$X_c^2 = \frac{784(|186 \times 360 - 55 \times 283| - \frac{1}{2} \times 784)^2}{141 \times 643 \times 369 \times 415} = 12.7.$$

The decrease from an X^2 of 13.4 makes little practical difference. Were smaller samples involved, the difference would be larger.

There is some controversy concerning the advisability of using X_c^2 rather than X^2. Cochran (1954) recommends use of X_c^2 rather than X^2. He recommends use of the exact test when some estimated expected value $n_i{}^A n_j{}^B/N$ is less than 5, provided suitable tables are available. For dissenting opinions, see Pearson (1947), Plackett (1964), and Grizzle (1967).

Two-by-Two Tables and Larger Tables

Two-by-two contingency tables are much simpler than larger two-way tables. Since $X^2 = r_{11}^2 = r_{12}^2 = r_{21}^2 = r_{22}^2$, residual analysis is trivial. This situation only arises in two-by-two tables. Unlike more complex tables, the exact significance level of X^2 can be computed in a feasible manner and a continuity correction can be used to improve the usual chi-square approximation for the significance level.

Many common measures of association for two-by-two tables are based on the log cross-product ratio $\tau_{(12)(12)}^{AB}$. In larger tables, no single parameter provides a comparable basis for measures of association.

In two-by-two tables, only two conditional probabilities (or conditional logits) need be compared, so that simultaneous inference problems are not normally important. In larger tables, multiple comparisons arise which do require simultaneous inferences.

Despite the unusual simplicity of two-by-two tables, they do illustrate many of the general procedures used in two-way tables. In the rest of this chapter, marginal probabilities, conditional probabilities, logits, chi-square statistics, additive log–linear models, adjusted residuals, and log cross-product ratios will continue to be important.

2.2   A 2 × s Table: Ethnicity and Women's Role Attitudes

Contingency tables formed by cross-classifying observations from a dichotomous and a polytomous variable retain a few of the simplifying features of two-by-two tables. There are some simplifications available in computation of the Pearson chi-square, and some adjusted residuals are redundant. Thus some special treatment of 2 × s tables is useful. To illustrate analysis of such tables, data are used from Dowdall (1974). She uses surveys conducted in 1968 and 1969 in Rhode Island to assess the effect of ethnic background on role attitudes of women between the ages of 15 and 64. As part of the investigation, responses were tabulated for the question, "Do you think it is all right for a married woman to have a job instead of only taking care of the house and the children while her husband provides for the family." Results are tabulated in Table 2.2 in terms of reported national origin of father for those white native-born women who responded to the question. The surveys were conducted by the Population Research Laboratory of Brown University. Dowdall (1974) discusses the extent to which this geographically limited sample is representative of the United States as a whole. She also notes that the survey preceded the increased public attention recently given to women's roles.

To determine whether ethnic origin and response are related, consider a probability model in which (A_h, B_h), $1 \leqslant h \leqslant N = 599$, are independent and identically distributed pairs of random variables. Let $A_h = 1$ if subject h has a favorable attitude and let $A_h = 2$ if subject h has an unfavorable attitude. Let B_h correspond to the ethnic origin of subject h. Given the arrangement of the table, it is reasonable to let $B_h = 1$ correspond to Italian, $B_h = 2$ correspond

Table 2.2

*Native-Born White Rhode Island Women Cross-Classified by Ethnic Origin
and Attitude toward Married Women Having Jobs[a]*

	Ethnic origin								
Attitude	Ital.	North. Eur.[b]	Other Eur.[c]	Engl.	Irish	French-Can.	French	Port.	Total
Favorable	78	56	43	53	43	36	42	29	380
Unfavorable	47	29	29	32	30	22	23	7	219
Total	125	85	72	85	73	58	65	36	599

[a] Computed from numbers in Dowdall (1974).
[b] Northern European excludes English, French, and Irish.
[c] Other European consists of Polish, Russian, and mixed background.

to northern European, etc. Let i, $1 \leqslant i \leqslant 2$, correspond to one of the two possible responses, and let j, $1 \leqslant j \leqslant 8$, correspond to one of the eight possible ethnic origins. Let $p_{ij} > 0$ be the probability that $A_h = i$ and $B_h = j$, and let n_{ij} be the number of subjects h such that $A_h = i$ and $B_h = j$. Note that ethnic origin and response are unrelated if

$$p_{ij} = p_i^A p_j^B, \qquad 1 \leqslant i \leqslant 2, \quad 1 \leqslant j \leqslant s = 8.$$

Maximum-Likelihood Estimates

As in Section 2.1, the independence model corresponds to a log–linear model of the form

$$\log m_{ij} = \lambda + \lambda_i^A + \lambda_j^B,$$

where

$$\sum \lambda_i^A = \sum \lambda_j^B = 0.$$

This model will be examined in detail in Section 2.6. As is well known, the maximum-likelihood estimate $\hat{m}_{ij}$ of m_{ij} satisfies the equation

$$\hat{m}_{ij} = n_i^A n_j^B / N.$$

For example,

$$\hat{m}_{11} = 380 \times 125/599 = 79.3$$

and

$$\hat{m}_{12} = 380 \times 85/599 = 53.9.$$

For other estimates, see Table 2.3.

Chi-Square Tests

As noted by Pearson (1911),

$$X^2 = \sum_i \sum_j (n_{ij} - n_i^A n_j^B / N)^2 / (n_i^A n_j^B / N)$$

$$= \sum_j \left(\frac{n_1^A n_2^A}{n_j^B} \right) \left(\frac{n_{1j}}{n_j^B} - \frac{n_{2j}}{n_j^B} \right)^2.$$

The likelihood-ratio chi-square of Wilks (1935) is

$$L^2 = 2 \sum_i \sum_j n_{ij} \log \left(\frac{n_{ij} N}{n_i^A n_j^B} \right).$$

The degrees of freedom are $s - 1 = 7$. This formula will be derived in Section 2.6. In this example, $X^2 = 6.03$ and $L^2 = 6.48$. Since X^2 and L^2 have approximate chi-square distributions with 7 degrees of freedom if the independence model holds, there is little evidence of ethnic differences in responses.

Adjusted Residuals

A weak hint exists that responses of women of Portuguese origin may differ from those of other women surveyed. The adjusted residuals

$$r_{ij} = \frac{n_{ij} - (n_i{}^A n_j{}^B/N)}{\{n_i{}^A n_j{}^B [1 - (n_i{}^A/N)][1 - (n_j{}^B/N)]/N\}^{1/2}}$$

have approximate standard normal distributions under independence. They are shown in Table 2.3. Since $r_{1j} + r_{2j} = 0$, it suffices to confine attention to r_{1j}, $1 \leqslant j \leqslant s = 8$. The largest adjusted residual in magnitude is

$$r_{18} = \frac{29 - (380 \times 36)/599}{[380 \times 36 \times (1 - 380/599) \times (1 - 36/599)/599]^{1/2}} = 2.20.$$

If the model of independence holds, than r_{18} has an approximate $N(0, 1)$ distribution. The probability that $|r_{18}| \geqslant 2.20$ is then 0.014. By itself, this result is fairly strong; however, no *a priori* reason exists to examine r_{18}. Under the independence hypothesis, $8(0.014) = 0.11$ is the expected number of ethnic origins j for which $|r_{ij}| \geqslant 2.20$. Thus the observed adjusted residual for women of Portuguese origin is not that surprising.

One may conclude that the data do not provide much evidence for a relationship between response and ethnic origin. Whether a larger sample might provide such evidence is unknown. It should be noted that Dowdall

Table 2.3

Observed Cell Counts, Estimated Expected Cell Counts, and Adjusted Residuals for the Independence Model for Table 2.2[a]

Attitude	Ital.	North. Eur.	Other Eur.	Engl.	Irish	French-Can.	French	Port.
Favorable	78	56	43	53	43	36	42	29
	79.3	53.9	45.7	53.9	46.3	36.8	41.2	22.8
	−0.27	0.50	−0.70	−0.22	−0.86	−0.23	0.21	2.20
Unfavorable	47	29	29	32	30	22	23	7
	45.7	31.1	26.3	31.1	26.7	21.2	23.8	13.2
	0.27	−0.50	0.70	0.22	0.86	0.23	−0.21	−2.20

Ethnic origin (column group header spanning Ital. through Port.)

[a] The first line is the observed count n_{ij}, the second line is the estimated expected count $\hat{m}_{ij}$, and the third line is the adjusted residual r_{ij}.

(1974) does report significant effects of ethnicity in connection with other questions and in connection with scales based on all questions asked.

The $2 \times s$ table is an intermediate case. It involves more complication than the two-by-two table, primarily because chi-square tests for independence have more than one degree of freedom. Thus analysis of residuals is non-trivial, and continuity corrections and exact tests disappear. On the other hand, in cross-classifications of two polytomous variables, more adjusted residuals appear and more problems arise in describing any associations that may be present. The next section provides an example of such difficulties.

2.3   An $r \times s$ Table: Psychiatric Therapy and the Psychoses

When observations are cross-classified from a polytomous variable with $r \geqslant 3$ categories and a polytomous variable with $s \geqslant 3$ categories, an $r \times s$ contingency table results. As in simpler two-way tables, the hypothesis of independence corresponds to an additive log–linear model, and chi-square tests and adjusted residuals may be used to test this hypothesis. If the hypothesis is untenable, then description of deviations from independence becomes a difficult but important problem. This section emphasizes use of log cross-product ratios to study the dependence of polytomous variables rather than measures of association which use a single numerical measure to describe the size of the dependence.

Data used in this section come from Hollingshead and Redlich's (1967 [1958]) exploration of relationships between social class and mental illness. As a preliminary step in their investigation of psychotic patients in the New Haven area, they checked the relationship existing in 1950 between diagnosis and type of therapy. Results are summarized in Table 2.4.

Testing Independence

To see if a relationship exists between diagnostic group and treatment, let (A_h, B_h), $1 \leqslant h \leqslant N = 1442$, be independent pairs with identical distributions. Let A_h be i if subject h is in diagnostic group i, where group 1 is affective psychosis, group 2 is alcoholic psychoses, etc. Let $B_h = j$ represent the type of therapy given subject h, so that $B_h = 1$ if psychotherapy is used, $B_h = 2$ if organic therapy is used, and $B_h = 3$ if custodial care is the only treatment. Let $p_{ij} > 0$ be the probability that $A_h = i$ and $B_h = j$, and let n_{ij} be the number of subjects who are in diagnostic group i and receive therapy j. If group and

Table 2.4

Number of Psychotic Patients in Specified Diagnostic Groups Receiving
a Principal Type of Psychiatric Therapy[a]

Diagnostic group	Type of therapy			Total
	Psychotherapy	Organic therapy	Custodial care	
Affective psychoses	30	102	280	160
Alcoholic psychoses	48	23	20	91
Organic psychoses	19	80	75	174
Schizophrenic psychoses	121	344	382	847
Senile psychoses	18	11	141	170
Total	236	560	646	1442

[a] Computed from Table 35 in Hollingshead and Redlich (1967 [1958], page 288).

therapy are independent, then

$$p_{ij} = p_i^A p_j^B, \qquad 1 \leqslant i \leqslant r = 5, \quad 1 \leqslant j \leqslant s = 3.$$

Equivalently, the log–linear model

$$\log m_{ij} = \lambda + \lambda_i^A + \lambda_j^B$$

is satisfied, where

$$\sum \lambda_i^A = \sum \lambda_j^B = 0.$$

As noted in Section 2.6, the maximum-likelihood estimate

$$\hat{m}_{ij} = n_i^A n_j^B / N,$$

so that the chi-square statistics are

$$X^2 = \sum_i \sum_j (n_{ij} - n_i^A n_j^B / N)^2 / (n_i^A n_j^B / N)$$

$$= N \sum_i (1/n_i^A) \sum_j (1/n_j^B) n_{ij}^2 - N$$

and

$$L^2 = 2 \sum_i \sum_j n_{ij} \log[(n_{ij} N)/(n_i^A n_j^B)]$$

$$= 2 \left(\sum_i \sum_j n_{ij} \log n_{ij} - \sum_i n_i^A \log n_i^A - \sum_j n_j^B \log n_j^B + N \log N \right).$$

There are

$$(r - 1)(s - 1) = (5 - 1)(3 - 1) = 8$$

degrees of freedom.

These formulas for $\hat{m}_{ij}$, X^2, and L^2 are all rather old. Neyman and Pearson (1928) showed that $\hat{m}_{ij}$ is the maximum-likelihood estimate of m_{ij}, Pearson (1904) developed the first formula for X^2. Leslie (1951) developed the second formula for X^2 for use with calculators. The formulas for L^2 are found in Wilks (1935). The second formula is sometimes used with calculators. The formula for degrees of freedom is due to Fisher (1922).

In this example, $X^2 = 254.3$ and $L^2 = 248.2$, so the independence model clearly fails. (Hollingshead and Redlich's X^2 of 228.5 on page 288 appears to be in error.) As an aid in checking calculations, the $\hat{m}_{ij}$ and the adjusted residuals

$$r_{ij} = \frac{n_{ij} - (n_i{}^A n_j{}^B / N)}{\{n_i{}^A n_j{}^B [1 - (n_i{}^A / N)][1 - (n_j{}^B / N)] / N\}^{1/2}}$$

are shown in Table 2.5. Note that the failure does not involve only a few isolated cells. Most adjusted residuals are very large.

Cross-Product Ratios

The relationship between diagnosis and treatment may be explored through examination of the cross-product ratios

$$q^{AB}_{(ii')(jj')} = \frac{m_{ij} m_{i'j'}}{m_{ij'} m_{i'j}} = \frac{p_{ij} p_{i'j'}}{p_{ij'} p_{i'j}}$$

or their logarithms

$$\tau^{AB}_{(ii')(jj')} = \log q^{AB}_{(ii')(jj')}$$
$$= \log m_{ij} - \log m_{ij'} - \log m_{i'j} + \log m_{i'j'}$$
$$= \log p_{ij} - \log p_{ij'} - \log p_{i'j} + \log p_{i'j'}.$$

There are only eight that are not redundant cross-product ratios, for there are many algebraic restrictions on the $\tau^{AB}_{(ii')(jj')}$. For instance, note that

$$\tau^{AB}_{(ii')(jj')} = -\tau^{AB}_{(ii')(j'j)} = -\tau^{AB}_{(i'i)(jj')} = \tau^{AB}_{(i'i)(j'j)}$$

and

$$\tau^{AB}_{(ii)(jj')} = \tau^{AB}_{(ii')(jj)} = 0.$$

Table 2.5

Observed Cell Counts, Estimated Expected Cell Counts, and Adjusted Residuals
for the Independence Model for Table 2.3

Diagnostic group psychoses	Type of therapy[a]		
	Psychotherapy	Organic therapy	Custodial care
Affective	30	102	280
	57.4	136.2	218.4
	−4.19	−3.97	5.95
Alcoholic	48	23	20
	12.7	30.1	48.2
	11.21	−1.71	−5.65
Organic	19	80	75
	24.2	57.5	92.2
	−1.24	4.04	−2.58
Schizophrenic	121	344	382
	118.0	280.0	449.0
	0.47	7.61	−6.63
Senile	18	11	141
	23.7	56.2	90.1
	−1.36	−8.21	7.68

[a] The first line is the observed count n_{ij}, the second line is the estimated expected count $\hat{m}_{ij}$, and the third line is the adjusted residual r_{ij}.

In this example, the log cross-product ratios $\tau_{(i4)(12)}^{AB}$ and $\tau_{(i4)(23)}^{AB} = -\tau_{(i4)(32)}^{AB}$ are examined. This decision partly reflects the fact to be noted in Section 2.6 that if i' and j' are fixed, then the additive model holds if and only if

$$\tau_{(ii')(jj')}^{AB} = 0, \qquad i \neq i', \quad j \neq j'.$$

It also follows from results to be presented in Section 2.6 that

$$\tau_{(ii')(12)}^{AB} = \tau_{(i4)(12)}^{AB} - \tau_{(i'4)(12)}^{AB},$$
$$\tau_{(ii')(23)}^{AB} = \tau_{(i4)(23)}^{AB} - \tau_{(i'4)(23)}^{AB},$$

and

$$\tau_{(ii')(13)}^{AB} = \tau_{(i4)(12)}^{AB} + \tau_{(i4)(23)}^{AB} - \tau_{(i'4)(12)}^{AB} - \tau_{(i'4)(23)}^{AB}.$$

Thus the log cross-product ratios $\tau_{(i4)(12)}^{AB}$ and $\tau_{(i4)(23)}^{AB} = -\tau_{(i4)(32)}^{AB}$ determine all the $\tau_{(ii')(jj')}^{AB}$. The choice of group 4, the schizophrenic group, as a base of comparison arises since most patients are in this group. The use of $\tau_{(i4)(12)}^{AB}$

and $\tau_{(i4)(23)}^{AB}$ also reflects the fact that choices are often made between psychotherapy (treatment 1) and organic therapy (treatment 2) or between organic therapy and custodial care (treatment 3).

To understand the significance of these ratios, consider $q_{(34)(12)}^{AB}$. The ratio $m_{31}/m_{32} = p_{31}/p_{32}$ is the relative odds that a subject in the organic diagnostic group will receive psychotherapy rather than organic therapy. The ratio $m_{41}/m_{42} = p_{41}/p_{42}$ is the relative odds that a subject in the schizophrenic diagnostic group will receive psychotherapy rather than organic therapy. The coefficient $q_{(34)(12)}^{AB}$ compares these two ratios.

If $q_{(34)(12)}^{AB} = 1$, then the relative odds that a subject receives psychotherapy rather than organic therapy are the same for the schizophrenic and organic groups. If $q_{(34)(12)}^{AB} > 1$, then the relative odds are higher in the organic group.

To estimate the $\tau_{(ii')(jj')}^{AB}$, let

$$x_{ij} = n_{ij} + \tfrac{1}{2}$$

and

$$\hat{\tau}_{(ii')(jj')}^{AB} = \log\left(\frac{x_{ij}x_{i'j'}}{x_{ij'}x_{i'j}}\right).$$

The asymptotic standard deviation may be estimated by

$$s(\hat{\tau}_{(ii')(jj')}^{AB}) = \left(\frac{1}{x_{ij}} + \frac{1}{x_{ij'}} + \frac{1}{x_{i'j}} + \frac{1}{x_{i'j'}}\right)^{1/2}.$$

Following Gart and Zweifel (1967), the $\frac{1}{2}$'s have been added to the n_{ij} to reduce asymptotic bias. For related formulas, see Plackett (1962) and Goodman (1964a).

As an example of computations, consider $\hat{\tau}_{(14)(12)}^{AB}$. Here

$$\hat{\tau}_{(14)(12)}^{AB} = \log\left(\frac{30.5 \times 344.5}{102.5 \times 121.5}\right) = -0.17$$

and

$$s(\hat{\tau}_{(14)(12)}^{AB}) = \left(\frac{1}{30.5} + \frac{1}{102.5} + \frac{1}{121.5} + \frac{1}{344.5}\right)^{1/2} = 0.23.$$

Of the eight estimated coefficients which can differ from 0, four are very large relative to their asymptotic standard errors and four are modest in size. The following conclusions can be reached. There is no clear evidence that the distribution of treatment types for organic psychoses differ from those for schizophrenic psychoses, although the asymptotic standard errors are large enough so that confidence intervals for $\tau_{(34)(12)}^{AB}$ and $\tau_{(34)(23)}^{AB}$ are wide. Thus the data are consistent with large differences in treatment distribution for the two psychoses. For example, the lower 95 percent confidence bound for $\tau^{AB}_{(34)(12)}$ of

$$-0.38 - 1.96(0.27) = -0.91$$

corresponds to an odds ratio of $e^{-0.91} = 0.40$.

Custodial care is much more common relative to organic therapy in affective psychoses than in schizophrenic psychoses. The estimated cross-product ratio, $e^{-0.90} = 0.41$, is not well determined; the 95 percent confidence interval has lower bound

$$\exp[-0.90 - 1.96(0.14)] = 0.31$$

and upper bound

$$\exp[-0.90 + 1.96(0.14)] = 0.53.$$

Similarly, custodial care is much more common relative to organic therapy with senile psychoses than with schizophrenic psychoses. Psychotherapy is much more common relative to organic therapy in the case of alcoholic psychoses or senile psychoses than in the case of schizophrenic psychoses. It is an exercise for the reader (Exercise 2.3) to quantify these conclusions by construction of confidence intervals for the corresponding cross-product ratios. A result of this exercise is the observation that no cross-product ratio is determined with much accuracy.

One issue should be mentioned in connection with the many confidence intervals which can be constructed from Table 2.6. A similar problem has been discussed in Section 1.4. Since there are eight degrees of freedom for

Table 2.6

Estimated Log Cross-Product Ratios $\hat{\tau}^{AB}_{(i4)(12)}$ and $\hat{\tau}^{AB}_{(i4)(23)}$
for the Data in Table 2.4

Diagnostic group psychoses	i	Type of therapy contrasted[a]	
		Psychotherapy versus organic therapy	Organic therapy versus custodial care
Affective	1	−0.17 (0.23)	−0.90 (0.14)
Alcoholic	2	1.77 (0.27)	0.24 (0.31)
Organic	3	−0.38 (0.27)	0.17 (0.18)
Schizophrenic	4	0.00 (0.00)	0.00 (0.00)
Senile	5	1.52 (0.39)	−2.41 (0.32)

[a] Estimated asymptotic standard deviations are in parentheses.

chi-square statistics, eight nontrivial approximate 95 percent confidence intervals for $\tau^{AB}_{(i4)(12)}$ or $\tau^{AB}_{(i4)(23)}$ can be constructed.

For each individual interval, there is an approximate 0.05 probability that the parameter is outside its confidence interval. The expected number of parameters outside their respective confidence intervals is about $8(0.05) = 0.4$. Thus there is a substantial probability that some parameter is not within its confidence interval.

For a price, one can ensure that 0.05 is at least as great as the approximate probability that any of the eight confidence intervals does not contain its corresponding parameter. Instead of intervals with bounds

$$\hat{\tau}^{AB}_{(i4)(jj')} - 1.96s(\hat{\tau}^{AB}_{(i4)(jj')})$$

and

$$\hat{\tau}^{AB}_{(i4)(jj')} + 1.96s(\hat{\tau}^{AB}_{(i4)(jj')})$$

for $j = 1$ and $j' = 2$ or $j = 2$ and $j' = 3$, intervals with bounds

$$\hat{\tau}^{AB}_{(i4)(jj')} - 2.74s(\hat{\tau}^{AB}_{(i4)(jj')})$$

and

$$\hat{\tau}^{AB}_{(i4)(jj')} + 2.74s(\hat{\tau}^{AB}_{(i4)(jj')})$$

may be used. Here 2.74 is $Z_{0.025/8}$. The intervals are substantially wider now, for $2.74/1.96$ is about 1.4; however, protection is obtained against human ingenuity in selection of interesting confidence intervals to compute after examination of the data. For further details, see Goodman (1964a). Exercise 2.3 is also relevant.

In this example, the confidence intervals considered are of enough interest on an individual basis for it to appear preferable to use $Z_{\alpha/2}$ rather than $Z_{\alpha/16}$. If confidence intervals are used to establish whether any coefficients at all are definitely different from 0, the simultaneous confidence interval approach appears desirable.

Measures of Association

In the preceding example, no attempt has been made to use a single number to describe the relationship between diagnostic group and type of treatment. No obvious purpose would be served by such an endeavor. However, applications do arise in which the strength of the relationship between two polytomous variables must be described by a single number, called a measure of association. Such applications generally involve use of one variable to predict the other.

Measures of association will receive relatively little attention in this book; a detailed exposition and helpful references are provided in the Goodman and Kruskal (1954, 1959, 1963, 1972) series on measures of association.

2.4 Comparison of Samples

So far in this chapter, the two-way tables examined have been cross-classifications of observations on two polytomous variables. The probability model used assumes that N independent and identically distributed pairs (A_h, B_h) $1 \leqslant h \leqslant N$, are observed, where for $1 \leqslant i \leqslant r$, $1 \leqslant j \leqslant s$, the probability is $p_{ij} > 0$ that $A_h = i$ and $B_h = j$. The observed count n_{ij} is the number of h, $1 \leqslant h \leqslant N$, for which $A_h = i$ and $B_h = j$. The hypothesis under study is that of independence. Under independence,

$$p_{ij} = p_i^A p_j^B,$$

where for $1 \leqslant i \leqslant r$, p_i^A is the probability that $A_h = i$, and for $1 \leqslant j \leqslant s$, p_j^B is the probability that $B_h = j$. As will be shown in Section 2.6, this independence hypothesis is equivalent to an additive log–linear model.

Another probability model for two-way tables can be considered. In this model, s independent samples are compared. Just as in the models of Sections 2.1 to 2.3, pairs (A_h, B_h), $1 \leqslant h \leqslant N$, are observed. It is still the case that $1 \leqslant A_h \leqslant r$, $1 \leqslant B_h \leqslant s$, and n_{ij}, $1 \leqslant i \leqslant r$, $1 \leqslant j \leqslant s$, is still the number of h such that $A_h = i$ and $B_h = j$. However, B_h is now a fixed variable indicating the sample from which A_h is taken. It is assumed that given the B_h, $1 \leqslant h \leqslant N$, the A_h are independently distributed. If $B_h = j$, then the probability is $p_{i \cdot j}^{A \cdot B} > 0$ that $A_h = i$. The hypothesis under study is that of homogeneity. Under homogeneity, the distribution of A_h is independent of the sample B_h from which it is drawn, so that

$$p_{i \cdot j}^{A \cdot B} = p_{i \cdot j'}^{A \cdot B}, \qquad 1 \leqslant i \leqslant r, \quad 1 \leqslant j \leqslant s, \quad 1 \leqslant j' \leqslant s.$$

Although the homogeneity hypothesis arises from a different probability model than does the independence hypothesis, they both will be shown in Section 2.6 to be equivalent to the same additive log–linear model. Consequently, maximum-likelihood estimates, chi-square tests, and adjusted residuals for the homogeneity model are the same as those for the independence model. This situation will often arise in this book; many different probability models often lead to the same log–linear model and to the same estimation and testing procedures.

To illustrate use of homogeneity, the fact is used that the General Social Survey asks many of the same questions from year to year. As an example, from 1972 to 1975, subjects have been asked whether courts were sufficiently harsh with criminals. Responses are reported in Table 2.7. Given these responses, overall changes in attitudes toward the courts can be investigated. Although the formal situation is different, the methods of the preceding section can still be used for statistical analysis.

Let A_h represent the response of subject h, $1 \leqslant h \leqslant N = 5360$. Let $A_h = 1$ if the response is "too harshly," let $A_h = 2$ if the response is "not harshly enough," etc. Let B_h represent the year of interview of subject h, so that $B_h = 1$ corresponds to 1972, $B_h = 2$ corresponds to 1973, etc. Let n_{ij} be the number of subjects h for whom $A_h = i$ and $B_h = j$. For example, $n_{11} = 105$ and $n_{12} = 68$. Given the B_h, $1 \leqslant h \leqslant N$, let the A_h be independent random variables with conditional probability $p_{i \cdot j}^{A \cdot B} > 0$ that $A_h = i$ given that $B_h = j$. For example, $p_{1 \cdot 1}^{A \cdot B}$ is the probability that a subject responds "too harshly" if interviewed in 1972. If attitudes are not changing over time, when the homogeneity hypothesis

$$p_{i \cdot j}^{A \cdot B} = p_{i \cdot 1}^{A \cdot B}, \qquad 1 \leqslant i \leqslant 5, \quad 1 \leqslant j \leqslant 4,$$

holds.

The homogeneity model corresponds to the same log–linear model used with the independence model. Given that n_j^B subjects h respond in year

Table 2.7

Cross-Classification of Respondents in 1972–1975 General Social Surveys by Attitude to Treatment of Criminals by Courts and by Year of Survey[a]

	Year of survey				
Response[b]	1972	1973	1974	1975	Total
Too harshly	105	68	42	61	276
Not harshly enough	1066	1092	580	1174	3912
About right	265	196	72	144	677
Don't know	173	138	51	104	466
No answer	4	10	8	7	29
Total	1613	1504	753	1490	5360

[a] From National Opinion Research Center (1972, page 29; 1973, page 57; 1974, page 55; 1975, page 58).
[b] The question asked is, "In general, do you think the courts in this area deal too harshly or not harshly enough with criminals?" The reduced 1974 sample results from an experimental change in the question used with half the sample.

$B_h = j$, then expected value m_{ij} of n_{ij} is $n_j^B p_{i \cdot j}^{A \cdot B}$. The homogeneity model holds if and only if

$$\log m_{ij} = \lambda + \lambda_i^A + \lambda_j^B$$

for some λ, λ_i^A, and λ_j^B such that

$$\sum \lambda_i^A = \sum \lambda_j^B = 0.$$

Chi-Square Tests and Adjusted Residuals

Once again, the maximum-likelihood estimate $\hat{m}_{ij}$ of m_{ij} is $n_i^A n_j^B / N$ and the chi-square statistics

$$X^2 = \sum_i \sum_j (n_{ij} - n_i^A n_j^B / N)^2 / (n_i^A n_j^B / N)$$

and

$$L^2 = 2 \sum_i \sum_j n_{ij} \log \left(\frac{n_{ij} N}{n_i^A n_j^B} \right)$$

have $(r-1)(s-1) = 12$ degrees of freedom. Use of X^2 for testing homogeneity is found in Pearson (1911) with the wrong degrees of freedom and in Fisher (1922) with the right degrees of freedom. The use of L^2 for testing homogeneity appears in Wilks (1935).

The adjusted residuals r_{ij} are still given by the formula

$$r_{ij} = \frac{n_{ij} - (n_i^A n_j^B / N)}{\{n_i^A n_j^B [1 - (n_i^A / N)][1 - (n_j^B / N)]/N\}^{1/2}}.$$

The same large-sample approximations used in Sections 2.1 to 2.3 also apply here, so that under homogeneity, r_{ij} has an approximate standard normal distribution.

Results of use of the homogeneity model are summarized in Table 2.8. Since $X^2 = 87.4$ and $L^2 = 87.0$, the homogeneity model clearly fails. Thus the distribution of attitudes toward the courts appears to have changed in the period of time under study. The adjusted residuals suggest an increasingly common belief that courts are not harsh enough with criminals. The very large value of 5.94 for r_{24} and the very small value -7.46 for r_{21} fit into this pattern. The decreasing use of the categories "too harshly," "about right," and "don't know" are suggested by residuals such as r_{11}, r_{14}, r_{31}, r_{34}, r_{41}, and r_{44}.

Table 2.8

Observed Counts, Estimated Expected Counts, and Adjusted Residuals
for the Homogeneity Model for Table 2.7[a]

	Year of survey			
Response	1972	1973	1974	1975
Too harshly	105	68	42	61
	83.1	77.4	38.8	76.7
	2.96	-1.30	0.57	-2.17
Not harshly enough	1066	1092	580	1174
	1177.2	1097.7	549.6	1087.5
	-7.46	0.39	2.69	5.94
About right	265	196	72	144
	203.7	190.0	95.1	188.2
	5.49	0.55	-2.73	-4.06
Don't know	173	138	51	104
	140.2	130.8	65.5	129.5
	3.46	0.78	-2.02	-2.76
No answer	4	10	8	7
	8.7	8.1	4.1	8.1
	-1.92	0.77	2.10	-0.44

[a] The first line is the observed count n_{ij}, the second line is the estimated
expected count $\hat{m}_{ij}$, and the third line is the adjusted residual r_{ij}.

Linear Trends

The clearest trend shown by the table is a tendency for an increasing
fraction n_{2j}/n_j^B of subjects to use the response "not harshly enough." This
fraction increases from 0.66 in 1972 to 0.73 in 1973, 0.77 in 1974, and 0.79
in 1975. A formal test of this pattern may be obtained as in Cochran (1954)
or Armitage (1955). Let year j have a score $2j - 5$. Consider estimation of
$\delta = \sum(2j - 5)p_2^{A\,:B}_{\cdot\,j}$. Note that under homogeneity

$$d = \frac{\sum(2j - 5)n_{2j}}{n_j^B} = -\frac{3n_{21}}{n_1^B} - \frac{n_{22}}{n_2^B} - \frac{n_{23}}{n_3^B} + \frac{3n_{24}}{n_4^B}$$

has mean 0 and variance

$$c = \frac{9k(1 - k)}{n_1^B} + \frac{k(1 - k)}{n_2^B} + \frac{k(1 - k)}{n_3^B} + \frac{9k(1 - k)}{n_4^B}$$

$$= k(1 - k)\left(\frac{9}{n_1^B} + \frac{1}{n_2^B} + \frac{1}{n_3^B} + \frac{9}{n_4^B}\right),$$

where

$$k = p_{2 \cdot 1}^{A \cdot B}.$$

To verify this formula, note that the n_{2j}/n_j^B are independent and have mean $p_{2 \cdot j}^{A \cdot B} = k$ and variances

$$\frac{1}{n_j^B} p_{2 \cdot j}^{A \cdot B}(1 - p_{2 \cdot j}^{A \cdot B}) = \frac{1}{n_j^B} k(1 - k).$$

The maximum-likelihood estimate of k is n_2^A/N, as noted in Section 2.6. Thus the estimated standard deviation of d is

$$\hat{c}^{1/2} = \left[\frac{n_2^A}{N} \left(1 - \frac{n_2^A}{N}\right) \left(\frac{9}{n_1^B} + \frac{1}{n_2^B} + \frac{1}{n_3^B} + \frac{9}{n_4^B}\right) \right]^{1/2}$$

$$= \left[\frac{3912}{5360} \left(1 - \frac{3912}{5360}\right) \left(\frac{9}{1613} + \frac{1}{1504} + \frac{1}{753} + \frac{9}{1490}\right) \right]^{1/2}$$

$$= 0.0518.$$

Since

$$d = -\frac{3 \times 1066}{1613} - \frac{1092}{1504} + \frac{580}{753} + \frac{3 \times 1174}{1490} = 0.425,$$

the standardized value

$$t = d/\hat{c}^{1/2} = 8.21.$$

Since this standardized value has an approximate standard normal distribution under the homogeneity model, the observed value of t is very inconsistent with such a model.

Given that a response other than "not harshly enough" is used, the distribution of responses has not changed as dramatically. Consider the test of homogeneity for the 4 × 4 table obtained from Table 2.7 by deletion of the second row. The Pearson chi-square drops to 16.87, and the likelihood-ratio chi-square drops to 16.57. The degrees of freedom is now $(4 - 1)(4 - 1) = 9$, so the approximate significance level of the test statistics is about 5 percent. If the "no answer" category is also removed, leaving a 3 × 4 table, then the chi-square statistics drop to an X^2 of 5.97 and an L^2 of 5.77. The degrees of freedom are now $(3 - 1)(4 - 1) = 6$, so a satisfactory fit is achieved.

More subtle analyses which employ log–linear models explicitly designed for ordered classifications like year of survey or harshness of courts are considered in Chapter 6. Further analysis of this table is deferred until then.

2.5   Additive Log–Linear Models for Poisson Observations

Rasch (1960, pages 13–23) considers a model for independent Poisson observations $n_{ij}, 1 \leqslant i \leqslant r, 1 \leqslant j \leqslant s$, with respective means $m_{ij} > 0, 1 \leqslant i \leqslant r, 1 \leqslant j \leqslant s$, in which

$$\log m_{ij} = \lambda + \lambda_i^A + \lambda_j^B,$$

where

$$\sum \lambda_i^A = \sum \lambda_j^B = 0.$$

In Rasch's case, the index $i, 1 \leqslant i \leqslant r$, represents a subject, while $j, 1 \leqslant j \leqslant s$, represents a passage the subject reads. The number of reading errors made when subject i reads passage j is n_{ij}. The coefficient λ_i^A provides a measure of the relative proneness to errors of subject i, while λ_j^B provides a measure of the relative difficulty of passage j. The additive log–linear model has the following important feature. Given that n_j^B errors are made in passage j, $1 \leqslant j \leqslant s$, the distribution of the $n_{ij}, 1 \leqslant i \leqslant r$, is multinomial. The sample size is n_j^B, and the probability for subject i is

$$p_i = g \exp \lambda_i^A,$$

where

$$g = \frac{1}{\sum \exp \lambda_i^A}.$$

This probability p_i is the same for all passages j, and it depends only on the ability measures λ_i^A. Thus conditional on the total number of errors made in each passage, inferences concerning subject abilities are independent of passage difficulties. A similar argument shows that conditional on the total number of errors made by each subject, inferences about passage difficulties are independent of subject abilities. Rasch regards this separation of inferences about subject abilities from passage difficulties as a fundamental requirement for assessment of them.

For now, it suffices to note that maximum-likelihood estimates $\hat{m}_{ij}$, chi-square statistics X^2 and L^2, degrees of freedom, and adjusted residuals r_{ij} for this model are the same as those for the independence model.

Some complications may arise, however, in the more general Poisson model in which

$$\log(m_{ij}/z_{ij}) = \lambda + \lambda_i^A + \lambda_j^B,$$

$$\sum \lambda_i^A = \sum \lambda_j^B = 0,$$

where the $z_{ij} > 0$ are known scale factors. The formula $\hat{m}_{ij} = n_i^A n_j^B / N$ only applies if for some $a_i > 0, 1 \leqslant i \leqslant r$, and some $b_j > 0, 1 \leqslant j \leqslant s$,

$$z_{ij} = a_i b_j.$$

Otherwise, $\log[n_i^A n_j^B/(N z_{ij})]$ is not expressible as a sum $\hat{\lambda} + \hat{\lambda}_i^A + \hat{\lambda}_j^B$, where $\sum \hat{\lambda}_i^A = \sum \hat{\lambda}_j^B = 0$. In addition, if $\hat{m}_{ij}$ is not $n_i^A n_j^B/N$, then the formula for r_{ij} used in earlier sections does not apply. Consequently, iterative computations are needed. Two approaches are available. The Newton–Raphson algorithm may be used, just as in Chapter 1, or the iterative proportional fitting algorithm of Deming and Stephan (1940) may be tried. The latter algorithm is quite easy to use, but it does not provide estimates of asymptotic variances and does not aid in computation of adjusted residuals. In this section, both the Newton–Raphson and Deming–Stephan algorithms will be used.

As an example, consider Table 1.8 and Table 1.10. Together, these tables yield a 12×3 table of counts n_{ij}, where n_{ij} is the number of reported suicides in month i of year $1967 + j$. If z_{ij} is the number of days in month i of year $1967 + j$, then

$$f_{ij} = m_{ij}/z_{ij}$$

is the expected rate per day of reported suicides in month i of year $1967 + j$. If

$$\log f_{ij} = \lambda + \lambda_i^A + \lambda_j^B,$$

then the ratio $f_{ij}/f_{i'j}$ of the expected rate in month i of $1967 + j$ and the expected rate in month i' of $1967 + j$ is independent of j. Thus the relative rates of suicides in different months are the same from year to year.

Since February has 29 days in 1962 and 28 days in 1969 and 1970, z_{ij} cannot be expressed as a product $a_i b_j$. Thus maximum-likelihood estimates $\hat{m}_{ij}$ of the mean m_{ij} cannot be expressed in closed form as $n_i^A n_j^B/N$; however, iterative algorithms are available for such computations. If adjusted residuals are not needed, then the iterative proportional fitting algorithm of Deming and Stephan (1940) is probably the simplest to apply to this example. Otherwise, a version of the Newton–Raphson algorithm may be tried. Here the Deming–Stephan algorithm will be employed first.

The Deming–Stephan Algorithm

The Deming–Stephan algorithm finds $\hat{m}_{ij}$ such that the maximum-likelihood equations

$$\hat{m}_i^A = \sum_j \hat{m}_{ij} = n_i^A,$$

$$\hat{m}_j^B = \sum_i \hat{m}_{ij} = n_j^B,$$

$$\log(\hat{m}_{ij}/z_{ij}) = \hat{\lambda} + \hat{\lambda}_i^A + \hat{\lambda}_j^B,$$

and

$$\sum_i \hat{\lambda}_i^A = \sum_j \hat{\lambda}_j^B = 0$$

are satisfied.

The estimates $\hat{\lambda}$, $\hat{\lambda}_i^A$, and $\hat{\lambda}_j^B$ do not explicitly appear in the algorithm. To use the procedure, let

$$m_{ij0} = z_{ij}.$$

Note that

$$\log(m_{ij0}/z_{ij}) = \lambda_0 + \lambda_{i0}^A + \lambda_{j0}^B$$

if $\lambda_0 = \lambda_{i0}^A = \lambda_{j0}^B = 0$. Then successive approximations m_{ijt}, $t \geqslant 0$, are obtained for $\hat{m}_{ij}$ by the formulas

$$m_{ij(t+1)} = m_{ijt} n_i^A / m_{it}^A, \qquad t \quad \text{even}$$
$$= m_{ijt} n_j^B / m_{jt}^B, \qquad t \quad \text{odd}.$$

If t is even, then

$$\sum_j m_{ij(t+1)} = \frac{n_i^A}{m_{it}^A} \sum_j m_{ijt} = \frac{n_i^A}{m_{it}^A} m_{it}^A = n_i^A,$$

so that one maximum-likelihood equation holds. If t is odd, then

$$\sum_i m_{ij(t+1)} = \frac{n_j^B}{m_{jt}^B} \sum_i m_{ijt} = \frac{n_j^B}{m_{jt}^B} m_{jt}^B = n_j^B,$$

so that another maximum-likelihood equation holds.

The term iterative proportional fitting arises since each iteration involves a proportional adjustment of a row or column such that either the row total m_{it}^A equals n_i^A or the column total m_{jt}^B equals n_j^B. Table 2.9 illustrates computations. Standard computer packages such as the C-TAB program described in Haberman (1973c) or the ECTA program of Fay and Goodman (1975) can perform the necessary calculations. The algorithm of Haberman (1972) may also be employed. Further discussion of the algorithm appears in Chapters 3 and 4 of this volume and in Chapters 7, 8, and 9 of Volume 2.

The approximations m_{ij2} for $\hat{m}_{ij}$ are probably accurate enough for practical purposes. Certainly m_{ij3} is quite accurate. Note that $m_{i3}^A = n_i^A$ by definition and m_{j3}^B and n_j^B agree to five significant figures for $1 \leqslant j \leqslant 3$.

The additive model does not provide an adequate fit, for $X^2 = 44.2$ and $L^2 = 44.1$. There are $(12 - 1)(3 - 1) = 22$ degrees of freedom, so the chi-square statistics are significant at about the 0.0035 level. Comparison of

Table 2.9

Calculation of Maximum-Likelihood Estimates by the Deming–Stephan Algorithm in Tables 1.8 and 1.10

Month	i	1968 ($j = 1$)	1969 ($j = 2$)	1970 ($j = 3$)	n_i^A	$m_{ij0} = z_{ij}$ 1968	1969	1970	m_{i0}^A
		Observed count n_{ij}							
Jan.	1	1720	1831	1867	5418	31	31	31	93
Feb.	2	1712	1609	1789	5110	29	28	28	85
Mar.	3	1924	1973	1944	5841	31	31	31	93
Apr.	4	1882	1944	2094	5920	30	30	30	90
May	5	1870	2003	2097	5970	31	31	31	93
June	6	1680	1774	1981	5435	30	30	30	90
July	7	1868	1811	1887	5566	31	31	31	93
Aug.	8	1801	1873	2024	5698	31	31	31	93
Sept.	9	1756	1862	1928	5546	30	30	30	90
Oct.	10	1760	1897	2032	5689	31	31	31	93
Nov.	11	1666	1866	1978	5510	30	30	30	90
Dec.	12	1733	1921	1859	5513	31	31	31	93
Total		21,372	22,364	23,480	67,216	366	365	365	1096

Month	$m_{ij1} = m_{ij0} n_i^A / m_{i0}^A$ 1968	1969	1970	$m_{ij2} = m_{ij1} n_j^B / m_{j2}^B$ 1968	1969	1970	m_{i2}^A
Jan.	1806.0	1806.0	1806.0	1719.6	1804.3	1894.3	5418.2
Feb.	1743.4	1683.3	1683.3	1660.0	1681.7	1765.6	5107.3
Mar	1947.0	1947.0	1947.0	1853.9	1945.1	2042.2	5841.2
Apr.	1973.3	1973.3	1973.3	1879.0	1971.5	2069.8	5920.2
May	1990.0	1990.0	1990.0	1894.8	1988.1	2087.3	5970.2
June	1811.7	1811.7	1811.7	1725.0	1806.9	1900.3	5435.2
July	1855.3	1855.3	1855.3	1766.6	1853.6	1946.1	5566.2
Aug.	1899.3	1899.3	1899.3	1808.5	1897.5	1992.2	5698.2
Sept.	1848.7	1848.7	1848.7	1760.3	1846.9	1939.1	5546.2
Oct.	1896.3	1896.3	1896.3	1805.6	1894.5	1989.1	5689.2
Nov.	1836.7	1836.7	1836.7	1748.8	1834.9	1926.5	5510.2
Dec.	1837.7	1837.7	1837.7	1749.8	1835.9	1927.5	5513.2
Total	22,445.	22,385.	22,385.	21,372.	23,364.	23,480.	67,216.

Month	$m_{ij3} = m_{ij2} n_i^A / m_{i2}^B$ 1968	1969	1970
Jan.	1719.6	1804.2	1894.2
Feb.	1660.9	1682.6	1766.5
Mar.	1853.8	1945.1	2042.1
Apr.	1878.9	1971.4	2069.7
May	1894.8	1988.0	2087.2
June	1725.0	1809.9	1900.2
July	1766.5	1853.5	1946.0
Aug.	1808.4	1897.4	1992.1
Sept.	1760.2	1846.8	1939.0
Oct.	1805.6	1894.4	1989.0
Nov.	1748.8	1834.8	1926.4
Dec.	1749.7	1835.8	1927.5
Total	22,372.	23,364.	23,480.

n_{ij} and m_{ij3} in Table 2.9 shows nothing dramatic. As noted in Chapter 1, there appear to be factors affecting monthly suicide rates that have not been examined here. These effects appear modest in size. The largest ratio $n_{ij}/\hat{m}_{ij}$ is

$$\frac{n_{71}}{\hat{m}_{71}} = \frac{1868}{1766.5} = 1.06.$$

The smallest ratio $n_{ij}/\hat{m}_{ij}$ is

$$\frac{n_{33}}{\hat{m}_{33}} = \frac{1944}{2042.1} = 0.95.$$

The Newton–Raphson Algorithm

As usual, the Newton-Raphson algorithm reduces to a series of weighted least squares problems. To begin computations, let $m_{ij0} = n_{ij} + \frac{1}{2}$ be an initial approximation for $\hat{m}_{ij}$ and let

$$y_{ij0} = \log(m_{ij0}/z_{ij})$$

be an empirical logarithm. Weighted-least-squares estimates λ_{i0}^A for λ_i^A and λ_{j0}^B for λ_j^B are found as in a regression model

$$y_{ij0} = \lambda + \lambda_i^A + \lambda_j^B + \varepsilon_{ij}$$

in which each ε_{ij} is an independent random variable with mean 0 and variance m_{ij0}^{-1} and $\sum \lambda_i^A = \sum \lambda_j^B = 0$. To write normal equations for this regression model, a reparametrization to be described in Section 2.6 will be used. Here β_k, $1 \leqslant k \leqslant 13$, and x_{ijk}, $1 \leqslant k \leqslant 13$, are defined so that

$$
\begin{aligned}
\beta_k &= \lambda_k^A, & 1 &\leqslant k \leqslant 11, \\
&= \lambda_{k-11}^B, & 12 &\leqslant k \leqslant 13, \\
x_{ijk} &= 1, & 1 &\leqslant i = k \leqslant 11, \\
&= -1, & i &= 12, \quad 1 \leqslant k \leqslant 11, \\
&= 0, & i &\neq k, \quad 1 \leqslant i \leqslant 11, \quad 1 \leqslant k \leqslant 11, \\
&= 1, & k &- 11 = j, \quad 12 \leqslant k \leqslant 13, \\
&= -1, & j &= 3, \quad 12 \leqslant k \leqslant 13, \\
&= 0, & j &\neq k - 11, \quad 12 \leqslant k \leqslant 13.
\end{aligned}
$$

Then the regression model can be written

$$y_{ij0} = \lambda + \sum_{k=1}^{13} \beta_k x_{ijk} + \varepsilon_{ij}.$$

The normal equations for the β_k become

$$\sum_l S_{kl0}\beta_{l0} = w_{k0},$$

$$S_{kl0} = \sum_i \sum_j (x_{ijk} - \theta_{k0})(x_{ijl} - \theta_{l0})m_{ij0},$$

$$w_{k0} = \sum_i \sum_j (x_{ijk} - \theta_{k0})y_{ij0}m_{ij0},$$

$$\theta_{k0} = \sum_i \sum_j x_{ijk}m_{ij0} \Big/ \sum_i \sum_j m_{ij0}.$$

New estimated means m_{ij1} are then found from the equations

$$v_{ij0} = \sum_k \beta_{k0}x_{ijk},$$

$$g_0 = N \Big/ \sum_i \sum_j z_{ij}\exp v_{ij0},$$

$$m_{ij1} = z_{ij}g_0 \exp v_{ij0}.$$

Here, N is the sum $\sum\sum n_{ij}$ of all suicides from 1968 to 1970.

In this example, the estimates m_{ij1} shown in Table 2.10 are very similar to the estimates m_{ij3} in Table 2.9. Nonetheless, an added iteration is included in Table 2.10 for completeness. The estimate m_{ij2} is defined by the series of equations

$$\sum_l S_{kl1}\delta_{l1} = a_{k0},$$

$$S_{kl1} = \sum_i \sum_j (x_{ijk} - \theta_{k1})(x_{ijl} - \theta_{l1})m_{ij1},$$

$$a_{k1} = \sum_i \sum_j x_{ijk}(n_{ij} - m_{ij1}),$$

$$\theta_{k1} = \sum_i \sum_j x_{ijk}m_{ij1} \Big/ \sum_i \sum_j m_{ij1},$$

$$\beta_{k1} = \beta_{k0} + \delta_{k1},$$

$$v_{ij1} = \sum_k \beta_{k1}x_{ijk},$$

$$g_1 = N \Big/ \sum_i \sum_j z_{ij}\exp v_{ij1},$$

$$m_{ij2} = z_{ij}g_1 \exp v_{ij1}.$$

Note the very close agreement between m_{ij3} in Table 2.9 and m_{ij2} in Table 2.10.

Table 2.10

Computation of Maximum-Likelihood Estimates for Tables 1.8 and 1.10
by the Newton–Raphson Algorithm

Month	λ_{i0}^A	λ_{i1}^A	$s(\hat{\lambda}_i^A)$	1968		1969		1970	
				m_{i11}	m_{i12}	m_{i21}	m_{i22}	m_{i31}	m_{i32}
Jan.	−0.05115	−0.05090	0.01299	1719.3	1719.6	1803.6	1804.2	1893.8	1894.2
Feb.	−0.01873	−0.01892	0.01334	1661.3	1660.9	1682.8	1682.6	1766.9	1766.5
Mar.	0.02460	0.02428	0.01255	1854.6	1853.8	1945.6	1945.1	2042.8	2042.1
Apr.	0.07021	0.07050	0.01248	1878.5	1878.9	1970.6	1971.4	2069.2	2069.7
May	0.04581	0.04612	0.01243	1894.3	1894.7	1987.3	1988.0	2086.6	2087.2
June	−0.01482	−0.01498	0.01297	1725.4	1725.0	1810.0	1809.9	1900.5	1900.2
July	−0.02352	−0.02395	0.01283	1767.4	1766.5	1854.2	1853.5	1946.8	1946.0
Aug.	−0.00077	−0.00051	0.01270	1808.1	1808.4	1896.8	1897.4	1991.6	1992.1
Sept.	0.00493	0.00524	0.01285	1759.8	1760.2	1846.1	1846.8	1938.4	1939.0
Oct.	−0.00224	−0.00209	0.01270	1805.4	1805.6	1894.0	1894.4	1988.7	1989.0
Nov.	−0.00107	−0.00127	0.01289	1749.2	1748.8	1835.1	1834.8	1926.8	1926.4
Dec.	−0.03325	−0.03352	0.01289	1750.3	1749.7	1836.2	1835.8	1928.0	1927.5

	1968	1969	1970
$\lambda_{\cdot j1}^B$	−0.04820	−0.00029	0.04849
$\lambda_{\cdot j2}^B$	−0.04827	−0.00021	0.04848
$s(\lambda_{j2}^B)$	0.00552	0.00546	0.00539

Adjusted Residuals

Given that the Newton–Raphson algorithm has been used, adjusted residuals are easily computed, just as in Chapter 1. Here

$$r_{ij} = (n_{ij} - \hat{m}_{ij})/\hat{c}_{ij}^{1/2}$$

and

$$\hat{c}_{ij} = \hat{m}_{ij}\left[1 - \hat{m}_{ij}/N - \hat{m}_{ij}\sum_k \sum_l (x_{ijk} - \bar{T}_k)(x_{ijl} - \bar{T}_l)S^{kl}\right],$$

where

$$\bar{T}_k = \frac{1}{N}\sum_i \sum_j x_{ijk}n_{ij} = \frac{1}{N}\sum_i \sum_j x_{ijk}\hat{m}_{ij},$$

$$\hat{S}_{kl} = \sum_i \sum_j (x_{ijk} - \bar{T}_k)(x_{ijl} - \bar{T}_l)\hat{m}_{ij},$$

and $\hat{S}^{-1}$, the matrix with elements $\hat{S}^{kl}$, $1 \leq k \leq 13$, $1 \leq l \leq 13$, is the inverse of the matrix S with elements S_{kl}, $1 \leq k \leq 13$, $1 \leq l \leq 13$. Results are shown in Table 2.11. As usual, the r_{ij} should be approximately standard normal

Table 2.11

Observed Counts, Estimated Expected Counts, and Adjusted Residuals for the Additive
Log–Linear Model for Tables 1.8 and 1.10

	1968			1969			1970		
Month	Obs.	Exp.	Res.	Obs.	Exp.	Res.	Obs.	Exp.	Res.
Jan.	1720	1719.6	0.01	1831	1804.2	0.81	1867	1894.2	−0.81
Feb.	1712	1660.9	1.59	1609	1682.6	−2.28	1789	1766.5	0.69
Mar.	1924	1853.8	2.06	1973	1945.1	0.81	1944	2042.1	−2.82
Apr.	1882	1878.9	0.09	1944	1971.4	−0.79	2094	2069.7	0.69
May	1870	1894.7	−0.72	2003	1988.0	0.43	2097	2087.2	0.28
June	1680	1725.0	−1.37	1774	1809.9	−1.08	1981	1900.2	2.40
July	1868	1766.5	3.05	1811	1853.5	−1.26	1887	1946.0	−1.73
Aug.	1801	1808.4	−0.22	1873	1897.4	−0.72	2024	1992.1	0.93
Sept.	1756	1760.2	−0.13	1862	1846.8	0.45	1928	1939.0	−0.32
Oct.	1760	1805.6	−1.36	1897	1894.4	0.08	2032	1989.0	1.25
Nov.	1666	1748.8	−2.50	1866	1834.8	0.93	1978	1926.4	1.52
Dec.	1733	1749.7	−0.50	1921	1835.8	2.54	1859	1927.5	−2.02

deviates if the model under study holds. There are no clear patterns to the adjusted residuals; however, the fact that eight of 36 adjusted residuals exceed 2 in magnitude and one exceeds 3 certainly casts doubt on the addictive log–linear model. Note that $36(0.0454) = 1.63$ is approximately the expected number of adjusted residuals to exceed 2 in magnitude under the additive model, while the corresponding expected number exceeding 3 in magnitude is about $36(0.00270) = 0.10$.

Parameter Estimates

Estimated parameters $\hat{\lambda}_i^A$, $1 \leqslant i \leqslant 11$, and $\hat{\lambda}_j^B$, $1 \leqslant j \leqslant 2$, are immediate by-products of computations, as can be seen in Table 2.10. The remaining parameter estimates

$$\hat{\lambda}_{12}^A = -\sum_{i=1}^{11} \hat{\lambda}_i^A$$

and

$$\hat{\lambda}_3^B = -\hat{\lambda}_1^B - \hat{\lambda}_2^B$$

follow immediately. The value of these parameter estimates is somewhat reduced since the fit is unsatisfactory; nonetheless, it is important to note that they are easily found if needed. It also should be noted that if the model fits, $\hat{\beta}$ has estimated asymptotic covariance matrix $\hat{S}^{-1}$, so that

$$s^2(\hat{\lambda}_i^A) = \hat{S}^{ii}, \qquad\qquad 1 \leqslant i \leqslant 11,$$

$$= \sum_{i'=1}^{11} \sum_{j'=1}^{11} \hat{S}^{i'j'}, \qquad i = 12,$$

$$s^2(\hat{\lambda}_j^B) = \hat{S}^{(j+11)(j+11)}, \qquad 1 \leqslant j \leqslant 2,$$

$$= \sum_{k=12}^{13} \sum_{l=12}^{13} \hat{S}^{kl}, \qquad j = 3.$$

The additive model is considerably better than the model

$$\log f_{ij} = \lambda + \lambda_j^B,$$

$$\sum \lambda_j^B = 0,$$

in which the expected rate f_{ij} per day depends on the year j but not the month i. In this model,

$$f_{ij} = k_j$$

for some k_j,

$$\lambda = \tfrac{1}{3}(\log k_1 + \log k_2 + \log k_3),$$

and

$$\lambda_j^B = \log k_j - \lambda.$$

Using the same arguments used in Section 1.3, one finds that

$$\hat{k}_j = n_j^B / z_j^B,$$

so that

$$\hat{k}_1 = \frac{21{,}372}{366} = 58.393,$$

$$\hat{k}_2 = \frac{22{,}364}{365} = 61.271,$$

$$\hat{k}_3 = \frac{23{,}480}{365} = 64.329.$$

Thus $\hat{k}_j$ is the observed average rate per day of suicide in $1967 + j$. Since $\hat{m}_{ij} = z_{ij}\hat{k}_j$,

$$X^2 = \sum_i \sum_j \frac{(n_{ij} - \hat{m}_{ij})^2}{\hat{m}_{ij}} = 116.0$$

and

$$L^2 = 2 \sum_i \sum_j n_{ij} \log(n_{ij}/\hat{m}_{ij}) = 115.7.$$

The degrees of freedom are $36 - 3 = 33$, so the fit is very poor. The R^2 statistic for comparing the additive model to the model ignoring seasonal effects is

$$1 - \frac{44.1}{115.7} = 0.62.$$

Thus much of the observed variations of f_{ij} within year j is explained by the additive model.

2.6 Additive Log–Linear Models

In the preceding sections of this chapter, an $r \times s$ table **n** with elements n_{ij}, $1 \leq i \leq r$, $1 \leq j \leq s$, has been considered, where $r \geq 2$ and $s \geq 2$. An additive model has been tried in which m_{ij}, the expected value of n_{ij}, is assumed to satisfy an equation

$$\log(m_{ij}/z_{ij}) = \lambda + \lambda_i^A + \lambda_j^B. \tag{2.1}$$

In this equation $z_{ij} > 0$ is known, λ, λ_i^A, and λ_j^B are unknown, and

$$\sum \lambda_i^A = \sum \lambda_j^B = 0. \tag{2.2}$$

This additive model is an example of a more general class of log–linear models for two-way tables. As in Draper and Smith (1966, page 258), (2.1) holds if and only if

$$\log(m_{ij}/z_{ij}) = \lambda + \sum_{i'=1}^{r-1} \lambda_{i'}^A x_{ii'}^A + \sum_{j=1}^{j-1} \lambda_{j'}^B x_{jj'}^B, \tag{2.3}$$

where for $1 \leq i' \leq r - 1$,

$$\begin{aligned} x_{ii'}^A &= 1, & i = i', \\ &= -1, & i \neq i', \quad i < r, \\ &= 0, & i = r, \end{aligned}$$

and for $1 \leq j' \leq s - 1$,

$$\begin{aligned} x_{jj'}^B &= 1, & j = j', \\ &= -1, & j \neq j', \quad j < s, \\ &= 0, & j = s. \end{aligned}$$

For example, if $i = j = 1$, then $x_{i1}^A = x_{j1}^B = 1$ and the other $x_{ii'}^A$ and $x_{jj'}^B$ are zero. Thus

$$\log(m_{11}/z_{11}) = \lambda + \lambda_1{}^A + \lambda_1{}^B.$$

In general, this book considers log–linear models in which

$$\log(m_{ij}/z_{ij}) = \alpha + \sum_{k=1}^{q} \beta_k x_{ijk} \qquad (2.4)$$

for some known $z_{ij} > 0$, $1 \leqslant i \leqslant r$, $1 \leqslant j \leqslant s$, and x_{ijk}, $1 \leqslant i \leqslant r$, $1 \leqslant j \leqslant s$, $1 \leqslant k \leqslant q$, and some unknown α and β_k, $1 \leqslant k \leqslant q$. Note that (2.4) is the same as (1.6) except for the subscripts used. To relate the general case of (2.4) to the special case (2.3), let $x_{ijk} = x_{ik}^A$ for $1 \leqslant k \leqslant r - 1$, let $x_{ijk} = x_{j(k-r+1)}^B$ for $r \leqslant k \leqslant r + s - 2$, let $q = r + s - 2$, let $\alpha = \lambda$, let $\beta_k = \lambda_k{}^A$ for $1 \leqslant k \leqslant r - 1$, and let $\beta_k = \lambda_{k-r+1}^B$ for $r \leqslant k \leqslant r + s - 2$.

Sampling Procedures

Four sampling procedures are commonly encountered when two-way tables are examined. These procedures lead to the Poisson model, the multinomial model, the row-multinomial model or the column-multinomial model.

In the Poisson model, each count n_{ij} has an independent Poisson distribution with mean $m_{ij} > 0$. All results in this section apply to this model. This model appears in this chapter in Section 2.5.

In the multinomial model, the complete table of counts n_{ij}, $1 \leqslant i \leqslant r$, $1 \leqslant j \leqslant s$, has a multinomial distribution with sample size $N > 0$ and probabilities $p_{ij} > 0$, $1 \leqslant i \leqslant r$, $1 \leqslant j \leqslant s$, so that each n_{ij} has mean Np_{ij}. The model applies if the following conditions hold:

(a)   Independent and identically distributed pairs (A_h, B_h), $1 \leqslant h \leqslant N$, are observed.

(b)   The A_h can have integral values from 1 to r.

(c)   The B_h can have integral values from 1 to s.

(d)   The count n_{ij} is the number of pairs (A_h, B_h), $1 \leqslant h \leqslant N$, such that $A_h = i$ and $B_h = j$.

(e)   For $1 \leqslant h \leqslant N$, $1 \leqslant i \leqslant r$, and $1 \leqslant j \leqslant s$, the probability is $p_{ij} > 0$ that $A_h = i$ and $B_h = j$.

As in the case of the Poisson model, all results of this section apply if the multinomial sampling model is used. This model has been used in Sections 2.1, 2.2, and 2.3.

In the row-multinomial model, each row of counts n_{ij}, $1 \leqslant j \leqslant s$, has an independent multinomial distribution with sample size $n_i^A > 0$ and probabilities $p_{j \cdot i}^{B \cdot A} > 0, 1 \leqslant j \leqslant s$. Thus each n_{ij} has mean $m_{ij} = n_i^A p_{j \cdot i}^{B \cdot A}$. This model holds if the following conditions are satisfied:

(a) Independent pairs (A_h, B_h), $1 \leqslant h \leqslant N$, are observed.

(b) The A_h can have integral values from 1 to r.

(c) The B_h can have integral values from 1 to s.

(d) The count n_{ij} is the number of pairs (A_h, B_h), $1 \leqslant h \leqslant N$, such that $A_h = i$ and $B_h = j$.

(e) The A_h, $1 \leqslant h \leqslant N$, are fixed, so that for $1 \leqslant i \leqslant r$, n_i^A of the A_h have value i.

(f) For $1 \leqslant h \leqslant N, 1 \leqslant i \leqslant r, 1 \leqslant j \leqslant s$, $p_{j \cdot i}^{B \cdot A} > 0$ is the conditional probability that $B_h = j$ given that $A_h = i$.

If row-multinomial sampling is present, then results of this section only apply to a log–linear model defined by (2.4) when for some α_{i*}, $1 \leqslant i^* \leqslant r$, and β_{i*k}, $1 \leqslant i^* \leqslant r, 1 \leqslant k \leqslant q$,

$$\alpha_{i*} + \sum_k \beta_{i*k} x_{ijk} = 1, \qquad i = i^*,$$

$$(2.5)$$

$$= 0, \qquad i \neq i^*.$$

In particular, results of this section apply to the additive log–linear model. The verify this claim, define λ_{i*}, $\lambda_{\cdot i*i}^A$, $1 \leqslant i \leqslant r$, and $\lambda_{\cdot j*j}^B$, $1 \leqslant j \leqslant s$, for $1 \leqslant i \leqslant r$ so that

$$\lambda_{i*} = \frac{1}{r}, \qquad 1 \leqslant i^* \leqslant r,$$

$$\lambda_{\cdot i*i}^A = \frac{r-1}{r}, \qquad i^* = i,$$

$$= -\frac{1}{r}, \qquad i^* \neq i,$$

and

$$\lambda_{\cdot i*j}^B = 0, \qquad 1 \leqslant j \leqslant s.$$

Then substitution of the λ_{i*}, $\lambda_{\cdot i*i}^A$, and $\lambda_{\cdot j*j}^B$ into (2.3) or (2.1) leads to

$$\lambda_{i*} + \sum_{i'} \lambda_{\cdot i*i'}^A x_{ii'}^A + \sum_{j'} \lambda_{\cdot i*j'}^B x_{jj'}^B = \lambda_{i*} + \lambda_{\cdot i*i}^A = 1, \qquad i = i^*,$$

$$= 0, \qquad i \neq i^*.$$

Row-multinomial sampling has not been used in this chapter; however, column multinomial sampling appears in Section 2.4. In column-multinomial sampling, each column of counts n_{ij}, $1 \leqslant i \leqslant r$, has an independent multinomial distribution with sample size $n_j^B > 0$ and probabilities $p_{i \cdot j}^{A \cdot B} > 0$,

$1 \leqslant i \leqslant r$. Thus each n_{ij} has mean $m_{ij} = n_j^B p_{i \cdot j}^{A \cdot B}$. This model holds when the following conditions are satisfied:

(a) Independent pairs (A_h, B_h), $1 \leqslant h \leqslant N$, are observed.

(b) The A_h can have integral values from 1 to r.

(c) The B_h can have integral values from 1 to s.

(d) The count n_{ij} is the number of pairs (A_h, B_h), $1 \leqslant h \leqslant N$, such that $A_h = i$ and $B_h = j$.

(e) The B_h, $1 \leqslant h \leqslant N$, are fixed so that for $1 \leqslant j \leqslant s$, n_j^B of the B_h have value j.

(f) For $1 \leqslant h \leqslant N$, $1 \leqslant i \leqslant r$, $1 \leqslant j \leqslant s$, $p_{i \cdot j}^{A \cdot B} > 0$ is the conditional probability that $B_h = i$ given that $A_h = j$.

If column-multinomial sampling is used, then results of this section hold for a log–linear model defined by (2.4) if for some α_{j*}, $1 \leqslant j \leqslant s$, and β_{j*k}, $1 \leqslant j^* \leqslant s$, $1 \leqslant k \leqslant q$,

$$\alpha_{j*} + \sum_k \beta_{j*k} x_{ijk} = 1, \qquad j = j^*,$$

$$\hspace{5.5cm} = 0, \qquad j \neq j^*. \tag{2.6}$$

In particular, results of this section do apply to the additive log–linear model. The argument to verify this assertion is similar to that used to show that results apply to the additive log–linear model under row-multinomial sampling.

Maximum-Likelihood Equations

Under any of the sampling procedures under study, the maximum-likelihood equations for a log–linear model defined by (2.4) are

$$\hat{M} = \sum_i \sum_j \hat{m}_{ij} = N = \sum_i \sum_j n_{ij}, \tag{2.7}$$

$$\sum_i \sum_j x_{ijk} \hat{m}_{ij} = \sum_i \sum_j x_{ijk} n_{ij}, \qquad 1 \leqslant k \leqslant q, \tag{2.8}$$

$$\log(\hat{m}_{ij}/z_{ij}) = \hat{\alpha} + \sum_{k=1}^q \hat{\beta}_k x_{ijk}. \tag{2.9}$$

These results follow from Birch (1963). They are similar to (1.7) and (1.9). The new feature in this chapter is their application to row-multinomial and column-multinomial sampling. Under row-multinomial sampling, note that (2.5) and (2.8) imply that $\hat{m}_i^A = n_i^A$, $1 \leqslant i \leqslant r$. Thus the probability estimates $\hat{p}_{j\cdot i}^{A\cdot B} = \hat{m}_{ij}/n_i^A$ satisfy the constraints

$$\sum_i \hat{p}_{i\cdot j}^{A\cdot B} = 1, \qquad 1 \leqslant j \leqslant s.$$

Here one also has

$$\sum_i p_{i\cdot j}^{A\cdot B} = 1, \qquad 1 \leqslant j \leqslant s.$$

It should also be noted that the condition (2.5) for row-multinomial sampling holds if (2.8) implies that $\hat{m}_i^A = n_i^A$, $1 \leqslant i \leqslant r$. Similarly, (2.6) holds if (2.8) implies that $\hat{m}_j^B = n_j^B$, $1 \leqslant j \leqslant s$.

For the additive model, the maximum-likelihood equations can be written as

$$\hat{M} = \sum_i \sum_j \hat{m}_{ij} = N = \sum_i \sum_j n_{ij}, \tag{2.10}$$

$$\hat{m}_{i'}^A - \hat{m}_r^A = \sum_i \sum_j x_{ii'}^A \hat{m}_{ij} = \sum_i \sum_j x_{ii'}^A n_{ij}$$
$$= n_{i'}^A - n_r^A, \qquad 1 \leqslant i' \leqslant r-1, \tag{2.11}$$

$$\hat{m}_{j'}^B - \hat{m}_s^B = \sum_i \sum_j x_{jj'}^B \hat{m}_{ij} = \sum_i \sum_j x_{jj'}^B n_{ij}$$
$$= n_{j'}^B - n_s^B, \qquad 1 \leqslant j' \leqslant s-1, \tag{2.12}$$

and

$$\log(\hat{m}_{ij}/z_{ij}) = \hat{\lambda} + \sum_{i'=1}^{r-1} \hat{\lambda}_{i'}^A x_{ii'}^A + \sum_{j'=1}^{s-1} \hat{\lambda}_{j'}^B x_{jj'}^B. \tag{2.13}$$

For example, let $z_{ij} = 1$ and $r = s = 2$. Then

$$\log \hat{m}_{ij} = \hat{\lambda} + \hat{\lambda}_1^A + \hat{\lambda}_1^B, \qquad i = j = 1,$$
$$= \hat{\lambda} - \hat{\lambda}_1^A - \hat{\lambda}_1^B, \qquad i = 2, \quad j = 1,$$
$$= \hat{\lambda} + \hat{\lambda}_1^A - \hat{\lambda}_1^B, \qquad i = 1, \quad j = 2,$$
$$= \hat{\lambda} - \hat{\lambda}_1^A - \hat{\lambda}_1^B, \qquad i = j = 2,$$

$$\hat{m}_{11} + \hat{m}_{12} + \hat{m}_{21} + \hat{m}_{22} = N = n_{11} + n_{12} + n_{21} + n_{22},$$

$$\hat{m}_1^A - \hat{m}_2^A = \hat{m}_{11} + \hat{m}_{12} - \hat{m}_{21} - \hat{m}_{22}$$
$$= n_{11} + n_{12} - n_{21} - n_{22} = n_1^A - n_2^A,$$

and

$$\hat{m}_1^B - \hat{m}_1^B = \hat{m}_{11} + \hat{m}_{21} - \hat{m}_{12} - \hat{m}_{22}$$
$$= n_{11} + n_{21} - n_{12} - n_{22} = n_1^B - n_2^B.$$

Eqs. (2.10)–(2.13) for the $\hat{m}_{ij}$ hold if and only if

$$\hat{m}_i^A = n_i^A, \qquad 1 \le i \le r, \tag{2.14}$$

$$\hat{m}_j^B = n_j^B, \qquad 1 \le j \le s, \tag{2.15}$$

$$\log(\hat{m}_{ij}/z_{ij}) = \hat{\lambda} + \hat{\lambda}_i^A + \hat{\lambda}_j^B, \tag{2.16}$$

and

$$\sum \hat{\lambda}_i = \sum \hat{\lambda}_j^B = 0. \tag{2.17}$$

It is easily verified that (2.14) implies (2.11) and (2.15) implies (2.12). To verify that (2.10) and (2.11) implies (2.14), and that (2.15) and (2.12) implies (2.15), see Exercise 2.7. Clearly (2.16) and (2.17) are equivalent to (2.13). Since

$$\hat{M} = \sum_i \hat{m}_i^A = \sum_j \hat{m}_j^B$$

and

$$N = \sum_i n_i^A = \sum_j n_j^B,$$

(2.14) or (2.15) implies (2.10). Given (2.14) and (2.15), it follows that the maximum-likelihood equations apply if the entire table has a multinomial distribution, if each row has an independent multinomial distribution, if each column has an independent multinomial distribution, or if each observation n_{ij} has an independent Poisson distribution. Thus these equations apply to all uses in this chapter of the additive log–linear model.

If each $z_{ij} = 1$, then (2.14)–(2.17) can be solved by setting

$$\hat{m}_{ij} = n_i^A n_j^B / N,$$

$$\hat{\lambda} = \frac{1}{r} \sum_{i'} \log n_{i'}^A + \frac{1}{s} \sum_{j'} \log n_{j'}^B - \log N,$$

$$\hat{\lambda}_i^A = \log n_i^A - \frac{1}{r} \sum_{i'} \log n_{i'}^A,$$

and

$$\hat{\lambda}_j^B = \log n_j^B - \frac{1}{s} \sum_{j'} \log n_{j'}^B.$$

Note that (2.14) holds since

$$\sum_j (n_i^A n_j^B / N) = (n_i^A / N) \sum_j n_j^B = (n_i^A / N) N = n_i^A.$$

Similarly, (2.15) holds since

$$\sum_i (n_i^A n_j^B / N) = (n_j^B / N) \sum_i n_i^A = (n_j^B / N)N = n_j^B.$$

Equation (2.17) holds since

$$\sum \hat{\lambda}_i^A = \sum_i \log n_i^A - r \frac{1}{r} \sum_{i'} \log n_{i'}^A = 0$$

and

$$\sum \hat{\lambda}_j^B = \sum_j \log n_j^B - s \frac{1}{s} \sum_{j'} \log n_{j'}^B = 0,$$

and (2.16) holds since

$$\hat{\lambda} + \hat{\lambda}_i^A + \hat{\lambda}_j^B = \log n_i^A + \log n_j^B - \log N = \log(n_i^A n_j^B / N).$$

The same formula for $\hat{m}_{ij}$ applies if for some $a_i > 0$, $1 \le i \le r$, and some $b_j > 0$, $1 \le j \le s$, $z_{ij} = a_i b_j$. For a proof, see Exercise 2.6. If z_{ij} cannot be expressed for all i, $1 \le i \le r$, and j, $1 \le j \le s$, as a product $a_i b_j$, then the iterative procedures described in Section 2.5 are required to compute the $\hat{m}_{ij}$. Note that the Newton–Raphson algorithm in that section applies to all models defined as in (2.4), while the iterative proportional fitting algorithm only applies to the additive log–linear model.

Independence and Additivity

The independence model holds if independent and identically distributed polytomous pairs (A_h, B_h), $1 \le h \le N$, are given such that each A_h and B_h are independent. If $1 \le A_h \le r$, $1 \le B_h \le s$, and $p_{ij} > 0$ is the probability that $A_h = i$ and $B_h = j$, then

$$p_{ij} = p_i^A p_j^B$$

and the mean m_{ij} of n_{ij} is $N p_{ij}$.

This model implies the additive model

$$\log(m_{ij} / z_{ij}) = \log m_{ij} = \log(N p_{ij}) = \lambda + \lambda_i^A + \lambda_j^B$$

holds with $z_{ij} = 1$,

$$\lambda = \log N + \frac{1}{r} \sum_{i'} \log p_{i'}^A + \frac{1}{s} \sum_{j'} \log p_{j'}^B,$$

$$\lambda_i^A = \log p_i^A - \frac{1}{r} \sum_{i'} \log p_{i'}^A,$$

and

$$\lambda_j{}^B = \log p_j{}^B - \frac{1}{s} \sum_{j'} \log p_{j'}^B.$$

Thus the $\lambda_i{}^A$ measure the relative size of the marginal probabilities $p_i{}^A$, and the $\lambda_j{}^B$ measure the relative size of the marginal probabilities $p_j{}^B$.

Conversely, additivity implies independence. Consider the log cross-product ratio

$$\log\left(\frac{p_{ij}p_{11}}{p_{i1}p_{1j}}\right) = \log\left(\frac{m_{ij}m_{11}}{m_{i1}m_{1j}}\right)$$

If the additive model holds, then

$$\begin{aligned}
\log\left(\frac{m_{ij}m_{11}}{m_{i1}m_{1j}}\right) &= \log m_{ij} - \log m_{i1} - \log m_{1j} + \log m_{11} \\
&= (\lambda + \lambda_i{}^A + \lambda_j{}^B) - (\lambda + \lambda_i{}^A + \lambda_1{}^B) - (\lambda + \lambda_1{}^A + \lambda_j{}^B) \\
&\quad + (\lambda + \lambda_1{}^A + \lambda_1{}^B) \\
&= 0.
\end{aligned}$$

Thus

$$\frac{p_{ij}p_{11}}{p_{i1}p_{1j}} = 1$$

and

$$p_{ij} = \frac{p_{i1}p_{1j}}{p_{11}}.$$

Note that

$$p_i{}^A = \sum_j \left(\frac{p_{i1}p_{1j}}{p_{11}}\right) = \frac{p_{i1}}{p_{11}} \sum_j p_{1j} = \frac{p_{i1}p_1{}^A}{p_{11}}.$$

Similarly,

$$p_j{}^B = \frac{p_1{}^B p_{1j}}{p_{11}}.$$

Since

$$\sum_i p_i{}^A = 1,$$

$$1 = \sum_i \left(\frac{p_{i1}p_1{}^A}{p_{11}}\right) = \frac{p_1{}^A}{p_{11}} \sum_i p_{i1} = \frac{p_1{}^A p_1{}^B}{p_{11}},$$

so that

$$p_{11} = p_1{}^A p_1{}^B.$$

More generally,

$$p_{ij} = \frac{p_{i1} p_{1j}}{p_{11}}$$

$$= \left(\frac{p_{i1} p_1{}^A}{p_{11}} \right) \left(\frac{p_1{}^B p_{1j}}{p_{11}} \right)$$

$$= p_i{}^A p_j{}^B.$$

Thus independence holds. For a different but related proof, see Bishop and Fienberg (1969).

Under the independence model, the maximum-likelihood equations imply that estimated marginal probabilities equal observed marginal relative frequencies, i.e.,

$$\hat{p}_i{}^A = \hat{m}_i{}^A / N = n_i{}^A / N$$

and

$$\hat{p}_j{}^B = n_j{}^B / N.$$

These equations are quite old. For example, see Neyman and Pearson (1928).

Homogeneity and Additivity

The (column) homogeneity model holds under the following conditions:

(a)   Independent observations (A_h, B_h), $1 \leqslant h \leqslant N$, are given such that each A_h can be any of the r integers from 1 to r and each B_h can be any of the integers from 1 to s.

(b)   Given the B_h, $1 \leqslant h \leqslant N$, the A_h are independent with conditional probability $p_{i \cdot j}^{A \cdot B} > 0$ that $A_h = i$ given that $B_h = j$.

(c)   The conditional probability $p_{i \cdot j}^{A \cdot B}$ is independent of j, so that

$$p_{i \cdot j}^{A \cdot B} = p_{i \cdot 1}^{A \cdot B}; \qquad 1 \leqslant i \leqslant r, \qquad 1 \leqslant j \leqslant s.$$

(d)   The number $n_j{}^B$ of $B_h = j$ is fixed.

A similar model may be defined for row homogeneity.

In the column homogeneity model, $m_{ij} = n_j{}^B p_{i \cdot 1}^{A \cdot B}$, so that

$$\log(m_{ij}/z_{ij}) = \log m_{ij} = \lambda + \lambda_i{}^A + \lambda_j{}^B,$$

where $z_{ij} = 1$,

$$\lambda = \frac{1}{s} \sum_{j'} \log n_{j'}^B + \frac{1}{r} \sum_{i'} \log p_{i' \cdot 1}^{A \cdot B},$$

$$\lambda_i^A = \log p_{i \cdot 1}^{A \cdot B} - \frac{1}{r} \sum_{i'} \log p_{i' \cdot 1}^{A \cdot B}$$

and

$$\lambda_j^B = \log n_j^B - \frac{1}{s} \sum_{j'} \log n_{j'}^B.$$

Here the λ_i^A measure the relative size of the common probabilities $p_{i \cdot 1}^{A \cdot B}$ The λ_j^B are of little interest since they just reflect sample sizes.

If the additive model holds, then the homogeneity model holds. The same argument used with the independence case shows that the log cross-product ratio

$$\log \left(\frac{p_{i \cdot j}^{A \cdot B} p_{1 \cdot 1}^{A \cdot B}}{p_{i \cdot 1}^{A \cdot B} p_{1 \cdot j}^{A \cdot B}} \right) = \log \left(\frac{m_{ij} m_{11}}{m_{i1} m_{1j}} \right)$$

satisfies the equation

$$p_{i \cdot j}^{A \cdot B} p_{1 \cdot 1}^{A \cdot B} = p_{i \cdot 1}^{A \cdot B} p_{1 \cdot j}^{A \cdot B}.$$

Since $\sum_i p_{i \cdot j}^{A \cdot B} = 1$,

$$p_{1 \cdot 1}^{A \cdot B} = \sum_i p_{i \cdot j}^{A \cdot B} p_{1 \cdot 1}^{A \cdot B} = \sum_i p_{i \cdot 1}^{A \cdot B} p_{1 \cdot j}^{A \cdot B} = p_{1 \cdot j}^{A \cdot B}.$$

More generally,

$$p_{i \cdot j}^{A \cdot B} = p_{i \cdot 1}^{A \cdot B}, \quad 1 \le i \le r, \quad 1 \le j \le s.$$

Thus the homogeneity model holds.

Under homogeneity, the maximum-likelihood estimate of the conditional probability $\hat{p}_{i \cdot j}^{A \cdot B} = \hat{p}_{i \cdot 1}^{A \cdot B}$ is the relative frequency n_i^A/N with which $A_h = i$, for

$$\hat{p}_{i \cdot 1}^{A \cdot B} = \hat{m}_{ij}/n_j^B = (n_i^A n_j^B/N)/n_j^B = n_i^A/N.$$

This equation is quite old. For example, see Neyman and Pearson (1928).

Parameter Estimation

Large-sample properties of parameter estimates in log-linear models defined as in (2.4) are quite similar to large-sample properties of parameter estimates in the log–linear models of Chapter 1; however, a few complications arise under row-multinomial sampling. Results discussed here are based on those of Haberman (1974a, pages 80–81).

Under Poisson sampling or multinomial sampling results are essentially the same as in Chapter 1. The analogous regression model to (2.4) has the form

$$Y_{ij} = \alpha + \sum_k \beta_k x_{ijk} + \varepsilon_{ij}.$$

Here the Y_{ij} are observed values of the dependent variable and the ε_{ij} are unobserved independent normal random variables with respective means of zero and variances of m_{ij}^{-1}. The weighted-least-squares estimate $\mathbf{b}$ of the vector $\boldsymbol{\beta}$ with coordinates β_k, $1 \leqslant k \leqslant q$, is determined by the simultaneous equations

$$\sum_l S_{kl} b_l = w_k, \qquad 1 \leqslant k \leqslant q,$$

where

$$w_k = \sum_i \sum_j (x_{ijk} - \theta_k) Y_{ij} m_{ij},$$

$$S_{kl} = \sum_i \sum_j (x_{ijk} - \theta_k)(x_{ijl} - \theta_l) m_{ij}, \qquad (2.18)$$

and

$$\theta_k = \sum_i \sum_j x_{ijk} m_{ij} \Big/ \sum_i \sum_j m_{ij}. \qquad (2.19)$$

If S has an inverse S^{-1}, then the weighted-least-squares estimate $\mathbf{b}$ has distribution $N(\boldsymbol{\beta}, S^{-1})$. Similarly, the maximum-likelihood estimate $\hat{\boldsymbol{\beta}}$ has an approximate distribution $N(\boldsymbol{\beta}, S^{-1})$. If c_k, $1 \leqslant k \leqslant q$, are known constants, then the linear combination

$$\hat{\tau} = \sum c_k \hat{\beta}_k \qquad (2.20)$$

has an approximate $N(\tau, \sigma^2(\hat{\tau}))$ distribution. Here

$$\tau = \sum c_k \beta_k, \qquad (2.21)$$

$$\sigma^2(\hat{\tau}) = \sum_k \sum_l c_k c_l S^{kl}, \qquad (2.22)$$

and S^{-1} has elements S^{kl}, $1 \leqslant k \leqslant q$, $1 \leqslant l \leqslant q$. The asymptotic covariance matrix S^{-1} has an estimate $\hat{S}^{-1}$. Here $\hat{S}^{-1}$ is the inverse of the matrix $\hat{S}$ with elements

$$\hat{S}_{kl} = \sum_i \sum_j (x_{ijk} - \bar{T}_k)(x_{ijl} - \bar{T}_l) m_{ij} \qquad (2.23)$$

and

$$\bar{T}_k = N^{-1} \sum_i \sum_j x_{ijk} \hat{m}_{ij} = N^{-1} \sum_i \sum_j x_{ijk} n_{ij}. \qquad (2.24)$$

The estimated asymptotic variance of $\hat{\tau}$ is then

$$s^2(\hat{\tau}) = \sum_k \sum_l c_k c_l \hat{S}^{kl}, \qquad (2.25)$$

where $\hat{S}^{kl}$, $1 \leqslant k \leqslant q$, $1 \leqslant l \leqslant q$, are the elements of $\hat{S}^{-1}$. Thus τ has an approximate level-$(1 - \alpha)$ confidence interval with lower bound

$$\hat{\tau} - Z_{\alpha/2} s(\hat{\tau})$$

and upper bound

$$\hat{\tau} + Z_{\alpha/2} s(\hat{\tau}).$$

As in Chapter 1, approximations become increasingly accurate as the means m_{ij} become large.

The normal approximation $N(\tau, \sigma^2(\hat{\tau}))$ for $\hat{\tau}$ holds under row-multinomial sampling if and only if τ is a function of the probabilities $p_{j \cdot i}^{B \cdot A}$, $1 \leqslant i \leqslant r$, $1 \leqslant j \leqslant s$. Similarly, the normal approximation for $\hat{\tau}$ holds under column-multinomial sampling if and only if τ is a function of the probabilities $p_{i \cdot j}^{A \cdot B}$, $1 \leqslant i \leqslant r$, $1 \leqslant j \leqslant s$.

The Saturated Model

For example, consider the saturated model

$$\log m_{ij} = \lambda + \lambda_i^A + \lambda_j^B + \lambda_{ij}^{AB},$$

where

$$\sum \lambda_i^A = \sum \lambda_j^B = \sum_i \lambda_{ij}^{AB} = \sum_j \lambda_{ij}^{AB} = 0.$$

The term "saturated" arises since, as Goodman (1970) and Bishop, Fienberg, and Holland (1975, pages 11–31) note, this log–linear model imposes no restrictions; for one may write

$$\lambda = \frac{1}{rs} \sum_i \sum_j \log m_{ij},$$

$$\lambda_i^A = \frac{1}{r} \sum_j \log m_{ij} - \lambda,$$

$$\lambda_j^B = \frac{1}{s} \sum_i \log m_{ij} - \lambda,$$

and

$$\lambda_{ij}^{AB} = \log m_{ij} - \lambda - \lambda_i{}^A - \lambda_j{}^B.$$

This model has been used implicitly in Sections 2.1 and 2.3, for estimates of log cross-product ratios have been derived without any restrictions being imposed on the logarithms of the means m_{ij}.

As in Draper and Smith (1966, page 258), one may write

$$\log m_{ij} = \lambda + \sum_{i'=1}^{r-1} \lambda_{i'}^A x_{ii'}^A + \sum_{j'=1}^{s-1} \lambda_{j'}^B x_{jj'}^B + \sum_{i'=1}^{r-1} \sum_{j'=1}^{s-1} \lambda_{i'j'}^{AB} x_{ii'}^A x_{jj'}^B.$$

The maximum likelihood equations

$$\sum_i \sum_j \hat{m}_{ij} = \sum_i \sum_j \hat{n}_{ij},$$

$$\sum_i \sum_j x_{ii'}^A \hat{m}_{ij} = \sum_i \sum_j x_{ii'}^A n_{ij}, \qquad 1 \leqslant i' \leqslant r-1,$$

$$\sum_i \sum_j x_{jj'}^B \hat{m}_{ij} = \sum_i \sum_j x_{jj'}^B n_{ij}, \qquad 1 \leqslant j' \leqslant s-1,$$

$$\sum_i \sum_j x_{ii'}^A x_{jj'}^B \hat{m}_{ij} = \sum_i \sum_j x_{ii'}^A x_{jj'}^B n_{ij}, \qquad 1 \leqslant i' \leqslant r-1, \quad 1 \leqslant j' \leqslant s-1,$$

and

$$\log \hat{m}_{i'j} = \hat{\lambda} + \sum_{i'=1}^{r-1} \hat{\lambda}_{i'}^A x_{ii'}^A + \sum_{j'=1}^{s} \hat{\lambda}_{j'}^B x_{jj'}^B + \sum_{i'=1}^{r-1} \sum_{j'=1}^{s-1} \hat{\lambda}_{i'j'}^{AB} x_{ii'}^A x_{jj'}^B$$

hold if $\hat{m}_{ij} = n_{ij}$,

$$\hat{\lambda} = \frac{1}{rs} \sum_i \sum_j \log n_{ij},$$

$$\hat{\lambda}_i{}^A = \frac{1}{r} \sum_j \log n_{ij} - \hat{\lambda},$$

$$\hat{\lambda}_j{}^B = \frac{1}{s} \sum_i \log n_{ij} - \hat{\lambda},$$

and

$$\hat{\lambda}_{ij}^{AB} = \log n_{ij} - \hat{\lambda} - \hat{\lambda}_i{}^A - \hat{\lambda}_j{}^B.$$

Consider constants $c_{ij}, 1 \leqslant i \leqslant r, 1 \leqslant j \leqslant s$, such that

$$\sum_i c_{ij} = 0, \qquad 1 \leqslant j \leqslant s,$$

$$\sum_j c_{ij} = 0, \qquad 1 \leqslant i \leqslant s.$$

Let

$$\tau = \sum_i \sum_j c_{ij} \lambda_{ij}^{AB}.$$

Then

$$\tau = \sum_i \sum_j c_{ij}(\log m_{ij} - \lambda - \lambda_1{}^A - \lambda_j{}^B)$$

$$= \sum_i \sum_j c_{ij} \log m_{ij}$$

and

$$\hat{\tau} = \sum_i \sum_j c_{ij} \log n_{ij}.$$

As in Section 1.5, it is easily verified that under Poisson or multinomial sampling, τ has asymptotic variance

$$\sigma^2(\hat{\tau}) = \sum_i \sum_j c_{ij}^2/m_{ij}.$$

This formula also applies under row-multinomial sampling, for then

$$\tau = \sum_i \sum_j c_{ij} \log(n_i{}^A p_{j\cdot i}^{A\cdot B})$$

$$= \sum_i \sum_j c_{ij} \log n_i{}^A + \sum_i \sum_j c_{ij} \log p_{j\cdot i}^{B\cdot A}$$

$$= \sum_i \sum_j c_{ij} \log p_{j\cdot i}^{B\cdot A}$$

is a function of the probabilities $p_{j\cdot i}^{A\cdot B}$. Similarly, this formula applies under column-multinomial sampling, for then

$$\tau = \sum_i \sum_j c_{ij} \log(n_j{}^B \log p_{i\cdot j}^{A\cdot B}) = \sum_i \sum_j c_{ij} \log p_{i\cdot j}^{A\cdot B}$$

is a function of the probabilities $p_{i\cdot j}^{A\cdot B}$.

An important application of this result involves the log cross-product ratio

$$\tau_{(i'i'')(j'j'')}^{AB} = \log m_{i'j'} - \log m_{i'j''} - \log m_{i''j'} + \log m_{i''j''}$$

$$= \lambda_{i'j'}^{AB} - \lambda_{i'j''}^{AB} - \lambda_{i''j'}^{AB} + \lambda_{i''j''}^{AB},$$

where $i' \neq i''$ and $j' \neq j''$. Let

$$c_{ij} = 1, \quad i = i' \text{ and } j = j' \text{ or } i = i'' \text{ and } j = j'',$$

$$= -1, \quad i = i' \text{ and } j = j'' \text{ or } i = i'' \text{ and } j = j',$$

$$= 0, \quad \text{otherwise.}$$

Then

$$\tau^{AB}_{(i'i'')(j'j'')} = \sum_i \sum_j c_{ij} \log m_{ij},$$

$$\sum_i c_{ij} = 0, \qquad 1 \leqslant j \leqslant s,$$

and

$$\sum_j c_{ij} = 0, \qquad 1 \leqslant i \leqslant r.$$

Thus the estimated log cross-product ratio

$$\hat{\tau}^{AB}_{(i'i'')(j'j'')} = \log n_{i'j'} - \log n_{i'j''} - \log n_{i''j'} + \log n_{i''j''}$$

has asymptotic variance

$$\sigma^2(\hat{\tau}^{AB}_{(i'i'')(j'j'')}) = \frac{1}{m_{i'j'}} + \frac{1}{m_{i'j''}} + \frac{1}{m_{i''j'}} + \frac{1}{m_{i''j''}}.$$

Cross Product Ratios

The emphasis in this chapter on asymptotic properties of estimates of log cross-product ratios reflects the role of these parameters in describing deviations from the additive model in the case $z_{ij} = 1$. The additive model holds in this case if and only if

$$\tau^{AB}_{(ii')(jj')} = \log\left(\frac{m_{ij}m_{i'j'}}{m_{ij'}m_{i'j}}\right) = \log m_{ij} - \log m_{ij'} - \log m_{i'j} + \log m_{i'j'}$$

$$= 0, \qquad i \neq i', \quad j \neq j'. \tag{2.26}$$

(Note that if $i = i'$ or $j = j'$, then $\tau^{AB}_{(ii')(jj')} = 0$, whether or not the additive model holds.) To verify that (2.26) holds for the additive model, note that

$$\log m_{ij} - \log m_{ij'} - \log m_{i'j} + \log m_{i'j'}$$

is then equal to

$$(\lambda + \lambda_i^A + \lambda_j^B) - (\lambda + \lambda_i^A + \lambda_{j'}^B) - (\lambda + \lambda_{i'}^A + \lambda_j^B) + (\lambda + \lambda_{i'}^A + \lambda_{j'}^B) = 0.$$

On the other hand, (2.26) implies that for all i, $1 \leqslant i \leqslant r$, and j, $1 \leqslant j \leqslant s$,

$$\log m_{ij} = \log m_{ij'} + \log m_{i'j} - \log m_{i'j'}$$
$$= \lambda + \lambda_i^A + \lambda_j^B,$$

where

$$\lambda = \frac{1}{r}\sum_{i''}\log m_{i''j'} + \frac{1}{s}\sum_{j''}\log m_{i'j''} - \log m_{i'j'},$$

$$\lambda_i^A = \log m_{ij'} - \frac{1}{r}\sum_{i''}\log m_{i''j'},$$

and

$$\lambda_j^B = \log m_{i'j} - \frac{1}{s}\sum_{j''}\log m_{i'j''}.$$

Thus (2.26) does imply the additive model.

Departures from the additive model can be described without use of all $\tau_{(ii')(jj')}^{AB}$. For any fixed i'' and j'', the $\tau_{(ii')(jj')}^{AB}$, $i \neq i''$, $j \neq j''$, determine all log cross-product ratios. This result is used in Section 2.3 with $i'' = 4$ and $j'' = 2$. As a proof, note that

$$\log\left[\frac{m_{ij}m_{i'j'}}{m_{ij'}m_{i'j}}\right] = \log\left(\frac{m_{ij}m_{i''j''}}{m_{ij''}m_{i''j}}\right) + \log\left(\frac{m_{ij''}m_{i'j'}}{m_{ij'}m_{i'j''}}\right)$$

$$+ \log\left(\frac{m_{i''j}m_{i'j''}}{m_{i''j''}m_{i'j}}\right) + \log\left(\frac{m_{i''j''}m_{i'j'}}{m_{i''j'}m_{i'j''}}\right).$$

In other words,

$$\tau_{(ii')(jj')}^{AB} = \tau_{(ii'')(jj'')}^{AB} + \tau_{(ii'')(j''j')}^{AB} + \tau_{(i''i')(jj'')}^{AB} + \tau_{(i''i')(j''j')}^{AB}$$

$$= \tau_{(ii'')(jj'')}^{AB} - \tau_{(ii'')(j'j'')}^{AB} - \tau_{(i'i'')(jj'')}^{AB} + \tau_{(i'i'')(j'j'')}^{AB}.$$

To explore departures from the additive log–linear model, the estimated log cross-product ratio

$$\hat{\tau}_{(ii')(jj')}^{AB} = \log n_{ij} - \log n_{ij'} - \log n_{i'j} + \log n_{i'j'} \qquad (2.27)$$

under the saturated model may be used, together with the corresponding estimated asymptotic variance

$$s^2(\hat{\tau}_{(ii')(jj')}^{AB}) = \frac{1}{n_{ij}} + \frac{1}{n_{ij'}} + \frac{1}{n_{i'j}} + \frac{1}{n_{i'j'}}. \qquad (2.28)$$

These estimates apply under all four sampling procedures considered in this chapter. For extensive discussion of use of $\hat{\tau}_{(ii')(jj')}^{AB}$ and $s^2(\hat{\tau}_{(ii')(jj')}^{AB})$, see Goodman (1964a). A slight modification of $\hat{\tau}_{(ii')(jj')}^{AB}$ and $s^2(\hat{\tau}_{(ii')(jj')}^{AB})$ can also be found in the literature in which $\frac{1}{2}$ is added to each cell count, so that $\hat{\tau}_{(ii')(jj')}^{AB}$

is estimated by

$$\bar{\tau}^{AB}_{(ii')(jj')} = \log(n_{ij} + \tfrac{1}{2}) - \log(n_{ij'} + \tfrac{1}{2}) - \log(n_{i'j} + \tfrac{1}{2}) + \log(n_{i'j'} + \tfrac{1}{2}) \qquad (2.29)$$

and $\sigma^2(\bar{\tau}^{AB}_{(ii')(jj')})$ is estimated by

$$s^2(\bar{\tau}^{AB}_{(ii')(jj')}) = \frac{1}{n_{ij} + \tfrac{1}{2}} + \frac{1}{n_{ij'} + \tfrac{1}{2}} + \frac{1}{n_{i'j} + \tfrac{1}{2}} + \frac{1}{n_{i'j'} + \tfrac{1}{2}}. \qquad (2.30)$$

The recommendation that $\tfrac{1}{2}$ be added to frequencies is provided in Goodman (1970). A discussion of the basis for this recommendation is found in Gart and Zweifel (1967), who provide a number of references to earlier work. The earliest use of this adjustment with logarithms of frequencies appears to be by Haldane (1955).

Chi-Square Tests

If a log–linear model defined by (2.4) is satisfied, then the Pearson chi-square statistic

$$X^2 = \sum_i \sum_j (n_{ij} - \hat{m}_{ij})^2 / \hat{m}_{ij}$$

and the likelihood-ratio chi-square statistic

$$L^2 = 2 \sum_i \sum_j n_{ij} \log(n_{ij}/\hat{m}_{ij})$$

both have approximate chi-square distributions with $rs - q - 1$ degrees of freedom. The approximation requires that the means m_{ij} be sufficiently large and that the parameters α and β_k, $1 \leqslant k \leqslant q$, be uniquely determined by $\log(m_{ij}/z_{ij})$, $1 \leqslant i \leqslant r$, $1 \leqslant j \leqslant s$. This uniqueness condition holds if the matrix S of (2.18) has an inverse.

Under the additive log–linear model, these chi-square statistics have approximate chi-square distributions with $(r - 1)(s - 1)$ degrees of freedom. Note that there are rs counts n_{ij} and $(r + s - 1)$ λ-parameters in (2.3). Thus the degrees of freedom are $rs - (r + s - 1) = (r - 1)(s - 1)$.

The accuracy of the large-sample approximation for X^2 has been checked by Monte-Carlo sampling by Lewontin and Felsenstein (1965) in the case of $2 \times s$ tables with s equal to 5 or 10. They found a minimum value of m_{ij} of 1 to be quite adequate to ensure acceptable accuracy. Haldane (1939) provides general formulas for the mean and variance of X^2, conditional on the observed marginal totals $n_i{}^A$ and $n_j{}^B$. The variance formula is given in a simpler form by Dawson (1954). Margolin and Light (1974) make some comparisons of X^2 and L^2 for 3×2 tables in which columns are independent and $n_1{}^B = n_2{}^B$. They consider $N = 10$ and $N = 20$, and note that in their case, X^2 is never greater than L^2, and that the chi-square distribution better describes the distribution of X^2. However, their results are not conclusive.

Adjusted Residuals

If a log–linear model defined as in (2.4) holds, then the adjusted residuals $r_{ij} = (n_{ij} - \hat{m}_{ij})/\hat{c}_{ij}^{1/2}$ have approximate standard normal distributions, where

$$\hat{c}_{ij} = \hat{m}_{ij}[1 - (\hat{m}_{ij}/N) - \sum_k \sum_l (x_{ijk} - \bar{T}_k)(x_{ijl} - \bar{T}_l)\hat{S}^{kl}].$$

This formula has been applied in Section 2.5. It is essentially the same as the formula for adjusted residuals in Section 1.5, As usual, the approximation improves as the m_{ij} increases. If all values of m_{ij} exceed 25, then the approximation is fairly accurate.

An unusual simplification in the formula for r_{ij} is available under the additive model, provided the z_{ij} are all 1. Then as noted by Haberman (1973b, 1974a, page 141),

$$r_{ij} = \frac{n_{ij} - (n_i^A n_j^B / N)}{\{n_i^A n_j^B [1 - (n_i^A/N)][1 - (n_j^B/N)]/N\}^{1/2}}.$$

A more general residual is also easily computed under the additive log–linear model if the z_{ij} are 1. Consider row category scores t_i, $1 \leqslant i \leqslant r$, and column scores u_j, $1 \leqslant j \leqslant s$. Let

$$d = \sum_i \sum_j t_i u_j (n_{ij} - n_i^A n_j^B / N),$$

$$\bar{t}_i = N^{-1} \sum_i n_i^A t_i,$$

$$\bar{u}_j = N^{-1} \sum_j n_j^B u_j,$$

and

$$\hat{c} = N^{-1} \left[\sum_i n_i^A (t_i - \bar{t})^2 \right] \left[\sum_j n_j^B (u_j - \bar{u})^2 \right].$$

Assume that the t_i are not constant for all i and the u_j are not constant for all j. Then as noted in Haberman (1976), $d/\hat{c}^{1/2}$ has an approximate $N(0,1)$ distribution.

To interpret $d/\hat{c}^{1/2}$, consider independent and identically distributed pairs (A_h, B_h), $1 \leqslant h \leqslant N$, such that $1 \leqslant A_h \leqslant r$, $1 \leqslant B_h \leqslant s$. Let $T_h = t_i$ if $A_h = i$ and $U_h = u_j$ if $B_h = j$. Then d is the estimated covariance of the pairs (T_h, U_h), and $N^{1/2}d/\hat{c}^{1/2}$ is the estimated correlation of these pairs.

An example of use of row and column scores appears in Section 2.4. Here $t_2 = 1$, $t_1 = t_3 = t_4 = 0$, $u_1 = -3/n_1^B$, $u_2 = -1/n_2^B$, $u_3 = 1/n_3^B$, and $u_4 =$

$3/n_4{}^B$. Thus

$$d = -3\left(\frac{n_{21}}{n_1{}^B} - \frac{n_2{}^A}{N}\right) - \left(\frac{n_{22}}{n_2{}^B} - \frac{n_2{}^A}{N}\right) + \left(\frac{n_{23}}{n_3{}^B} - \frac{n_2{}^A}{N}\right) + 3\left(\frac{n_{24}}{n_4{}^B} - \frac{n_2{}^A}{N}\right)$$

$$= -3\frac{n_{21}}{n_1{}^B} - \frac{n_{22}}{n_2{}^B} + \frac{n_{23}}{n_3{}^B} + 3\frac{n_{24}}{n_4{}^B},$$

$$\bar{t} = \frac{n_2{}^A}{N},$$

$$\bar{u} = N^{-1}(-3 - 1 + 1 + 3) = 0,$$

$$n_1{}^A\left(1 - \frac{n_2{}^A}{N}\right)^2 + (n_1{}^A + n_3{}^A + n_4{}^A)\left(-\frac{n_2{}^A}{N}\right)^2$$

$$= n_2{}^A\left(1 - \frac{n_2{}^A}{N}\right)^2 + (N - n_2{}^A)\left(\frac{n_2{}^A}{N}\right)^2$$

$$= n_2{}^A\left(1 - \frac{n_2{}^A}{N}\right),$$

and

$$\hat{c} = N^{-1}n_2{}^A\left(1 - \frac{n_2{}^A}{N}\right)\left(\frac{9}{n_1{}^B} + \frac{1}{n_2{}^B} + \frac{1}{n_3{}^B} + \frac{9}{n_4{}^B}\right).$$

Adjusted residuals r_{ij} are obtained by the BMDP program for two-way contingency tables described in Dixon (1975).

EXERCISES

2.1 Lynd and Lynd (1956[1929], page 524) examine student attitudes both to traits desirable in fathers and traits desirable in mothers. Consider the data in Table 2.12. Find an approximate 95 per cent confidence interval for the cross-product ratio.

Solution

The estimated log cross-product ratio is

$$\hat{\tau} = \log\left(\frac{26.5 \times 407.5}{16.5 \times 343.5}\right) = 0.645$$

and

$$s(\hat{\tau}) = \left(\frac{1}{26.5} + \frac{1}{16.5} + \frac{1}{343.5} + \frac{1}{407.5}\right)^{1/2} = 0.322.$$

Table 2.12

High School Student Ratings of Traits Most Desirable
in a Mother[a]

Being a college graduate	Sex of respondent		Total
	Male	Female	
Mentioned	26	16	42
Not mentioned	343	407	750
Total	369	423	792

[a] From Lynd and Lynd (1956 [1929], page 524).

An approximate 95 percent confidence interval for the log cross-product ratio has bounds

$$0.645 - 1.96(0.322) = 0.013$$

and

$$0.645 + 1.96(0.322) = 1.28.$$

Corresponding bounds for the cross-product ratio are

$$e^{0.013} = 1.01$$

and

$$e^{1.28} = 3.58.$$

The difference between male and female respondents is less clear here than in Table 2.1.

2.2 In the test for independence in a 2×2 table, show that

$$X^2 = \frac{N(n_{11}n_{22} - n_{12}n_{21})^2}{n_1{}^A n_2{}^A n_1{}^B n_2{}^B}.$$

Solution

The important observation is that if $d_{ij} = n_{ij} - n_i{}^A n_j{}^B / N$, then

$$d_{11} = -d_{12} = -d_{21} = d_{22}.$$

For example,

$$
\begin{aligned}
d_{11} + d_{12} &= n_{11} - n_1{}^A n_1{}^B / N + n_{12} - n_1{}^A n_2{}^B / N \\
&= (n_{11} + n_{12}) - n_1{}^A (n_1{}^B + n_2{}^B) / N \\
&= n_1{}^A - n_1{}^A \\
&= 0.
\end{aligned}
$$

It now follows that

$$X^2 = d_{11}^2 \sum_i \sum_j 1/(n_i^A n_j^B/N)$$

$$= d_{11}^2 \left(\frac{N}{n_1^A n_1^B} + \frac{N}{n_1^A n_2^B} + \frac{N}{n_2^A n_1^B} + \frac{N}{n_2^A n_2^B} \right)$$

$$= d_{11}^2 \left[\frac{N(n_1^B + n_2^B)}{n_1^A n_1^B n_2^B} + \frac{N(n_1^B + n_2^B)}{n_2^A n_1^B n_2^B} \right]$$

$$= \frac{d_{11}^2 N^2}{n_1^B n_2^B} \left(\frac{1}{n_1^A} + \frac{1}{n_2^A} \right)$$

$$= \frac{d_{11}^2 N^2}{n_1^B n_2^B} \left(\frac{n_1^A + n_2^A}{n_1^A n_2^A} \right)$$

$$= \frac{d_{11}^2 N^3}{n_1^A n_2^A n_1^B n_2^B}.$$

Next note that

$$d_{11} = n_{11} - \frac{(n_{11} + n_{12})(n_{11} + n_{21})}{n_{11} + n_{21} + n_{12} + n_{22}}$$

$$= \frac{n_{11}^2 + n_{11}n_{21} + n_{11}n_{12} + n_{11}n_{22} - n_{11}^2 - n_{11}n_{21} - n_{12}n_{11} - n_{12}n_{21}}{n_{11} + n_{21} + n_{12} + n_{22}}$$

$$= \frac{n_{11}n_{22} - n_{12}n_{21}}{N}.$$

The formula for X^2 now follows without difficulty.

2.3 Construct individual approximate 95 percent confidence intervals for each $\tau_{(i4)(jj')}^{AB}$ of Section 2.3. Compare these approximate confidence intervals to approximate simultaneous 95 percent confidence intervals. Construct corresponding confidence intervals for the cross-product ratios $q_{(i4)(jj')}^{AB}$.

Solution

Results are summarized in Table 2.13. The individual intervals have bounds

$$\hat{\tau}_{(i4)(jj')}^{AB} - 1.96s(\hat{\tau}_{(i4)(jj')}^{AB})$$

and

$$\hat{\tau}_{(i4)(jj')}^{AB} + 1.96s(\hat{\tau}_{(i4)(jj')}^{AB})$$

and the simultaneous intervals have bounds

$$\hat{\tau}_{(i4)(jj')}^{AB} - 2.74s(\hat{\tau}_{(i4)(jj')}^{AB})$$

Table 2.13

Approximate Individual and Simultaneous 95 percent Confidence Intervals for Log Cross-Product Ratios and Cross-Product Ratios for Data in Table 2.4

Diagnostic group (i) compared to schizophrenic group (4)	Types of therapy (j and j') compared	Individual confidence bounds				Simultaneous confidence bounds			
		$\tau^{AB}_{(i4)(jj')}$		$q^{AB}_{(i4)(jj')}$		$\tau^{AB}_{(i4)(jj')}$		$q^{AB}_{(i4)(jj')}$	
		Lower bound	Upper bound	Lower bound	Upper bound	Lower bound	Upper bound	Lower bound	Upper bound
Affective (1)	Psych. and org. (1 and 2)	-0.62	0.28	0.54	1.53	-0.80	0.46	0.45	1.59
	Org. and cust. (2 and 3)	-1.17	-0.63	0.31	0.53	-1.28	-0.53	0.28	0.59
Alcoholic (2)	Psych. and org. (1 and 2)	1.23	2.30	3.43	9.98	1.02	2.51	2.18	12.34
	Org. and cust. (2 and 3)	-0.37	0.85	0.69	2.34	-0.61	1.09	0.54	2.98
Organic (3)	Psych. and org. (1 and 2)	-0.91	0.16	0.40	1.17	-1.12	0.37	0.32	1.45
	Org. and cust. (2 and 3)	-0.18	0.51	0.84	1.67	-0.31	0.65	0.73	1.92
Schizophrenic (4)	Psych. and org. (1 and 2)	0.00	0.00	1.00	1.00	0.00	0.00	1.00	1.00
	Org. and cust. (2 and 3)	0.00	0.00	1.00	1.00	0.00	0.00	1.00	1.00
Senile (5)	Psych. and org. (1 and 2)	0.75	2.28	2.12	9.80	0.45	2.58	1.57	13.26
	Org. and cust. (2 and 3)	-3.02	-1.79	0.05	0.17	-3.27	-1.54	0.04	0.21

and

$$\hat{\tau}^{AB}_{(i4)(jj')} + 2.74s(\hat{\tau}^{AB}_{(i4)(jj')})$$

for $j = 1$ and $j' = 2$ or $j = 2$ and $j' = 3$. The confidence intervals for $q^{AB}_{(i4)(jj')}$ are obtained by exponentiation. For example, the lower bound for the ordinary interval for $q^{AB}_{(i4)(12)}$ is

$$\exp[\hat{\tau}^{AB}_{(i4)(12)} - 1.96s(\hat{\tau}^{AB}_{(i4)(12)})].$$

2.4 Hollingshead and Redlich (1967 [1958], page 260), examine the relationship between diagnostic group and type of psychiatric therapy among neurotic patients. Is there any evidence of a relationship? If so, describe this relationship. (See Table 2.14.)

Table 2.14

Number of Neurotic Patients in Specified Diagnostic Groups Receiving
a Principal Type of Psychiatric Therapy[a]

	Type of therapy			
Diagnostic group	Psychotherapy	Organic therapy	Custodial care	Total
Character disorders	91	13	8	112
Phobic-anxiety reactions	87	9	6	102
Antisocial reactions	85	7	5	97
Depressive reactions	38	8	3	49
Somatization reactions	35	8	3	46
Obsessive-compulsive reactions	18	3	1	22
Hysterical reactions	18	2	1	21
Total	372	50	27	449

[a] Hollingshead and Redlich (1967 [1958], page 260). The percentages listed in the source are not consistent with the reported marginal totals. The totals have been changed in Table 2.4 to provide counts consistent with the reported percentages.

Solution

There are $(7 - 1)(3 - 1) = 12$ degrees of freedom. The X^2 statistic is 6.16, and L^2 is 6.00, so there is no indication that treatment and diagnostic group are related. The estimated expected values and adjusted residuals of Table 2.15 are consistent with this conclusion. The cell counts are rather small, so the quality of the chi-square approximation is open to some question under the Cochran (1954) rule that no more than 20 percent of the cells have expected counts less than 5. Nonetheless, X^2 and L^2 are so small that modest deviations from large-sample theory cannot affect the inference.

2.5 The question on social class in Table 1.6 appears in the General Social Survey from 1972 to 1975. Responses are summarized in Table 2.16. Have there been any changes over time in the distribution of responses?

Solution

The homogeneity model yields a Pearson chi-square 14.7 and the likelihood-ratio chi-square is 14.6. There are 9 degrees of freedom, so the significance level of these statistics is about 10 percent. These statistics provide weak evidence for some lack of homogeneity. A glance at the adjusted residual $r_{11} = 3.19$ in Table 2.17 suggests something anomalous occurred in 1972 with the classification of lower class.

Table 2.15

Observed Counts, Estimated Expected Counts, and Adjusted Residuals for the Independence Model for Table 2.14[a]

Diagnostic group	Psychotherapy	Organic therapy	Custodial care
	Type of therapy[a]		
Character disorders	91	13	8
	92.8	12.5	6.7
	−0.52	0.18	0.58
Phobic-anxiety reactions	87	9	6
	84.5	11.4	6.1
	0.74	−0.84	−0.06
Antisocial reactions	85	7	5
	80.4	10.8	5.8
	1.41	−1.38	−0.40
Depressive reactions	38	8	3
	40.6	5.6	2.9
	−1.04	1.22	0.03
Somatization reactions	35	8	3
	38.1	5.1	2.8
	−1.28	1.42	0.15
Obsessive-compulsive reactions	18	3	1
	18.2	2.4	1.3
	−0.13	0.38	−0.30
Hysterical reactions	18	2	1
	17.4	2.3	1.3
	0.36	−0.24	−0.25

[a] The first line gives the observed count n_{ij}, the second line the estimated expected count $\hat{m}_{ij}$, and the third line is the adjusted residual r_{ij}.

Table 2.16

Subjects in the 1972–1975 General Social Surveys Cross-Classified by Reported Social Class and Year of Survey[a]

Response[b]	1972	1973	1974	1975	Total
		Year			
Lower class	104	27	64	72	267
Working class	760	358	689	714	2521
Middle class	704	343	676	655	2377
Upper class	36	21	46	41	144
Total	1604	748	1475	1482	5309

[a] National Opinion Research Center (1972, page 45, 1973, page 29, 1974, page 29, 1975, page 41).

[b] There are 31 excluded responses of "don't know" or "no answer." Only about half the subjects in 1973 were asked the question described in Table 1.6.

Table 2.17

Observed Counts, Estimated Expected Counts, and Adjusted
Residuals for the Homogeneity Model for Table 2.16[a]

| | Year | | | |
Response	1972	1973	1974	1975
Lower class	104	27	64	72
	80.7	37.6	74.2	74.5
	3.19	-1.92	-1.43	-0.35
Working class	760	358	689	714
	761.7	355.2	700.4	703.7
	-0.10	0.22	-0.70	0.63
Middle class	704	342	676	655
	718.2	334.9	660.4	663.5
	-0.85	0.56	0.96	-0.52
Upper class	36	21	46	41
	43.5	20.3	40.0	40.2
	-1.38	0.17	1.13	0.15

[a] The first line gives the observed count n_{ij}, the second line
the estimated expected count $\hat{m}_{ij}$, and the third line is the
adjusted residual.

2.6 Show that under the additive model of Section 2.6, if $z_{ij} = a_i b_j$, $1 \leq i \leq r$, $1 \leq j \leq s$, for some $a_i > 0$, $1 \leq i \leq r$, and $b_j > 0$, $1 \leq j \leq s$, then

$$\hat{m}_{ij} = n_i^A n_j^B / N.$$

Show that if the Deming–Stephan algorithm is used, then

$$m_{ij2} = \hat{m}_{ij}.$$

Thus only 2 steps of the algorithm are required.

Solution

To show that $\hat{m}_{ij} = n_i^A n_j^B / N$, note that it is already known that

$$\sum_j (n_i^A n_j^B / N) = n_i^A$$

and

$$\sum_i (n_i^A n_j^B / N) = n_j^B.$$

Let $w_i = z_{ij}$,

$$\hat{\lambda} = \frac{1}{r} \sum_{i'} \log(n_{i'}^A / a_{i'}) + \frac{1}{s} \sum_{j'} \log(n_j^A / b_{j'}) - \log N,$$

$$\hat{\lambda}_i^A = \log(n_i^A / a_i) - \frac{1}{r} \sum_{i'} \log(n_{i'}^A / a_{i'}),$$

and

$$\hat{\lambda}_j^B = \log(n_j^B/b_j) - \frac{1}{s}\sum_{j'} \log(n_{j'}^B/b_{j'}).$$

Note that

$$\sum \hat{\lambda}_i^A = 0,$$
$$\sum \hat{\lambda}_j^B = 0,$$

and

$$\log\left(\frac{n_i^A n_j^B/N}{z_{ij}}\right) = \log(n_i^A/a_i) + \log(n_j^B/b_j) - \log N$$

$$= \hat{\lambda} + \hat{\lambda}_i^A + \hat{\lambda}_j^B.$$

To show that $m_{ij2} = \hat{m}_{ij}$, note that

$$m_{ij1} = m_{ij0}n_i^A/m_{i0}^A = a_i b_j n_i^A \bigg/ \left(a_i \sum_{j'} b_{j'}\right) = n_i^A b_j \bigg/ \sum_{j'} b_{j'}$$

and

$$m_{ij2} = m_{ij0}n_j^B/m_{j0}^B = \left(n_i^A b_j \bigg/ \sum_{j'} b_{j'}\right)n_j^B \bigg/ \left(Nb_j \bigg/ \sum_{j'} b_{j'}\right) = n_i^A n_j^B/N.$$

2.7   Verify that (2.10), (2.11), and (2.12) imply (2.14) and (2.15).

Solution

It suffices to consider (2.14). Addition of (2.11) over i' yields

$$\sum_{i'=1}^{r-1} (\hat{m}_{i'}^A - m_r^A) = \sum_{i'=1}^{r-1} (n_{i'}^A - n_r^A).$$

Note that

$$\sum_{i'=1}^{r-1} \hat{m}_{i'}^A = \hat{M} - \hat{m}_r^A$$

and

$$\sum_{i'=1}^{r-1} n_{i'}^A = N - n_r^A.$$

Thus

$$\sum_{i'=1}^{r-1} (\hat{m}_{i'}^A - \hat{m}_r^A) = \hat{M} - r\hat{m}_r^A$$

and

$$\sum_{i'=1}^{r-1} (n_{i'}^A - n_r{}^A) = N - rn_r{}^A.$$

Since

$$\hat{M} = N$$

and

$$\sum_{i'=1}^{r-1} (\hat{m}_{i'}^A - \hat{m}_r{}^A) = \sum_{i'=1}^{r-1} (n_{i'}^A - n_r{}^A),$$

it follows that $\hat{m}_r{}^A = n_r{}^A$. By (2.11), $\hat{m}_i{}^A = n_i{}^A$, $1 \leqslant i \leqslant r - 1$. Thus (2.14) holds for $1 \leqslant i \leqslant r$.

3 *Complete Three-Way Tables*

When three polytomous variables are cross-classified, a large number of relationships between the variables may be considered. In this chapter, a class of log-linear models called hierarchical models is used to describe such relationships. These models are not the only log–linear models appropriate for three-way tables; others are described in Chapters 5 to 8.

Nonetheless, hierarchical models do provide a general tool for examination of contingency tables in which two or more variables are cross-classified. Such models are based on a general method of parametrization commonly encountered in the analysis of variance of factorial tables. In the case of log–linear models for a three-way table, let counts n_{ijk}, $1 \leqslant i \leqslant r$, $1 \leqslant j \leqslant s$, $1 \leqslant k \leqslant t$, be arranged in a three-way table. Assume each n_{ijk} has a positive expected value m_{ijk}. Then for some unique λ, λ_i^A, $1 \leqslant i \leqslant r$, λ_j^B, $1 \leqslant j \leqslant s$, λ_k^C, $1 \leqslant k \leqslant t$, λ_{ij}^{AB}, $1 \leqslant i \leqslant r$, $1 \leqslant j \leqslant s$, λ_{ik}^{AC}, $1 \leqslant i \leqslant r$, $1 \leqslant k \leqslant t$, λ_{jk}^{BC}, $1 \leqslant j \leqslant s$, $1 \leqslant k \leqslant t$, and λ_{ijk}^{ABC}, $1 \leqslant i \leqslant r$, $1 \leqslant j \leqslant s$, $1 \leqslant k \leqslant t$,

$$\log m_{ijk} = \lambda + \lambda_i^A + \lambda_j^B + \lambda_k^C + \lambda_{ij}^{AB} + \lambda_{ik}^{AC} + \lambda_{jk}^{BC} + \lambda_{ijk}^{ABC} \tag{3.1}$$

and

$$\sum \lambda_i^A = \sum \lambda_j^B = \sum \lambda_k^C = \sum_i \lambda_{ij}^{AB} = \sum_j \lambda_{ij}^{AB} = \sum_i \lambda_{ik}^{AC} = \sum_k \lambda_{ik}^{AC}$$

$$= \sum_j \lambda_{jk}^{BC} = \sum_k \lambda_{jk}^{BC} = \sum_i \lambda_{ijk}^{ABC} = \sum_j \lambda_{ijk}^{ABC} = \sum_k \lambda_{ijk}^{ABC} = 0. \tag{3.2}$$

As in the analysis of variance, the λ_i^A parameters are called the main effects for variable A, the λ_{ij}^{AB} are $A \times B$ interactions, and the λ_{ijk}^{ABC} are $A \times B \times C$ interactions. The λ_{ij}^{AB}, λ_{ik}^{AC}, and λ_{jk}^{BC} are the two-factor interactions, while the λ_{ijk}^{ABC} are the three-factor interactions. If no restrictions are imposed on the

λ-parameters, then in the language of Goodman (1970), (3.1) and (3.2) specify a saturated model. The saturated model is one example of a hierarchical model.

In other hierarchical models, some λ-parameters in (3.1) and (3.2) are set to 0. The hierarchy restriction is followed that if any λ-parameter with superscript S is set to 0, then any λ-parameter of the same or higher order is set to 0. Here a λ-parameter has the same or higher order if its superscript contains each letter of S. For example, one may assume that each λ_{ijk}^{ABC} is 0, so that one has the model of no three-factor interaction

$$\log m_{ijk} = \lambda + \lambda_i^A + \lambda_j^B + \lambda_k^C + \lambda_{ij}^{AB} + \lambda_{ik}^{AC} + \lambda_{jk}^{BC}.$$

One may also assume that each λ_{ij}^{AB} and each λ_{ijk}^{ABC} are 0, so that

$$\log m_{ijk} = \lambda + \lambda_i^A + \lambda_j^B + \lambda_k^C + \lambda_{ik}^{AC} + \lambda_{jk}^{BC}.$$

On the other hand, the assumption

$$\log m_{ijk} = \lambda + \lambda_{ij}^{AB} + \lambda_k^C$$

does not yield a hierarchical log–linear model. Since the λ_i^A are assumed 0, the λ_{ij}^{AB} n. 'st also be 0.

Several factors make hierarchical models of special interest. As Birch (1963) and Goodman (1970) note, except for the model of no three-factor interaction, each hierarchical model for a three-way table is equivalent to one or more hypotheses of independence, conditional independence, or equiprobability, and the model of no three-factor interaction can be interpreted in terms of comparison of cross-product ratios in several two-way tables. Thus hierarchical models have simple interpretations. Some computational convenience is associated with hierarchical models. Maximum-likelihood estimates and adjusted residuals for hierarchical models can be found without recourse to iterative computation, except for the model of no three-factor interaction. In the latter model the iterative proportional fitting algorithm provides an alternative to the Newton–Raphson algorithm for computation of maximum-likelihood estimates. Even if the Newton–Raphson algorithm is used, special simplifications are sometimes available. Thus hierarchical models are an attractive family of models for three-way tables. They have the further virtue of easy generalization to higher-way tables, as will be seen in Chapter 4.

Section 3.1 provides an example of the simplest possible three-way table, a $2 \times 2 \times 2$ contingency table. This table can be analyzed by comparing log odds ratios and cross-product ratios similar to those encountered in the 2×2 table of Section 2.1. Complications arise only when a log–linear model is used in which log cross-product ratios for distinct subtables are assumed equal. The resulting log–linear model, called a model of no three-factor interaction, has maximum-likelihood estimates which can only be obtained by iterative computation. For this purpose, a version of the Newton–Raphson algorithm is introduced which has a much different appearance than those found in earlier chapters, although the new version is algebraically equivalent to the earlier examples. Chi-square tests for the model of no three-factor interaction are introduced, and implications of the model are discussed for estimation of log odds and log cross-product ratios. The relationship of the model of no three-factor interaction to the general class of hierarchical models is also explored.

Section 3.2 examines a slightly more complex $2 \times 3 \times 2$ table. This table is used to introduce the concepts of marginal and partial association. Chi-square tests and adjusted residuals are introduced to test whether these forms of association are present. Measurement of these forms by association by log cross-product ratios is also considered. The model of no three-factor interaction is discussed in terms of its relationship to measurement of partial association. In this section, iterative proportional fitting is applied to this model.

Section 3.3 provides an enumeration of all possible hierarchical models. Their interpretations are given in terms of hypotheses of independence, conditional independence, equiprobability, or constant partial association.

Section 3.4 provides a general discussion of procedures for computation of maximum-likelihood estimates and estimated asymptotic variances. The Newton-Raphson algorithm and the iterative proportional fitting algorithm are applied to both these tasks.

Section 3.5 considers problems of model selection. Chi-square statistics, partitions of likelihood-ratio chi-squares, adjusted residuals, and standardized parameters are all described.

3.1   Race, Sex, and Homicide Weapons—A $2 \times 2 \times 2$ Table

The $2 \times 2 \times 2$ table has a position relative to general $r \times s \times t$ tables analogous to the position of the 2×2 table relative to $r \times s$ tables. The $2 \times 2 \times 2$ table is somewhat simpler to describe and analyze than a general $r \times s \times t$ table; however, the basic principles involved are not much different.

The principal gain with $2 \times 2 \times 2$ tables is the ability to describe the table by use of a limited number of log odds and log cross-product ratios. An added gain is associated with the model of no three-factor interaction introduced in this section. A very simple version of the Newton–Raphson algorithm is available for $2 \times 2 \times 2$ tables. The algorithm for $r \times s \times t$ tables is much less convenient.

To illustrate analysis of $2 \times 2 \times 2$ tables, consider Table 3.1. This table contains a cross-classification of homicide victims in 1970 by race, sex, and type of assault. To simplify analysis, only homicides involving firearms and explosives or cutting and piercing instruments have been included and racial groups other than whites and blacks are excluded. The other types of assault such as legal interventions or hangings and strangulations are less common and have a somewhat different relationship to sex and race than do the types of assault used in Table 3.1. Relatively few homicides involve racial groups other than whites and blacks. In all, 82 percent of all reported homicides are included in Table 3.1.

The principal question considered here is the extent to which type of assault depends on either the race or the sex of the victim. To begin an investigation of this problem, let n_{ijk} be the number of reported 1970 homicide victims of race i and sex j killed by an assault of k. For instance, n_{221} is the number of black women reported killed by firearms and explosives. As a rough approximation, assume that each n_{ijk} has an independent Poisson distribution with mean m_{ijk}.

Comparison of Relative Risks

The relative risk of homicide by firearms and explosives rather than by cutting and piercing instruments is m_{ij1}/m_{ij2} for persons of race i and sex j. The log relative risk $\log(m_{ij1}/m_{ij2})$ may be denoted by $\tau_{12 \cdot ij}^{C \cdot AB}$. Here the 1 and

Tables 3.1

*Reported Homicides in 1970, Classified by Race and Sex of Victim
and by Type of Assault*[a]

| | | Type of assault[b] | | |
| | | Firearms and explosives | Cutting and piercing instruments | Total |
Race	Sex			
White	Male	3910	808	4718
	Female	1050	234	1284
Black	Male	5218	1385	6603
	Female	929	298	1227
Total		11,107	2725	13,832

[a] From National Center for Health Statistics (1970, pages 1–183).

[b] Homicides are only included if the assault is by one of the two specified methods, and other races than white and black are excluded. Of 16,848 reported homicides in 1970, 3016 are eliminated by these restrictions.

2 indicate a comparison of relative risks for values 1 (firearms and explosives) and 2 (cutting and piercing instruments) of variable C (type of assault) given value i of variable A (race of victim) and value j of variable B (sex of victim). The estimated value of $\tau_{12 \cdot ij}^{C \cdot AB}$ is

$$\hat{\tau}_{12 \cdot ij}^{C \cdot AB} = \log(n_{ij1}/n_{ij2}),$$

which has an approximate normal distribution with asymptotic mean $\tau_{12 \cdot ij}^{C \cdot AB}$ and asymptotic variance

$$\sigma^2(\hat{\tau}_{12 \cdot ij}^{C \cdot AB}) = \frac{1}{m_{ij1}} + \frac{1}{m_{ij2}}.$$

The estimated asymptotic variance (EASD) is then

$$s^2(\hat{\tau}_{12 \cdot ij}^{C \cdot AB}) = \frac{1}{n_{ij1}} + \frac{1}{n_{ij2}}.$$

Halves may be added to the n_{ijk} to improve approximations; however, the effect is negligible given the large values of n_{ijk} observed. Note the similarity of formulas for $\hat{\tau}_{12 \cdot ij}^{C \cdot AB}$ here and corresponding formulas in Section 2.1 for $\hat{\tau}_{12 \cdot j}^{A \cdot B}$.

Estimates $\log(n_{ij1}/n_{ij2})$ and estimated asymptotic standard deviations

$$\left(\frac{1}{n_{ij1}} + \frac{1}{n_{ij2}} \right)^{1/2}$$

are presented in Table 3.2. Variations involving both race and sex appear present, with variation due to race larger than variation due to sex.

Variations involving race are considered first. Note that given that a victim is of sex j, the difference

$$\tau_{(12)(12) \cdot j}^{AC \cdot B} = \log\left(\frac{m_{1j1} m_{2j2}}{m_{1j2} m_{2j1}} \right)$$

$$= \log\left(\frac{m_{1j1}}{m_{1j2}} \right) - \log\left(\frac{m_{2j1}}{m_{2j2}} \right)$$

$$= \tau_{12 \cdot 1j}^{C \cdot AB} - \tau_{12 \cdot 2j}^{C \cdot AB}$$

in log relative risk for blacks and whites has an estimate

$$\hat{\tau}_{(12)(12) \cdot j}^{AC \cdot B} = \log\left(\frac{n_{1j1} n_{2j2}}{n_{1j2} n_{2j1}} \right)$$

$$= \log\left(\frac{n_{1j1}}{n_{1j2}} \right) - \log\left(\frac{n_{2j1}}{n_{2j2}} \right)$$

$$= \hat{\tau}_{12 \cdot 1j}^{C \cdot AB} - \hat{\tau}_{12 \cdot 2j}^{C \cdot AB}.$$

Table 3.2

Estimated Log Relative Risks of Assault by Firearms and
Explosives Rather than Cutting and Piercing Instruments,
Based on Table 3.1

Race	Sex	Estimated log relative risk	Estimate asymptotic standard deviation
White	Male	1.577	0.039
	Female	1.501	0.072
Black	Male	1.326	0.030
	Female	1.137	0.067

This estimate has an approximate normal distribution with asymptotic mean $\tau^{AC \cdot B}_{(12)(12) \cdot j}$ and asymptotic variance

$$\sigma^2(\hat{\tau}^{AC \cdot B}_{(12)(12) \cdot j}) = \frac{1}{m_{1j1}} + \frac{1}{m_{1j2}} + \frac{1}{m_{2j1}} + \frac{1}{m_{2j2}}$$

$$= \sigma^2(\hat{\tau}^{C \cdot AB}_{12 \cdot 1j}) + \sigma^2(\hat{\tau}^{C \cdot AB}_{12 \cdot 2j}).$$

The symbol $\tau^{AC \cdot B}_{(12)(12) \cdot j}$ indicates that a log cross-product ratio has been taken involving categories 1 and 2 of variable A and categories 1 and 2 of variable C. The log cross-product ratio has been taken conditional on value j of variable B. Note that $\hat{\tau}^{AC \cdot B}_{(12)(12) \cdot j}$ is the estimated log cross-product ratio of the 2 × 2 table of victims of sex j in which race and type of assault are cross-classified.

In the case of male victims, $\hat{\tau}^{AC \cdot B}_{(12)(12) \cdot 1} = 1.577 - 1.326 = 0.250$. (The discrepancy in the last decimal place is due to rounding.) The EASD is

$$s(\hat{\tau}^{AC \cdot B}_{(12)(12) \cdot 1}) = \left(\frac{1}{3910} + \frac{1}{808} + \frac{1}{5218} + \frac{1}{1385} \right)^{1/2} = 0.049.$$

If the relative risk $\tau^{C \cdot AB}_{12 \cdot 11}$ for white males is equal to the relative risk $\tau^{C \cdot AB}_{12 \cdot 21}$ for black males, then $\tau^{AC \cdot B}_{(12)(12) \cdot 1} = 0$. Thus the standardized value

$$\hat{\tau}^{AC \cdot B}_{(12)(12) \cdot 2} / s(\hat{\tau}^{AC \cdot B}_{(12)(12) \cdot 2})$$

has an approximate $N(0, 1)$ distribution. Since in Table 3.1, this standardized value is

$$0.250/0.049 = 5.10,$$

there is very strong evidence of different relative risks for white and black males. The magnitude of the difference is somewhat uncertain. An approximate 95 percent confidence interval has lower bound

$$0.250 - 1.96(0.049) = 0.154$$

and upper bound

$$0.250 + 1.96(0.049) = 0.346;$$

nonetheless, some difference is clearly present.

Somewhat similar results appear for female victims. The estimate $\hat{\tau}^{AC \cdot B}_{(12)(12) \cdot 2}$ is

$$1.501 - 1.137 = 0.364$$

and

$$s(\hat{\tau}^{AC \cdot B}_{(12)(12) \cdot 2}) = \left(\frac{1}{1050} + \frac{1}{234} + \frac{1}{929} + \frac{1}{298} \right)^{1/2} = 0.098.$$

The standardized value

$$0.364/0.098 = 3.71$$

is also highly significant. Due to the smaller number of homicide victims who are female, the EASD for females is about twice that for males. Thus the difference in relative risk is rather poorly determined. The corresponding approximate 95 percent confidence bounds for females are

$$0.364 - 1.96(0.098) = 0.172$$

and

$$0.364 + 1.96(0.098) = 0.557.$$

To explore contrasts between males and females, consider the difference

$$\tau^{BC \cdot A}_{(12)(12) \cdot i} = \log\left(\frac{m_{i11} m_{i22}}{m_{i21} m_{i12}} \right)$$

$$= \log\left(\frac{m_{i11}}{m_{i12}} \right) - \log\left(\frac{m_{i21}}{m_{i22}} \right)$$

$$= \tau^{C \cdot AB}_{12 \cdot i1} - \tau^{C \cdot AB}_{12 \cdot i2}$$

in log relative risk for males and females of race i. The estimate

$$\hat{\tau}^{BC \cdot A}_{(12)(12) \cdot i} = \log\left(\frac{n_{i11} n_{i22}}{n_{i21} n_{i12}} \right)$$

$$= \log\left(\frac{n_{i11}}{n_{i12}} \right) - \log\left(\frac{n_{i21}}{n_{i22}} \right)$$

$$= \hat{\tau}^{C \cdot AB}_{12 \cdot i1} - \hat{\tau}^{C \cdot AB}_{12 \cdot i2}$$

has an approximate normal distribution with asymptotic mean $\tau^{BC \cdot A}_{(12)(12) \cdot i}$ and asymptotic variance

$$\sigma^2(\hat{\tau}^{BC \cdot A}_{(12)(12) \cdot i}) = \frac{1}{m_{i11}} + \frac{1}{m_{i21}} + \frac{1}{m_{i12}} + \frac{1}{m_{i22}}$$

$$= \sigma^2(\hat{\tau}^{C \cdot AB}_{12 \cdot i1}) + \sigma^2(\hat{\tau}^{C \cdot AB}_{12 \cdot i2}).$$

The estimated asymptotic variance is

$$s^2(\hat{\tau}^{BC \cdot A}_{(12)(12) \cdot i}) = \frac{1}{n_{i11}} + \frac{1}{n_{i21}} + \frac{1}{n_{i12}} + \frac{1}{n_{i22}}.$$

Given that the victim is white, the estimated difference is

$$\hat{\tau}^{BC \cdot A}_{(12)(12) \cdot 1} = 1.577 - 1.501 = 0.076.$$

The corresponding EASD is

$$\left(\frac{1}{3910} + \frac{1}{808} + \frac{1}{1050} + \frac{1}{234} \right)^{1/2} = 0.082.$$

There is no real evidence here that the relative risks for white male and white female victims differ. Note that the standardized value $0.076/0.082$ is only 0.92. On the other hand, this difference need not be negligible. The upper bound of a 95 percent confidence interval for $\tau^{BC \cdot A}_{(12)(12) \cdot 1}$ is

$$0.076 + 1.96(0.082) = 0.236.$$

The situation with black victims is somewhat clearer. The estimated difference $\hat{\tau}^{BC \cdot A}_{(12)(12) \cdot 2}$ in log relative risk is

$$1.326 - 1.137 = 0.189.$$

The EASD is

$$\left(\frac{1}{5218} + \frac{1}{1385} + \frac{1}{929} + \frac{1}{298} \right)^{1/2} = 0.073,$$

so the standardized value is

$$0.189/0.073 = 2.59.$$

The probability that a standard normal deviate exceeds 2.59 in magnitude is only about 1 percent.

So far, the following conclusions can be reached. Given either sex, white homicide victims are relatively more likely than blacks to be killed by firearms and explosives rather than by cutting or piercing instruments. Black male homicide victims are relatively more likely than black female homicide victims to be killed by firearms or explosives rather than by cutting or piercing instruments.

It should be noted that these statements of relative risk are conditional on a person being a victim. Since there are far fewer blacks in the general population than whites and about the same number of men and women in the population, blacks are much more likely than whites to be homicide victims and men are much more likely than women to be homicide victims.

A more subtle issue to consider is whether the difference $\tau^{AC \cdot B}_{(12)(12) \cdot 1}$ between white male and black male relative risks is the same as the difference $\tau^{AC \cdot B}_{(12)(12) \cdot 2}$ between white female and black female relative risks. Equality permits a single number to be used to express the difference between whites and blacks in relative risk. This single expression can be estimated more efficiently than can the differences between whites and blacks that are conditional on sex. Thus there is an added gain if $\tau^{AC \cdot B}_{(12)(12) \cdot 1} = \tau^{AC \cdot B}_{(12)(12) \cdot 2}$. It is then the case that

$$\tau^{ABC}_{(12)(12)(12)} = \log\left(\frac{m_{111}m_{212}m_{122}m_{221}}{m_{112}m_{211}m_{121}m_{222}}\right) = 0.$$

Since

$$\tau^{BC \cdot A}_{(12)(12) \cdot 1} - \tau^{BC \cdot A}_{(12)(12) \cdot 2} = \tau^{ABC}_{(12)(12)(12)},$$

the difference $\tau^{BC \cdot A}_{(12)(12) \cdot i}$ in log relative risk for males and females of race i is the same for $i = 1$ (whites) or $i = 2$ (blacks). Thus a single number can express the difference in log relative risk for males and females.

The contrast $\hat{\tau}^{ABC}_{(12)(12)(12)}$ has an estimate

$$\hat{\tau}^{ABC}_{(12)(12)(12)} = \log\left(\frac{n_{111}n_{212}n_{122}n_{221}}{n_{112}n_{211}n_{121}n_{222}}\right)$$

$$= \hat{\tau}^{AC \cdot B}_{(12)(12) \cdot 1} - \hat{\tau}^{AC \cdot B}_{(12)(12) \cdot 2}$$

$$= \hat{\tau}^{BC \cdot A}_{(12)(12) \cdot 1} - \hat{\tau}^{BC \cdot A}_{(12)(12) \cdot 2}.$$

The estimate is approximately normally distributed with asymptotic mean $\tau^{ABC}_{(12)(12)(12)}$ and asymptotic variance

$$\sigma^2(\hat{\tau}^{ABC}_{(12)(12)(12)}) = \sum_i \sum_j \sum_k \frac{1}{m_{ijk}}$$

$$= \frac{1}{m_{111}} + \frac{1}{m_{112}} + \frac{1}{m_{211}} + \frac{1}{m_{212}} + \frac{1}{m_{121}} + \frac{1}{m_{122}} + \frac{1}{m_{221}} + \frac{1}{m_{222}}$$

$$= \sigma^2(\hat{\tau}^{AC \cdot B}_{(12)(12) \cdot 1}) + \sigma^2(\hat{\tau}^{AC \cdot B}_{(12)(12) \cdot 2})$$

$$= \sigma^2(\hat{\tau}^{BC \cdot A}_{(12)(12) \cdot 1}) + \sigma^2(\hat{\tau}^{BC \cdot A}_{(12)(12) \cdot 2}).$$

The estimated asymptotic variance is

$$s^2(\hat{\tau}^{ABC}_{(12)(12)(12)}) = \sum_i \sum_j \sum_k \frac{1}{n_{ijk}}.$$

In Table 3.1,

$$\hat{\tau}^{ABC}_{(12)(12)(12)} = 0.250 - 0.364 = -0.114.$$

and

$$s(\hat{\tau}^{ABC}_{(12)(12)(12)}) = \left(\frac{1}{3910} + \frac{1}{5218} + \frac{1}{808} + \frac{1}{1385} + \frac{1}{1050} + \frac{1}{929} + \frac{1}{234} + \frac{1}{298} \right)^{1/2}$$

$$= 0.110.$$

The standardized value of $-0.114/0.110 = -1.04$ provides no indication that $\tau^{ABC}_{(12)(12)(12)}$ is not zero. Thus an attempt may be made to find a common estimate $\hat{\tau}^{AC}_{(12)(12)}$ for $\tau^{AC \cdot B}_{(12)(12) \cdot 1}$ and $\tau^{AC \cdot B}_{(12)(12) \cdot 2}$ and a common estimate $\hat{\tau}^{BC}_{(12)(12)}$ for $\tau^{BC \cdot A}_{(12)(12) \cdot 1}$ and $\tau^{BC \cdot A}_{(12)(12) \cdot 2}$.

The Model of No Three-Factor Interaction

There are two closely related procedures for estimation of the common differences $\tau^{AC}_{(12)(12)}$ and $\tau^{BC}_{(12)(12)}$. The method of maximum likelihood goes back to Bartlett (1935), although the computational procedures used here are somewhat newer. An approach that corresponds to use of the first cycle of the Newton–Raphson algorithm is due to Woolf (1955). The maximum-likelihood approach is considered here. See Exercise 3.1 for the Woolf (1955) approach.

The model that only assumes that the contrast $\tau^{ABC}_{(12)(12)(12)} = 0$ is called a log–linear model of no three-factor interaction. The model is a log–linear model since the logarithms $\mu_{ijk} = \log m_{ijk}$ of the expected values m_{ijk} are subject to the linear constraint.

$$\mu_{111} - \mu_{211} - \mu_{121} + \mu_{221} - \mu_{112} + \mu_{212} + \mu_{122} - \mu_{222} = 0.$$

The name of the model arises from a decomposition commonly encountered in analysis of variance. Explicit use of this decomposition in the analysis of contingency tables appears to be due to Birch (1963). Whenever all mean m_{ijk} are positive, one may write

$$\mu_{ijk} = \log m_{ijk} = \lambda + \lambda_i^A + \lambda_j^B + \lambda_k^C + \lambda_{ij}^{AB} + \lambda_{ik}^{AC} + \lambda_{jk}^{BC} + \lambda_{ijk}^{ABC},$$

where

$$\sum_i \lambda_i^A = \sum_j \lambda_j^B = \sum_k \lambda_k^C = \sum_j \lambda_{ij}^{AB} \sum_i \lambda_{ij}^{AB} = \sum_k \lambda_{ik}^{AC} = \sum_i \lambda_{ik}^{AC}$$

$$= \sum_k \lambda_{jk}^{BC} = \sum_j \lambda_{jk}^{BC} = \sum_k \lambda_{ijk}^{ABC} = \sum_j \lambda_{ijk}^{ABC} = \sum_i \lambda_{ijk}^{ABC} = 0.$$

As is well known, in a $2 \times 2 \times 2$ table,

$$\lambda = \tfrac{1}{8}(\mu_{111} + \mu_{211} + \mu_{121} + \mu_{221} + \mu_{112} + \mu_{212} + \mu_{122} + \mu_{222}),$$

$$\lambda_1{}^A = -\lambda_2{}^A = \tfrac{1}{8}(\mu_{111} - \mu_{211} + \mu_{121} - \mu_{221} + \mu_{112} - \mu_{212} + \mu_{122} - \mu_{222}),$$

$$\lambda_1{}^B = -\lambda_2{}^B = \tfrac{1}{8}(\mu_{111} + \mu_{211} - \mu_{121} - \mu_{221} + \mu_{112} + \mu_{212} - \mu_{122} - \mu_{222}),$$

$$\lambda_1{}^C = -\lambda_2{}^C = \tfrac{1}{8}(\mu_{111} + \mu_{211} + \mu_{121} + \mu_{221} - \mu_{112} - \mu_{212} - \mu_{122} - \mu_{222}),$$

$$\lambda_{11}^{AB} = -\lambda_{21}^{AB} = -\lambda_{12}^{AB} = \lambda_{22}^{AB}$$
$$= \tfrac{1}{8}(\mu_{111} - \mu_{211} - \mu_{121} + \mu_{221} + \mu_{112} - \mu_{212} - \mu_{122} + \mu_{222}),$$

$$\lambda_{11}^{AC} = -\lambda_{21}^{AC} = -\lambda_{12}^{AC} = \lambda_{22}^{AC}$$
$$= \tfrac{1}{8}(\mu_{111} - \mu_{211} + \mu_{121} - \mu_{221} - \mu_{112} + \mu_{212} - \mu_{122} + \mu_{222}),$$

$$\lambda_{11}^{BC} = -\lambda_{21}^{BC} = -\lambda_{12}^{BC} = \lambda_{22}^{BC}$$
$$= \tfrac{1}{8}(\mu_{111} + \mu_{211} - \mu_{121} - \mu_{221} - \mu_{112} - \mu_{212} + \mu_{122} + \mu_{222}),$$

and

$$\lambda_{111}^{ABC} = -\lambda_{211}^{ABC} = -\lambda_{121}^{ABC} = \lambda_{221}^{ABC}$$
$$= -\lambda_{112}^{ABC} = -\lambda_{212}^{ABC} = \lambda_{122}^{ABC} = -\lambda_{222}^{ABC}$$
$$= \tfrac{1}{8}(\mu_{111} - \mu_{211} - \mu_{121} + \mu_{221} - \mu_{112} + \mu_{212} + \mu_{122} - \mu_{222})$$
$$= \tfrac{1}{8}\tau_{(12)(12)(12)}^{ABC}.$$

For example, see Goodman (1970). In this decomposition, the λ_{ijk}^{ABC} are called three-factor interactions. Thus the assumption that $\tau_{(12)(12)(12)}^{ABC} = 0$ is equivalent to the assumption that the three-factor interactions λ_{ijk}^{ABC} are 0. Thus $\tau_{(12)(12)(12)}^{ABC} = 0$ if and only if

$$\mu_{ijk} = \lambda + \lambda_i{}^A + \lambda_j{}^B + \lambda_k{}^C + \lambda_{ij}^{AB} + \lambda_{ik}^{AC} + \lambda_{jk}^{BC}.$$

The Newton–Raphson Algorithm

The model of no three-factor interaction for a $2 \times 2 \times 2$ table is a log–linear model defined by a single constraint

$$\gamma = \sum_i \sum_j \sum_k c_{ijk}\mu_{ijk} = c_{111}\mu_{111} + c_{211}\mu_{211} + c_{121}\mu_{121} + c_{221}\mu_{221}$$

$$+ c_{112}\mu_{112} + c_{212}\mu_{212} + c_{122}\mu_{122} + c_{222}\mu_{222}$$

$$= 0$$

such that

$$\sum_i \sum_j \sum_k c_{ijk} = 0.$$

In the case under study, the c_{ijk} are defined as in Table 3.3. For models defined by single constraints, the Newton–Raphson algorithm has an un-

Table 3.3

Computation of Maximum-Likelihood Estimates in Table 3.1 Under the Model of No Three-Factor Interaction

Race	i	Sex	j	Type of assault[a]	k	Observed count $m_{ijk0} = n_{ijk}$	y_{ijk0}	c_{ijk}	$c_{ijk}y_{ijk0}$
White	1	Male	1	F. and e.	1	3910	8.2713	1	8.2713
	1		1	C. and p.	2	808	6.6946	−1	−6.6946
	1	Female	2	F. and e.	1	1050	6.9565	−1	−6.9565
	1		2	C. and p.	2	234	5.4553	1	5.4553
Black	2	Male	1	F. and e.	1	5218	8.5599	−1	−8.5599
	2		1	C. and p.	2	1385	7.2335	1	7.2335
	2	Female	2	F. and e.	1	929	6.8341	1	6.8341
	2		2	C. and p.	2	298	5.6971	−1	−5.6971
Total						13,832		0	−0.11389

[a] F. and e. is Firearms and explosives, and C. and p. is Cutting and piercing instruments.

i	j	k	$1000(c_{ijk}^2/m_{ijk0})$	$1000(c_{ijk}/m_{ijk0})$	μ_{ijk0}	$\exp \mu_{ijk0}$	m_{ijk2}
1	1	1	0.25575	0.25575	8.2737	3919.4	3919.3
1	1	2	1.2376	−1.2376	6.6829	798.61	798.58
1	2	1	0.95238	−0.95238	6.9476	1040.6	1040.6
1	2	2	0.42735	0.42735	5.4957	243.63	243.62
2	1	1	0.19164	−0.19164	8.5581	5208.6	5208.4
2	1	2	0.72202	0.72202	7.2403	1394.5	1394.4
2	2	1	1.0764	1.0764	6.8443	938.49	938.45
2	2	2	0.33557	−0.33557	5.6654	288.71	288.70
Total			12.065			13,833.	13,832.

$f_0 = -0.11389 \times 1000/12.06 = -9.4399$; $g_0 = 13,832/13,833. = 0.99996$; $\log g_0 = -0.00038744$

i	j	k	$n_{ijk} - m_{ijk1}$	$c_{ijk}(n_{ijk} - m_{ijk1})/m_{ijk1}$
1	1	1	−9.2974	−0.0023722
1	1	2	9.4165	−0.011792
1	2	1	9.4387	−0.0090708
1	2	2	−9.6235	−0.039501
2	1	1	9.6328	−0.0018495
2	1	2	−9.4110	−0.0067534
2	2	1	−9.4514	−0.010071
2	2	2	9.3032	−0.032225
Total			0.0000	−0.11363

Table 3.3 (continued)

i	j	k	c_{ijk}^2/m_{ijk1}	$n_{ijk} - m_{ijk1} - f_1 c_{ijk}$	μ_{ijk1}	$\exp \mu_{ijk2}$
1	1	1	0.25515	0.16303	8.2737	3919.5
1	1	2	1.2522	−0.043887	6.6827	798.54
1	2	1	0.96102	−0.021670	6.9475	1040.5
1	2	2	4.1047	−0.16308	5.4950	243.46
2	1	1	0.19200	0.17242	8.5581	5208.5
2	1	2	0.71715	0.043399	7.2403	1394.5
2	2	1	1.0656	0.0089750	6.8442	938.46
2	2	2	3.4638	−0.15717	5.6648	288.54
Total			12.012	0.0000		13,832.

$f_1 = -0.11363 \times 1000/12.012 = -9.4604$; $g_1 = 1.0000$.

i	j	k	m_{ijk2}	$n_{ijk} - m_{ijk2}$
1	1	1	3919.5	−9.4568
1	1	2	798.54	9.4607
1	2	1	1040.54	9.4607
1	2	2	243.46	−9.4604
2	1	1	5208.5	9.4648
2	1	2	1394.5	−9.4592
2	2	1	938.46	−9.4600
2	2	2	288.54	9.4604
Total			13,832.	0.0000

usually simple form described, except for slight variations, in Haberman (1974a, pages 50–54, 1974b). The algorithm is related in concept to Norton (1945).

To follow the development of the Newton–Raphson algorithm, observe that as Bartlett (1935) notes, maximum-likelihood estimates $\hat{m}_{ijk}$ of the means m_{ijk} satisfy the equations

$$n_{ijk} - \hat{m}_{ijk} = f c_{ijk}$$

and

$$\hat{\gamma} = \sum_i \sum_j \sum_k c_{ijk} \log \hat{m}_{ijk}$$
$$= \log \hat{m}_{111} - \log \hat{m}_{211} - \log \hat{m}_{121} - \log \hat{m}_{112}$$
$$+ \log \hat{m}_{221} + \log \hat{m}_{212} + \log \hat{m}_{122} - \log \hat{m}_{222}$$
$$= 0$$

for some real number f. See also Haberman (1974b). The Newton–Raphson algorithm uses a series of weighted-least-squares problems to approximate f.

As in Chapter 1, an initial estimate $m_{ijk0} > 0$ of $\hat{m}_{ijk}$ is used. In Table 3.3, $m_{ijk0} = n_{ijk}$. If some counts n_{ijk} are zero or are near zero, $m_{ijk0} = n_{ijk} + \frac{1}{2}$ appears preferable. Given m_{ijk0}, the empirical logarithm

$$y_{ijk0} = \log m_{ijk0}$$

may be defined. The weighted-leasted-squares estimates μ_{ijk0} of μ_{ijk} are then found for the linear model

$$y_{ijk0} = \mu_{ijk} + \varepsilon_{ijk},$$

where the μ_{ijk} are unknown parameters such that

$$\sum_i \sum_j \sum_k c_{ijk}\mu_{ijk} = \mu_{111} - \mu_{211} - \mu_{121} - \mu_{112} + \mu_{221} + \mu_{212} + \mu_{122} - \mu_{222}$$

$$= 0$$

and the ε_{ijk} are independent random variables with 0 means and respective variances m_{ijk0}^{-1}. This linear model has an unfamiliar form since no regression parameters appear. However, using standard arguments from linear algebra such as those in Rao (1973, page 11), one can show that this model is equivalent to the linear model

$$y_{ijk} = \alpha + \sum_{l=1}^{6} \beta_l x_{ijkl} + \varepsilon_{ijk}$$

in which

$$
\begin{aligned}
x_{ijk1} &= 1, & i &= 1,\\
&= -1, & i &= 2,\\
x_{ijk2} &= 1, & j &= 1,\\
&= -1, & j &= 2,\\
x_{ijk3} &= 1, & k &= 1,\\
&= -1, & k &= 2,\\
x_{ijk4} &= 1, & i &= j,\\
&= -1, & i &\neq j,\\
x_{ijk5} &= 1, & i &= k,\\
&= -1, & i &\neq k,\\
x_{ijk6} &= 1, & j &= k,\\
&= -1, & j &\neq k.
\end{aligned}
$$

The important feature in the algorithm under study is that the weighted-least-squares estimates μ_{ijk0} may be computed without first determining weighted-least-squares estimates β_{l0}, $1 \leqslant l \leqslant 6$. As noted by Haberman, the μ_{ijk0} may be found by the equation $\mu_{ijk0} = y_{ijk0} - f_0 c_{ijk}/m_{ijk0}$, where

$$f_0 = \gamma_0/h_0,$$

$$\gamma_0 = \sum_i \sum_j \sum_k c_{ijk} y_{ijk0} = y_{1110} - y_{2110} - y_{1210} - y_{1120}$$

$$+ y_{2210} + y_{2120} + y_{1220} - y_{2220},$$

and

$$h_0 = \sum_i \sum_j \sum_k (c_{ijk}^2/m_{ijk0}) = \sum_i \sum_j \sum_k (1/m_{ijk0}).$$

Note that

$$\sum_i \sum_j \sum_k c_{ijk}\mu_{ijk0} = \sum_i \sum_j \sum_k c_{ijk} y_{ijk0}$$

$$- (\gamma_0/h_0) \sum_i \sum_j \sum_k (c_{ijk}^2/m_{ijk})$$

$$= \gamma_0 - (\gamma_0/h_0)h_0$$

$$= 0.$$

Thus μ_{ijk0} satisfies the constraints on $\log \hat{m}_{ijk}$.

To obtain a new approximation m_{ijk1} for $\hat{m}_{ijk}$, let

$$g_0 = N \Big/ \sum_i \sum_j \sum_k \exp \mu_{ijk0},$$

where

$$N = \sum_i \sum_j \sum_k n_{ijk},$$

and let

$$m_{ijk1} = g_0 \exp \mu_{ijk0}.$$

In this example, the approximation m_{ijk1} for $\hat{m}_{ijk}$ is quite adequate for all practical purposes. If a further iteration is desired, then one may define a new working logarithm

$$y_{ijk1} = \log m_{ijk1} + (n_{ijk} - m_{ijk1})/m_{ijk1}$$

and proceed as in the regression model

$$y_{ijk1} = \mu_{ijk} + \varepsilon_{ijk},$$

where the μ_{ijk} are unknown parameters such that

$$\sum_i \sum_j \sum_k c_{ijk}\mu_{ijk} = \mu_{111} - \mu_{211} - \mu_{121} - \mu_{112} + \mu_{221} + \mu_{212}$$

$$+ \mu_{122} - \mu_{122}.$$
$$= 0$$

and the ε_{ijk} are random variables with means 0 and respective variances m_{ijk1}^{-1}. Thus μ_{ijk} is estimated by

$$\mu_{ijk1} = y_{ijk1} - (f_1 c_{ijk}/m_{ijk1})$$
$$= \mu_{ijk0} + \log g_0 + [(n_{ijk} - m_{ijk1} - f_1 c_{ijk})/m_{ijk1}],$$

where

$$f_1 = \gamma_1/h_1,$$

$$\gamma_1 = \sum_i \sum_j \sum_k c_{ijk} y_{ijk1}$$

$$= \sum_i \sum_j \sum_k [c_{ijk}(n_{ijk} - m_{ijk1})/m_{ijk1}],$$

and

$$h_1 = \sum_i \sum_j \sum_k (c_{ijk}^2/m_{ijk1}) = \sum_i \sum_j \sum_k (1/m_{ijk1}).$$

If

$$g_1 = N \Big/ \sum_i \sum_j \sum_k \exp \mu_{ijk1},$$

then

$$m_{ijk2} = g_1 \exp \mu_{ijk1}.$$

As can be seen in Table 3.3, this iteration hardly matters. Further iterations can be conducted in a similar manner. Normally, f_t approaches f, g_t approaches 1, μ_{ijkt} approaches $\hat{\mu}_{ijk} = \log \hat{m}_{ijk}$, m_{ijkt} approaches $\hat{m}_{ijk}$, and h_t approaches

$$h = \sum_i \sum_j \sum_k (c_{ijk}^2/\hat{m}_{ijk})$$

$$= \sum_i \sum_j \sum_k (1/\hat{m}_{ijk})$$

as t becomes large. As is apparent from this example, convergence can be very rapid.

Chi-Square Tests

The chi-square statistics for the model are

$$X^2 = \sum_i \sum_j \sum_k (n_{ijk} - \hat{m}_{ijk})^2/\hat{m}_{ijk}$$

$$= f^2 h$$

and

$$L^2 = 2 \sum_i \sum_j \sum_k n_{ijk} \log(n_{ijk}/\hat{m}_{ijk}).$$

Thus $X^2 = 1.08$ and $L^2 = 1.07$. Since one linear contrast is assumed zero, there is one degree of freedom. A more general formula for degrees of freedom for models of no three-factor interaction in $r \times s \times t$ tables will be given in Section 3.5. Note that X^2 and L^2 are very closely approximated by

$$X_0^{\ 2} = \left[\hat{\tau}^{ABC}_{(12)(12)(12)}/s(\hat{\tau}^{ABC}_{(12)(12)(12)})\right]^2 = f_0^{\ 2} h_0 = 1.08.$$

As noted in Goodman (1964b, 1965), this last statistic is algebraically equivalent to Woolf's (1955) test statistic for no three-factor interaction in a $2 \times 2 \times 2$ table. Using any of these test statistics, the model of no three-factor interaction appears quite tenable.

Estimation of Cross-Product Ratios

Given the model of no three-factor interaction, estimates for the log cross-product ratios

$$\tau^{AC(B)}_{(12)(12)} = \tfrac{1}{2}(\tau^{AC \cdot B}_{(12)(12) \cdot 1} + \tau^{AC \cdot B}_{(12)(12) \cdot 2}) = \tau^{AC \cdot B}_{(12)(12) \cdot 1} = \tau^{AC \cdot B}_{(12)(12) \cdot 2}$$

and

$$\tau^{BC(A)}_{(12)(12)} = \tfrac{1}{2}(\tau^{BC \cdot A}_{(12)(12) \cdot 1} + \tau^{BC \cdot A}_{(12)(12) \cdot 2}) = \tau^{BC \cdot A}_{(12)(12) \cdot 1} = \tau^{BC \cdot A}_{(12)(12) \cdot 2}$$

are readily obtained. Note that

$$\hat{\tau}^{AC(B)}_{(12)(12)} = \log\left(\frac{\hat{m}_{111}\hat{m}_{212}}{\hat{m}_{112}\hat{m}_{221}}\right) = \log\left(\frac{\hat{m}_{121}\hat{m}_{222}}{\hat{m}_{122}\hat{m}_{221}}\right) = 0.273$$

and

$$\hat{\tau}^{BC(A)}_{(12)(12)} = \log\left(\frac{\hat{m}_{111}\hat{m}_{122}}{\hat{m}_{112}\hat{m}_{121}}\right) = \log\left(\frac{\hat{m}_{211}\hat{m}_{222}}{\hat{m}_{212}\hat{m}_{221}}\right) = 0.138.$$

The asymptotic variance of $\hat{\tau}^{AC(B)}_{(12)(12)}$ is

$$\sigma^2(\hat{\tau}^{AC(B)}_{(12)(12)}) = \left[\sum_j \left(\frac{1}{m_{1j1}} + \frac{1}{m_{1j2}} + \frac{1}{m_{2j1}} + \frac{1}{m_{2j2}}\right)^{-1}\right]^{-1}$$

To interpret this curious expression, recall that

$$v_j = \frac{1}{m_{1j1}} + \frac{1}{m_{1j2}} + \frac{1}{m_{2j2}} + \frac{1}{m_{2j2}}$$

is the asymptotic variance of the log cross-product ratio

$$d_j = \log\left(\frac{n_{1j1}n_{2j2}}{n_{1j2}n_{2j2}}\right).$$

The d_j are independently distributed. Under the model of no three-factor interaction, d_j has asymptotic mean $\tau_{(12)(12)}^{AC(B)}$. The weighted average of d_1 and d_2 with the smallest asymptotic variance is

$$d = \frac{(d_1/v_1) + (d_2/v_2)}{(1/v_1) + (1/v_2)},$$

which has asymptotic mean $\tau_{(12)(12)}^{AC(B)}$ and asymptotic variance

$$\frac{(v_1/v_1{}^2) + (v_2/v_2{}^2)}{[(1/v_1) + (1/v_2)]^2} = \frac{1}{(1/v_1) + (1/v_2)} = \sigma^2(\hat{\tau}_{(12)(12)}^{AC(B)}).$$

Since it can be shown that $\hat{\tau}_{(12)(12)}^{AC(B)}$ has the same large-sample distribution as d, the asymptotic variance of $\sigma^2(\hat{\tau}_{(12)(12)}^{AC(B)})$ is the same as the asymptotic variance of the weighted average d. For further details, see Exercise 3.1.
The estimated asymptotic variance of $\hat{\tau}_{(12)(12)}^{AC(B)}$ is

$$s^2(\hat{\tau}_{(12)(12)}^{AB(C)}) = \left[\sum_j \left(\frac{1}{\hat{m}_{1j1}} + \frac{1}{\hat{m}_{2j1}} + \frac{1}{\hat{m}_{1j2}} + \frac{1}{\hat{m}_{2j2}}\right)^{-1}\right]^{-1}.$$

In this example,

$$\frac{1}{\hat{m}_{111}} + \frac{1}{\hat{m}_{211}} + \frac{1}{\hat{m}_{112}} + \frac{1}{\hat{m}_{212}} = \frac{1}{3919.5} + \frac{1}{5208.5} + \frac{1}{798.54} + \frac{1}{1394.5}$$

$$= 0.0024165$$

$$\frac{1}{\hat{m}_{121}} + \frac{1}{\hat{m}_{221}} + \frac{1}{\hat{m}_{122}} + \frac{1}{\hat{m}_{222}} = \frac{1}{1040.5} + \frac{1}{938.46} + \frac{1}{243.46} + \frac{1}{288.54}$$

$$= 0.0095998$$

and

$$s(\hat{\tau}_{(12)(12)}^{AC(B)}) = \left(\frac{1}{0.0024165} + \frac{1}{0.0095998}\right)^{-1/2} = 0.044.$$

Thus an approximate 95 percent confidence interval for $\tau_{(12)(12)}^{AC}$ has bounds

$$0.273 - 1.96(0.044) = 0.187$$

and

$$0.273 + 1.96(0.044) = 0.359.$$

Corresponding bounds for the cross-product ratio

$$q^{AC}_{(12)(12)} = \frac{m_{111}}{m_{112}} \frac{m_{212}}{m_{211}} = \frac{m_{121}}{m_{122}} \frac{m_{222}}{m_{222}}$$

are

$$\exp(0.187) = 1.206$$

and

$$\exp(0.359) = 1.432.$$

Thus controlling for sex of victim, the odds of assault by firearms and explosives rather than by cutting and piercing instruments are estimated to be from 1.21 to 1.43 times higher for whites than for blacks.

The confidence bounds for $\tau^{AC(B)}_{(12)(12)}$ are similar to those derived earlier for $\tau^{AC \cdot B}_{(12)(12) \cdot 1}$ and somewhat narrower than those obtained for $\tau^{AC \cdot B}_{(12)(12) \cdot 2}$. The result is predictable. Most information concerning an interaction of race and type of assault comes from male victims. Thus the estimate $\hat{\tau}^{AC(B)}_{(12)(12)}$ reflects the observed log cross-product ratio d_1 for male victims rather than the log ratio d_2 for female victims.

Similarly,

$$\sigma^2(\hat{\tau}^{BC(A)}_{(12)(12)}) = \left[\sum_j \left(\frac{1}{m_{i11}} + \frac{1}{m_{i21}} + \frac{1}{m_{i12}} + \frac{1}{m_{i22}} \right)^{-1} \right]^{-1}$$

and

$$\sigma^2(\hat{\tau}^{BC(A)}_{(12)(12)}) = \left[\sum_i \left(\frac{1}{\hat{m}_{i11}} + \frac{1}{\hat{m}_{i21}} + \frac{1}{\hat{m}_{i12}} + \frac{1}{\hat{m}_{i22}} \right)^{-1} \right]^{-1}.$$

Here

$$\frac{1}{\hat{m}_{111}} + \frac{1}{\hat{m}_{121}} + \frac{1}{\hat{m}_{112}} + \frac{1}{\hat{m}_{122}} = \frac{1}{3919.5} + \frac{1}{1040.5} + \frac{1}{798.54} + \frac{1}{243.46}$$

$$= 0.0065760,$$

$$\frac{1}{\hat{m}_{211}} + \frac{1}{\hat{m}_{221}} + \frac{1}{\hat{m}_{212}} + \frac{1}{\hat{m}_{222}} = \frac{1}{5208.5} + \frac{1}{938.46} + \frac{1}{1394.5} + \frac{1}{288.54}$$

$$= 0.0054421,$$

and

$$s^2(\hat{\tau}^{BC(A)}_{(12)(12)}) = \left(\frac{1}{0.0065760} + \frac{1}{0.0054421} \right)^{-1/2} = 0.054.$$

Thus an approximate 95 percent confidence interval for $\tau^{BC(A)}_{(12)(12)}$ has bounds

$$0.138 + 1.96(0.054) = 0.032$$

and

$$0.138 - 1.96(0.054) = 0.245.$$

The corresponding bounds for the cross-product ratio

$$q^{BC(A)}_{(12)(12)} = \frac{m_{111}m_{122}}{m_{112}m_{121}} = \frac{m_{211}m_{222}}{m_{212}m_{122}}$$

are $\exp(0.0315) = 1.032$ and $\exp(0.245) = 1.278$. Note that these bounds do not include 1, so that there is definite evidence that controlling for race, the odds of assault by firearms and explosives rather than by cutting and piercing instruments is higher for males than for females. The difference may well be negligible, given the lower bound of 1.032. The upper bound of 1.278 is comparable to the estimate $\hat{q}^{AC(B)}_{(12)(12)} = \exp(0.273) = 1.314$ of the cross-product ratio for race by assault given sex. Thus considerable uncertainty exists concerning the magnitude of the interaction between sex of victim and type of assault.

Association of Sex and Race

So far, comparisons have involved the log odds that an assault is by firearms and explosives rather than by cutting and piercing instruments. Another perspective is also possible. Controlling for type of assault, one may investigate the relationship of sex of victim to race of victim. Consider the log cross-product ratio

$$\tau^{AB \cdot C}_{(12)(12) \cdot k} = \log\left(\frac{m_{11k}m_{22k}}{m_{21k}m_{12k}}\right)$$

$$= \log m_{11k} - \log m_{21k} - \log m_{12k} + \log m_{22k}.$$

This ratio compares the odds for males and females that a victim of type of assault k is white rather than black. Under the model of no three-factor interaction,

$$\tau^{AB \cdot C}_{(12)(12) \cdot 1} = \tau^{AB \cdot C}_{(12)(12) \cdot 2} = \tau^{AB(C)}_{(12)(12)}.$$

The maximum-likelihood estimate of $\tau^{AB(C)}_{(12)(12)}$ is

$$\hat{\tau}^{AB(C)}_{(12)(12)} = \log\left(\frac{\hat{m}_{111}\hat{m}_{221}}{\hat{m}_{211}\hat{m}_{121}}\right) = \log\left(\frac{\hat{m}_{112}\hat{m}_{222}}{\hat{m}_{212}\hat{m}_{122}}\right) = -0.388.$$

The estimated asymptotic variance $s^2(\hat{\tau}^{AB(C)}_{(12)(12)})$

$$\left[\sum_k \left(\frac{1}{\hat{m}_{11k}} + \frac{1}{\hat{m}_{21k}} + \frac{1}{\hat{m}_{12k}} + \frac{1}{\hat{m}_{22k}}\right)^{-1}\right]^{-1}.$$

Since

$$\frac{1}{\hat{m}_{111}} + \frac{1}{\hat{m}_{211}} + \frac{1}{\hat{m}_{121}} + \frac{1}{\hat{m}_{221}} = \frac{1}{3919.5} + \frac{1}{5208.5} + \frac{1}{1040.5} + \frac{1}{938.46}$$

$$= 0.0024738$$

and

$$\frac{1}{\hat{m}_{112}} + \frac{1}{\hat{m}_{212}} + \frac{1}{\hat{m}_{122}} + \frac{1}{\hat{m}_{222}} = \frac{1}{798.54} + \frac{1}{1394.5} + \frac{1}{243.46} + \frac{1}{288.54}$$

$$= 0.0095426,$$

$$s(\hat{\tau}^{AB(C)}_{(12)(12)}) = \left(\frac{1}{0.0024738} + \frac{1}{0.0095426}\right)^{-1/2} = 0.044.$$

The standardized value $\hat{\tau}^{AB(C)}_{(12)(12)}/s(\hat{\tau}^{AB(C)}_{(12)(12)}) = -0.388/0.0044 = 8.74$ provides exceedingly strong evidence that $\tau^{AB(C)}_{(12)(12)}$ is not zero.

The approximate 95 percent confidence bounds are

$$0.388 - 1.96(0.044) = -0.475$$

and

$$0.388 + 1.96(0.044) = -0.301.$$

Corresponding bounds for

$$q^{AB(C)}_{(12)(12)} = \frac{m_{111}m_{221}}{m_{211}m_{121}} = \frac{m_{112}m_{222}}{m_{212}m_{122}}$$

are $\exp(-0.475) = 0.622$ and $\exp(-0.301) = 0.740$. Thus there is substantial interaction between race and sex of victim. It is estimated that controlling for type of assault, the odds that a victim is black rather than white are from 0.62 to 0.74 times as large for men as for women.

Log Cross-Product Ratios and λ-Parameters

Log cross-product ratios such as $\tau^{AB(C)}_{(12)(12)}$ or $\tau^{AB \cdot C}_{(12)(12) \cdot 1}$ are related to the λ-parameters obtained in the decomposition

$$\log m_{ijk} = \lambda + \lambda_i^A + \lambda_j^B + \lambda_k^C + \lambda_{ij}^{AB} + \lambda_{ik}^{AC} + \lambda_{jk}^{BC} + \lambda_{ijk}^{ABC}.$$

Note that

$$\lambda_{11}^{AB} - \lambda_{21}^{AB} - \lambda_{12}^{AB} + \lambda_{22}^{AB} = 4\lambda_{11}^{AB}$$
$$= \tfrac{1}{2}(\mu_{111} - \mu_{211} - \mu_{121} + \mu_{221} + \mu_{112}$$
$$- \mu_{212} - \mu_{122} + \mu_{222})$$
$$= \tfrac{1}{2}(\tau_{(12)(12)\cdot 1}^{AB\cdot C} + \tau_{(12)(12)\cdot 2}^{AB\cdot C})$$
$$= \tau_{(12)(12)}^{AB(C)}.$$

Similarly,

$$\lambda_{11}^{AC} - \lambda_{21}^{AC} - \lambda_{12}^{AC} + \lambda_{22}^{AC} = 4\lambda_{11}^{AC} = \tau_{(12)(12)}^{AC(B)}$$

and

$$\lambda_{11}^{BC} - \lambda_{21}^{BC} - \lambda_{12}^{BC} + \lambda_{22}^{BC} = 4\lambda_{11}^{BC} = \tau_{(12)(12)}^{BC(A)}.$$

There is very strong evidence that $\tau_{(12)(12)}^{AC(B)}$ and $\tau_{(12)(12)}^{AB(C)}$ are not zero, so no log–linear model can fit these data and assume that the λ_{ij}^{AB} or the λ_{ik}^{AC} are zero. Although smaller in magnitude than standardized values for $\tau_{(12)(12)}^{AB(C)}$ and $\tau_{(12)(12)}^{AB(B)}$, the standardized value

$$\hat{\tau}_{(12)(12)}^{BC(A)}/s(\hat{\tau}_{(12)(12)}^{BC(A)}) = 0.138/0.054 = 2.54$$

is still significant at about the 0.01 level. Thus it also appears unreasonable to assume that the λ_{jk}^{BC} are 0.

Possible Hierarchial Models

If attention is restricted to hierarchical models, then the simplest model that can fit the data is the model of no three-factor interaction

$$\log m_{ijk} = \lambda + \lambda_i^A + \lambda_j^B + \lambda_k^C + \lambda_{ij}^{AB} + \lambda_{ik}^{AC} + \lambda_{jk}^{BC}.$$

Since the λ_{ij}^{AB}, λ_{ik}^{AC}, and λ_{jk}^{BC} cannot be set to zero, no possible hierarchical model can set the lower-order terms λ, λ_i^A, λ_j^B, or λ_k^C to zero.

Although a few computational conveniences are associated with hierarchical models, the primary gain achieved by restriction to hierarchical models is ease of interpretation. As an example, assume that $\lambda_1^A = \lambda_2^A = 0$. Let $\tau_{12\cdot jk}^{A\cdot BC} = \log(m_{1jk}/m_{2jk}) = \mu_{1jk} - \mu_{2jk}$ denote the log odds that a victim of sex j killed by an assault of type k is white rather than black. Since

$$\lambda_1^A - \lambda_2^A = 2\lambda_1^A = \tfrac{1}{4}(\mu_{111} - \mu_{211} + \mu_{121} - \mu_{221} + \mu_{112} - \mu_{212} + \mu_{122} - \mu_{222})$$
$$= \tfrac{1}{4}(\tau_{12\cdot 11}^{A\cdot BC} + \tau_{12\cdot 21}^{A\cdot BC} + \tau_{12\cdot 12}^{A\cdot BC} + \tau_{12\cdot 22}^{A\cdot BC}),$$

$\lambda_1^A = \lambda_2^A = 0$ implies that the average of the log odds $\tau_{12\cdot jk}^{A\cdot BC}$, $1 \leqslant j \leqslant 2$, $1 \leqslant k \leqslant 2$, is 0. There is relatively little interest in the fact that this average is 0 unless the $\tau_{12\cdot jk}^{A\cdot BC}$ are all 0. The condition that all $\tau_{12\cdot jk}^{A\cdot BC}$ are equal can only hold if

$$
\begin{aligned}
8\lambda_{11}^{AB} &= \tau_{(12)(12)\cdot 1}^{AB\cdot C} + \tau_{(12)(12)\cdot 2}^{AB\cdot C} \\
&= \tau_{12\cdot 11}^{A\cdot BC} - \tau_{12\cdot 21}^{A\cdot BC} + \tau_{12\cdot 12}^{A\cdot BC} - \tau_{12\cdot 22}^{A\cdot BC} \\
&= 0.
\end{aligned}
$$

Thus a simple interpretation of the hypothesis that the λ_i^A are 0 is only possible if the λ_{ij}^{AB} are 0. Note that equality of the $\tau_{12\cdot jk}^{A\cdot BC}$ also requires that

$$
\lambda_{11}^{AC} = \tau_{12\cdot 11}^{A\cdot BC} + \tau_{12\cdot 21}^{A\cdot BC} - \tau_{12\cdot 12}^{A\cdot BC} - \tau_{12\cdot 22}^{A\cdot BC} = 0
$$

and

$$
\lambda_{111}^{ABC} = \tau_{12\cdot 11}^{A\cdot BC} - \tau_{12\cdot 21}^{A\cdot BC} - \tau_{12\cdot 12}^{A\cdot BC} + \tau_{12\cdot 22}^{A\cdot BC} = 0.
$$

Thus it is difficult to interpret a model with the λ_i^A set to 0 but the λ_{ij}^{AB}, λ_{ji}^{AC}, or λ_{ijk}^{ABC} not set to 0.

Estimation of Log Odds

The log odds ratios $\tau_{12\cdot jk}^{C\cdot AB}$ may be estimated under the model of no three-factor interaction. Results are rather similar to those obtained without restrictions on $\tau_{(12)(12)(12)}^{ABC}$. For details, see Exercise 3.2. In that exercise, log odds ratios $\tau_{12\cdot ik}^{B\cdot AC}$ and $\tau_{12\cdot jk}^{A\cdot BC}$ are also examined.

Special Features of the Model of No Three-Factor Interaction in 2 × 2 × 2 Tables

This section has explored analysis of a $2 \times 2 \times 2$ table by means of a hierarchical model called the model of no three-factor interaction. Several special features have arisen. Unlike the case in general $r \times s \times t$ tables, the model of no three-factor interaction can be reduced to a model of equality of cross-product ratios for two 2×2 tables. Thus the model of no three-factor interaction has an unusually simple interpretation in $2 \times 2 \times 2$ tables. The Newton–Raphson algorithm is unusually simple in $2 \times 2 \times 2$ tables. Since only one degree of freedom is associated with chi-square tests of this model in $2 \times 2 \times 2$ tables, residual analysis is trivial. The model of no three-factor interaction is unusual among hierarchical models for three-way tables since iterative calculations are required to find maximum-likelihood estimates. Thus this section has afforded an important but limited introduction to hierarchical models for three-way tables.

3.2 Woman's Place, Sex, and Education

In this section, a $2 \times 3 \times 2$ table rather than a $2 \times 2 \times 2$ table is examined, and attention is given to models of conditional and marginal independence as well as the model of no three-factor interaction. Relationships between these models are explored in terms of maximum-likelihood estimates of means and log cross-product ratios, chi-square tests, and adjusted residuals. Partitions of the likelihood-ratio chi-square and iterative proportional fitting are new features of this section, at least as far as three-way tables are concerned. General maximum-likelihood equations for hierarchical models are also introduced. To illustrate results, Table 3.4 was compiled from the 1975 General Social Survey. Similar data for 1974 are considered

Table 3.4

Subjects in the 1975 General Social Survey, Cross-Classified by Attitude toward Women Staying at Home, Sex of Respondent, and Education of Respondent[a]

| Sex of respondent | Education of respondent, yrs. | Response[b] | | | | | |
| | | Agree | | Disagree | | | |
		No.	Percent.	No.	Percent.	Total no.
Male	≤8	72	60.5	47	39.5	119
	9–12	110	35.9	196	64.1	306
	≥13	44	19.7	179	80.3	223
	Total	226	34.9	422	65.1	648
Female	≤8	86	69.4	38	30.6	124
	9–12	173	37.9	283	62.1	456
	≥13	28	13.0	187	87.0	215
	Total	287	36.1	508	53.9	795
Total	≤8	158	65.0	85	35.0	243
	9–12	283	37.1	479	62.9	762
	≥13	72	16.4	366	83.6	438
	Total	513	35.6	930	64.4	1443

[a] From National Opinion Research Center, 1975 General Social Survey, University of Chicago.

[b] Three subjects did not state their educational level, two others did not indicate their attitude, and 42 others were not sure whether they agreed with the statement asked. The question asked was the following: "Do you agree with this statement?—Women should take care of running their homes and leave running the country up to men." See National Opinion Research Center (1975, page 51).

in Exercise 3.3, and Example 4.1 compares results for the two years. The table permits a comparison of the relative importance of sex of respondent and education of respondent in determining attitudes toward roles for women. The question used asks whether "women should take care of running their homes and leave running the country up to men." For a fuller version of the question, see footnote b of Table 3.4.

A glance at the table indicates that education is a very important variable for prediction of response but sex is a relatively minor variable. For example, note that 60.5 percent of males with no more than eight years of education agreed with the statement but only 19.7% of males with at least 13 years of education agreed. Given a fixed level of education, the differences in percentages for males and females never exceed 10 percent.

To begin analysis of the table, consider the question whether sex and response are related. To examine this issue, let n_{ijk} be the number of subjects h of sex $S_h = i$, educational level $E_h = j$, and response $R_h = k$, so that $n_{111} = 72$, $n_{112} = 47$, $n_{121} = 110$, etc. There are some restrictions in the survey on the number of men and the number of women interviewed, so it appears that a realistic probability model should assume that the number n_i^S of $S_h = I$ is fixed (i.e., inferences are conditional on the observed number n_1^S of male subjects and the observed number n_2^S of female subjects). Given $S_h = i$, the probability is $p_{jk \cdot i}^{ER \cdot S} > 0$ that $E_h = j$ and $R_h = k$, and given the S_h, $1 \leqslant h \leqslant N = 1443$, the pairs (E_h, R_h), $1 \leqslant h \leqslant N$, are independently distributed. The expected value of n_{ijk} is then $n_i^S p_{jk \cdot i}^{ER \cdot S}$, and each table of counts n_{ijk}, $1 \leqslant j \leqslant 3$, $1 \leqslant k \leqslant 2$, has an independent multinomial distribution.

Marginal and Partial Association

The relationship between sex and response can be examined from two perspectives. The first perspective, that of marginal association, considers only the marginal probabilities $p_{k \cdot i}^{R \cdot S}$ of response k given sex i. The intervening variable education is ignored. The second perspective, that of partial association, considers the conditional probabilities $p_{k \cdot ij}^{R \cdot SE}$ of response k given sex i and education j.

The marginal association of sex and response may be measured by the cross-product ratio

$$q_{(12)(12)}^{SR} = \frac{m_{11}^{SR} m_{22}^{SR}}{m_{12}^{RS} m_{21}^{RS}} = \frac{N_1^S p_{1 \cdot 1}^{R \cdot S} N_2^S p_{2 \cdot 2}^{R \cdot S}}{N_1^S p_{1 \cdot 1}^{R \cdot S} N_2^S p_{2 \cdot 1}^{R \cdot S}} = \frac{p_{1 \cdot 1}^{R \cdot S} p_{2 \cdot 2}^{R \cdot S}}{p_{1 \cdot 2}^{R \cdot S} p_{2 \cdot 1}^{R \cdot S}}$$

or by its logarithm

$$\tau_{(12)(12)}^{SR} = \log q_{(12)(12)}^{SR} = \log\left(\frac{p_{1 \cdot 1}^{R \cdot S}}{p_{2 \cdot 1}^{R \cdot S}}\right) - \log\left(\frac{p_{1 \cdot 2}^{R \cdot S}}{p_{2 \cdot 2}^{R \cdot S}}\right).$$

Here m_{ik}^{SR}, the expected number of subjects with sex i and response k, is the sum of the means m_{ijk} for the educational levels j, $1 \leqslant j \leqslant 3$. Sex and response are marginally independent if

$$p_{i \cdot 1}^{R \cdot S} = p_{i \cdot 2}^{R \cdot S}, \quad 1 \leqslant i \leqslant 2,$$

in which case $q_{(12)(12)}^{SR} = 1$ and $\tau_{(12)(12)}^{SR} = 0$.

The partial association of sex and response given educational level j may be measured by the cross-product ratio

$$q_{(12)(12) \cdot j}^{SR \cdot E} = \frac{m_{1j1} m_{2j2}}{m_{2j2} m_{2j1}} = \frac{p_{1 \cdot 1j}^{R \cdot SE} p_{1 \cdot 2j}^{R \cdot SE}}{p_{1 \cdot 2j}^{R \cdot SE} p_{2 \cdot 1j}^{R \cdot SE}}$$

or its logarithm

$$\tau_{(12)(12) \cdot j}^{SR \cdot E} = \log q_{(12)(12) \cdot j}^{SR \cdot E} = \log\left(\frac{p_{1 \cdot 1j}^{R \cdot SE}}{p_{2 \cdot 1j}^{R \cdot SE}}\right) - \log\left(\frac{p_{2 \cdot 2j}^{R \cdot SE}}{p_{2 \cdot 2j}^{R \cdot SE}}\right).$$

Sex and response are conditionally independent given education if

$$p_{i \cdot 1j}^{R \cdot SE} = p_{i \cdot 2j}^{R \cdot SE}, \quad 1 \leqslant i \leqslant 2, \quad 1 \leqslant j \leqslant 3,$$

in which case each $q_{(12)(12) \cdot j}^{SR \cdot E} = 1$ and each $\tau_{(12)(12) \cdot j}^{SR \cdot E} = 0$. The notions of marginal and partial association considered here may be found in Yule (1900).

The relationship of the partial association measures $\tau_{(12)(12) \cdot j}^{SR \cdot E}$, $1 \leqslant j \leqslant 3$, to the marginal association measure $\tau_{(12)(12)}^{SR}$ is complex. Yule (1900), Simpson (1951), Goodman (1972), and Bishop, Fienberg, and Holland (1975, pages 39–42) all provide relevant discussions. The basic results to keep in mind are the following. The equation $\tau_{(12)(12) \cdot j}^{SR \cdot E} = \tau_{(12)(12)}^{SR}$ holds if either R_h and E_h are conditionally independent given S_h or S_h and E_h are conditionally independent given R_h (see Exercise 3.10). It is possible for any combination of $\tau_{(12)(12) \cdot 1}^{SR \cdot E}$, $\tau_{(12)(12) \cdot 2}^{SR \cdot E}$, $\tau_{(12)(12) \cdot 3}^{SR \cdot E}$, and $\tau_{(12)(12)}^{SR}$ to exist. Bizarre possibilities such as $\tau_{(12)(12) \cdot j}^{SR \cdot E} = -\tau_{(12)(12)}^{SR} \neq 0$, $1 \leqslant j \leqslant 3$, are possible in which the marginal and partial association coefficients are of opposite signs (see Exercise 3.11). In this section, both marginal and partial association are considered.

To test the hypothesis of marginal independence, the marginal table $\mathbf{n}^{SR}$ of counts

$$n_{ik}^{SR} = \sum_j n_{ijk} = n_{i1k} + n_{i2k}$$

is considered. Here n_{ik}^{SR} is the number of subjects with sex i and response k. Thus the table is a 2×2 contingency table with counts $n_{11}^{SR} = 226$, $n_{12}^{SR} = 422$, $n_{21}^{SR} = 287$, and $n_{22}^{SR} = 508$. To test for marginal independence, one may apply ordinary chi-square statistics for 2×2 tables to $\mathbf{n}^{SR}$.

Chi-Square Tests for Marginal Association

To define chi-square statistics, let $N = 1443$ be the number of subjects in the survey with tabulated responses, let n_i^S be the number of subjects of sex i, and let n_k^R be the number of subjects with response k. The Pearson chi-square statistic is then

$$
\begin{aligned}
X_{SR}^2 &= \sum_i \sum_k (n_{ik}^{SR} - n_i^S n_k^R / N)^2 / (n_i^S n_k^R / N) \\
&= \frac{(n_{11}^{SR} n_{22}^{SR} - n_{12}^{SR} n_{21}^{SR})^2 N}{n_1^S n_2^S n_1^R n_2^R} \\
&= \frac{(226 \times 508 - 422 \times 287)^2 1443}{648 \times 795 \times 513 \times 930} \\
&= 0.23,
\end{aligned}
$$

and the likelihood-ratio chi-square statistic is

$$
\begin{aligned}
L_{SR}^2 &= 2 \sum_i \sum_k n_{ik}^{SR} \log\left(\frac{n_{ik}^{SR} N}{n_i^S n_k^R}\right) \\
&= 2\left[226 \log\left(\frac{226 \times 1443}{648 \times 513}\right) + 422 \log\left(\frac{422 \times 1443}{648 \times 930}\right) \right. \\
&\quad \left. + 287 \log\left(\frac{287 \times 1443}{795 \times 573}\right) + 508 \log\left(\frac{508 \times 1443}{795 \times 930}\right) \right] \\
&= 0.23.
\end{aligned}
$$

There is one degree of freedom in a chi-square test of independence for a 2×2 table, so no evidence exists of a marginal association between sex and response. It is consistent with the data to assert that men and women are equally likely to agree that women should run their homes and let men run the country.

Conditional Independence

To interpret the conditional independence model in terms of log–linear models, note that the means have a unique representation

$$
\log m_{ijk} = \lambda + \lambda_i^S + \lambda_j^E + \lambda_k^R + \lambda_{ij}^{SE} + \lambda_{ik}^{SR} + \lambda_{jk}^{ER} + \lambda_{ijk}^{SER},
$$

where

$$
\begin{aligned}
\sum_i \lambda_i^S = \sum_j \lambda_j^E = \sum_k \lambda_k^E &= \sum_i \lambda_{ij}^{SE} = \sum_j \lambda_{ij}^{SE} = \cdots \\
&= \sum_i \lambda_{ijk}^{SER} = \sum_j \lambda_{ijk}^{SER} = \sum_k \lambda_{ijk}^{SER} = 0.
\end{aligned}
$$

The partial association coefficient is

$$\tau^{SR \cdot E}_{(12)(12) \cdot j} = \log m_{1j1} - \log m_{1j2} - \log m_{2j1} + \log m_{2j2}$$
$$= (\lambda^{SR}_{11} + \lambda^{SER}_{1j1}) - (\lambda^{SR}_{12} + \lambda^{SER}_{1j2}) - (\lambda^{SR}_{21} + \lambda^{SER}_{2j1}) + (\lambda^{SR}_{22} + \lambda^{SER}_{2j2})$$
$$= 4(\lambda^{SR}_{11} + \lambda^{SER}_{1j1}).$$

Here it is noted that

$$\lambda^{SR}_{11} = -\lambda^{SR}_{21} = -\lambda^{SR}_{12} = \lambda^{SR}_{22}$$

and

$$\lambda^{SER}_{1j1} = -\lambda^{SER}_{1j1} = -\lambda^{SER}_{1j2} = \lambda^{SER}_{2j2}.$$

Each log cross-product ratio $\tau^{SR \cdot E}_{(12)(12) \cdot j}$ is 0 if and only if the λ^{SR}_{ik} and the λ^{SER}_{ijk} are 0. Thus as Birch (1963) notes, conditional independence of S_h and E_h given R_h holds if and only if the hierarchical model

$$\log m_{ijk} = \lambda + \lambda^S_i + \lambda^E_j + \lambda^R_k + \lambda^{SE}_{ij} + \lambda^{ER}_{jk}$$

is satisfied, so that all λ-parameters involving both sex and response are 0.

Maximum-Likelihood Estimates and Chi-Square Tests for Conditional Independence

Chi-square tests and adjusted residuals are easily computed for the model of conditional independence of education and sex controlling for response. As is well known, the maximum-likelihood estimate $\hat{m}_{ijk}$ of m_{ijk} is

$$\hat{m}_{ijk} = n^{SE}_{ij} n^{ER}_{jk} / n^E_j,$$

where n^E_j is the number of subjects of education j, n^{SE}_{ij} is the number with sex i and education j, and n^{ER}_{jk} is the number with education j and response k. One thus obtains the familiar chi-square statistic

$$X^2_{SR \cdot E} = \sum_i \sum_j \sum_k (n_{ijk} - n^{SE}_{ij} n^{ER}_{jk} / n^E_j)^2 / (n^{SE}_{ij} n^{ER}_{jk} / n^E_j)$$

$$= \sum_j \frac{(n_{1j1} n_{2j2} - n_{1j2} n_{2j1})^2 n^E_j}{n^{SE}_{1j} n^{SE}_{2j} n^{ER}_{j1} n^{ER}_{j2}}$$

of Pearson (1916) or the likelihood-ratio statistic

$$L^2_{SR \cdot E} = 2 \sum_i \sum_j \sum_k n_{ijk} \log \frac{n_{ijk} n^E_j}{n^{SE}_{ij} n^{ER}_{jk}}$$

found in Kullback (1968 [1959], page 166). The corresponding degrees of freedom for these statistics are $(2 - 1)3(2 - 1) = 3$. For further details, see

Sections 3.3 and 3.5. In this example, $X^2_{SR \cdot E} = 5.99$ and $L^2_{SR \cdot E} = 6.02$. Results are significant at the 11 percent level. Thus evidence for partial association between sex and response given education appears to be rather weak. Nonetheless, there is somewhat more indication of partial association than of marginal association.

Adjusted Residuals for Conditional Independence

To obtain further insight into the possible existence of partial association, the adjusted residuals

$$r_{ijk} = \frac{n_{ijk} - n^{SE}_{ij} n^{ER}_{jk}/n^{E}_{j}}{[n^{SE}_{ij} n^{ER}_{jk}(1 - n^{SE}_{ij}/n^{E}_{j})(1 - n^{ER}_{jk}/n^{E}_{j})/n^{E}_{j}]^{1/2}}$$

may be examined. For related formulas, see Section 3.5. For now, it is helpful to note that

$$r_{1j1} = -r_{2j1} = -r_{1j2} = r_{2j2} = \frac{n_{1j1}n_{2j2} - n_{1j2}n_{2j1}}{(n^{SE}_{1j}n^{SE}_{2j}n^{ER}_{j1}n^{ER}_{j2}/n^{E}_{j})^{1/2}}.$$

For details, see Exercise 3.4 and Cochran (1954).

Given the constraints on the r_{ijk}, it suffices to examine r_{111}, r_{121}, and r_{131}. Note that

$$X^2 = r^2_{111} + r^2_{121} + r^2_{131},$$

so that the r^2_{1j1} terms may be regarded as components of X^2. Under the conditional independence model, each component has an approximate chi-square distribution with one degree of freedom, and each r_{1j1} has an approximate $N(0, 1)$ distribution. In Table 3.4,

$$r_{111} = \frac{72 \times 38 - 86 \times 47}{(119 \times 124 \times 158 \times 85/243)^{1/2}} = -1.45,$$

$$r_{121} = \frac{110 \times 283 - 173 \times 196}{(306 \times 456 \times 283 \times 479/762)^{1/2}} = -0.56,$$

and

$$r_{131} = \frac{44 \times 187 - 28 \times 179}{(223 \times 215 \times 72 \times 366/438)^{1/2}} = 1.89.$$

The residual r_{131} has an approximate significance level of $2[1 - \Phi(1.89)] = 0.06$; however, since each residual is computed from a distinct group of counts, this residual is one of three independent residuals. Thus the sizes of the observed residuals are not exceptionally large. The residuals r_{1j1} have an obvious tendency to increase with increasing educational level j.

This pattern suggests that the log cross-product ratios

$$\tau^{SR \cdot E}_{(12)(12) \cdot j} = \log\left(\frac{m_{1j1} m_{2j2}}{m_{1j2} m_{2j1}}\right),$$

may increase with increasing educational level j. In this case, the three-factor interaction λ^{SER}_{ijk} would not be 0. This possibility is formally checked later in this section by use of a model of no three-factor interaction.

Log Cross-Product Ratios

For now note both the uncertainty that any coefficients $\tau^{SR \cdot E}_{(12)(12) \cdot j}$ differ from 0 and the uncertainty that these coefficients are negligible. An approximate 95 percent confidence interval for $\tau^{SR \cdot E}_{(12)(12) \cdot j}$ has lower bound

$$\hat{\tau}^{SR \cdot E}_{(12)(12) \cdot j} - 1.96 s(\hat{\tau}^{SR \cdot E}_{(12)(12) \cdot j})$$

and upper bound

$$\hat{\tau}^{SR \cdot E}_{(12)(12) \cdot j} + 1.96 s(\hat{\tau}^{SR \cdot E}_{(12)(12) \cdot j}),$$

where

$$\hat{\tau}^{SR \cdot E}_{(12)(12) \cdot j} = \log\left(\frac{n_{1j1} n_{2j2}}{n_{1j2} n_{2j1}}\right)$$

and

$$s^2(\hat{\tau}^{SR \cdot E}_{(12)(12) \cdot j}) = \frac{1}{n_{1j1}} + \frac{1}{n_{1j2}} + \frac{1}{n_{2j1}} + \frac{1}{n_{2j2}}.$$

For the group with no more than eight years education ($j = 1$), the confidence bounds for $\tau^{SR \cdot E}_{(12)(12) \cdot 1}$ are -0.920 and 0.140. Corresponding bounds for $q^{SR \cdot E}_{(12)(12) \cdot 1}$ are

$$\exp(-0.920) = 0.398$$

and

$$\exp(0.140) = 1.150.$$

In the group with 9–12 years education ($j = 2$), the confidence bounds for $\tau^{SR \cdot E}_{(12)(12) \cdot 2}$ are -0.386 and 0.215. Corresponding bounds for $q^{SR \cdot E}_{(12)(12) \cdot 2}$ are 0.680 and 1.240. In the group with at least 13 years education ($j = 3$), we have confidence bounds of -0.021 and 1.012 for $\tau^{SR \cdot E}_{(12)(12) \cdot 3}$ and bounds of 0.980 and 2.751 for $q^{SR \cdot E}_{(12)(12) \cdot 3}$.

Thus it is quite possible that women with no more than eight years education are somewhat more likely to agree with the statement in question than are men with similar education, and it is also quite possible that women with at least 13 years education are somewhat more likely to disagree than are men with similar education.

Partial Association of Education and Response
Controlling for Sex

The partial association between education and response given sex may also be described by cross-product ratios and log–linear models. For $1 \leqslant j < j' \leqslant 3$, let

$$q^{ER \cdot S}_{(jj')(12) \cdot i} = \frac{m_{ij1} m_{ij'2}}{m_{ij2} m_{ij'1}} = \frac{p^{R \cdot SE}_{1 \cdot ij} p^{R \cdot SE}_{2 \cdot ij'}}{p^{R \cdot SE}_{2 \cdot ij} p^{R \cdot SE}_{1 \cdot ij'}}$$

compare the odds $p^{R \cdot SE}_{1 \cdot ij}/p^{R \cdot SE}_{2 \cdot ij}$ of agreement among persons of sex i and education j to the corresponding odds $p^{R \cdot SE}_{1 \cdot ij'}/p^{R \cdot SE}_{2 \cdot ij'}$ for persons of sex i and education j'. Let

$$\tau^{ER \cdot S}_{(jj')(12) \cdot i} = \log q^{ER \cdot S}_{(jj')(12) \cdot i},$$

so that

$$\tau^{ER \cdot S}_{(jj')(12) \cdot i} = (\lambda^{ER}_{j1} - \lambda^{ER}_{j2} - \lambda^{ER}_{j'1} + \lambda^{ER}_{j'2}) + (\lambda^{SER}_{ij1} - \lambda^{SER}_{ij2} - \lambda^{SER}_{ij'1} + \lambda^{SER}_{ij'2})$$
$$= 2(\lambda^{ER}_{j1} - \lambda^{ER}_{j'1}) + 2(\lambda^{SER}_{ij1} - \lambda^{SER}_{ij'1}).$$

If education and response are conditionally independent given sex, then the $q^{ER \cdot S}_{(jj')(12) \cdot i}$ are all 1, the $\tau^{ER \cdot S}_{(jj')(12) \cdot i}$ are all 0, and the λ^{ER}_{ij} and λ^{SER}_{ijk} are all 0.

The model of conditional independence of education and response given sex holds if and only if

$$\log m_{ijk} = \lambda + \lambda^S_i + \lambda^E_j + \lambda^R_k + \lambda^{SE}_{ij} + \lambda^{SR}_{ik}.$$

Maximum-likelihood estimates for this model may be found by the formula

$$\hat{m}_{ijk} = n^{SE}_{ij} n^{SR}_{ik}/n^S_i.$$

The associated chi-square statistics are

$$X^2_{ER \cdot S} = \sum_i \sum_j \sum_k (n_{ijk} - n^{SE}_{ij} n^{SR}_{ik}/n^S_i)^2/(n^{SE}_{ij} n^{SR}_{ik}/n^S_i)$$

$$= 166.8$$

and

$$L^2_{ER \cdot S} = 2 \sum_i \sum_j \sum_k n_{ijk} \log\left(\frac{n_{ijk} n^S_i}{n^{SE}_{ij} n^{SR}_{ik}}\right) = 172.5.$$

The degrees of freedom are $2(3 - 1)(2 - 1) = 4$, so that very clear evidence exists that education and response are partially associated given sex. Thus the model

$$\log m_{ijk} = \lambda + \lambda_i^S + \lambda_j^E + \lambda_k^R + \lambda_{ij}^{SE} + \lambda_{ik}^{SR}$$

is not tenable.

To obtain some impression of the size of the partial association of education and response, consider the estimates

$$\hat{\tau}_{(jj')(12) \cdot i}^{ER \cdot S} = \log\left(\frac{n_{ij1}n_{ij'2}}{n_{ij2}n_{ij'1}}\right),$$

$$s(\hat{\tau}_{(jj')(12) \cdot i}^{ER \cdot S}) = \left(\frac{1}{n_{ij1}} + \frac{1}{n_{ij2}} + \frac{1}{n_{ij'1}} + \frac{1}{n_{ij'2}}\right)^{1/2},$$

and

$$\hat{q}_{(jj')(12) \cdot i}^{ER \cdot S} = \frac{n_{ij1}n_{ij'2}}{n_{ij2}n_{ij'1}}$$

shown in Table 3.5. The coefficients $\hat{\tau}_{(jj')(12) \cdot i}^{ER \cdot S}$ are much larger in magnitude than the coefficients $\hat{\tau}_{(12)(12) \cdot j}^{SR \cdot E}$ observed earlier. The contrast between females with no more than eight years of schooling and females with at least 13 years of schooling is especially striking. The odds of agreement with the statement of women's proper place are estimated to be about 15 times larger in the former case than in the latter case.

The coefficient estimates $\hat{\tau}_{(12)(12) \cdot i}^{ER \cdot S}$ and $\hat{\tau}_{(23)(12) \cdot i}^{ER \cdot S}$ are rather similar in size. Models which assume that $\tau_{(12)(12) \cdot i}^{ER \cdot S} = \tau_{(23)(12) \cdot i}^{ER \cdot S}$ will be considered in Chapter 5. For now it suffices to note that the change from the lowest educational level to the intermediate level is comparable to the change from the intermediate level to the highest educational level.

Table 3.5

Estimated Partial Association Coefficients for Interaction in Table 3.14 of Education and Response Given Sex

Education levels compared	j	j'	Sex	i	$\hat{\tau}_{(jj')(12) \cdot i}$	$s(\hat{\tau}_{(jj')(12) \cdot i})$	$\hat{q}_{(jj')(12) \cdot i}$
$\leqslant 8, 9-12$	1	2	M	1	1.00	0.22	2.73
			F	2	1.31	0.22	3.70
$\leqslant 8, \geqslant 13$	1	3	M	1	1.83	0.25	6.23
			F	2	2.72	0.28	15.11
$9-12, \geqslant 13$	2	3	M	1	0.83	0.21	2.28
			F	2	1.41	0.22	4.08

If no three-factor interactions λ_{ijk}^{SER} are present, then

$$\tau_{(jj')(12)\cdot 1}^{ER\cdot S} = \tau_{(jj')(12)\cdot 2}^{ER\cdot S}.$$

Since each $\hat\tau_{(jj')(12)\cdot 1}^{ER\cdot S}$ is less than the corresponding $\hat\tau_{(jj')(12)\cdot 2}^{ER\cdot S}$, some doubts concerning this hypothesis are raised. These doubts also affect use of marginal association between education and response. The association appears stronger for females than for males. To investigate more thoroughly, a model of no three-factor interaction needs to be tested.

The Model of No Three-Factor Interaction

The model of no three-factor interaction is important for determination whether any partial association of sex and response given education exists and for determination whether the partial association coefficients

$$\tau_{(jj')(12)\cdot i}^{ER\cdot S}$$

for education and response given sex depend on sex. If this model fails to fit the data, then there is partial association of sex and response given education and the $\tau_{(jj')(12)\cdot i}^{ER\cdot S}$ do depend on sex.

Computation of maximum-likelihood estimates for the model of no three-factor interaction may be accomplished by use of the Newton–Raphson algorithm. For details, see Example 3.5. Here an alternate procedure will be used, the iterative proportional fitting algorithm of Deming and Stephan (1940) introduced for two-way tables in Chapter 2. This algorithm is used by Darroch (1962) to compute maximum-likelihood estimates for the model of no three-factor interaction. The version used here is based on that of Bishop (1969) and Bishop and Mosteller (1969). The algorithm always produces approximations m_{ijkv} of $\hat m_{ijk}$ that approach $\hat m_{ijk}$ as v becomes large. Rigorous discussions of convergence properties are given by Darroch and Ratcliff (1972) and Haberman (1974a, pages 65–67). As noted in Section 2.5, numerous computer programs for this algorithm are available. For example, see Haberman (1972, 1973b) and Fay and Goodman (1975).

The essential observation required for use of iterative proportional fitting is found in Birch (1963). In a hierarchical model, if λ-parameters with superscript T are not assumed 0, then the table of marginal totals $\hat{\mathbf{m}}^T$ are equal to the table of marginal totals $\mathbf{n}^T$. For example, in the model of no three-factor interaction, the λ_{ij}^{SE} are not set to 0. Therefore $\hat{\mathbf{m}}^{SE} = \mathbf{n}^{SE}$; i.e.,

$$\hat m_{ij}^{SE} = \sum_k \hat m_{ijk} = \sum_k n_{ijk} = n_{ij}^{SE}.$$

Similarly, the λ_{ik}^{SR} are not set to 0 and

$$\hat m_{ik}^{SR} = \sum_j \hat m_{ijk} = \sum_j n_{ijk} = n_{ik}^{SE}.$$

In addition, the λ_{jk}^{ER} are not 0 and

$$\hat{m}_{jk}^{ER} = \sum_i \hat{m}_{ijk} = \sum_i n_{ijk} = n_{jk}^{ER}.$$

The equation of $\hat{m}_{ij}^{SE}$ and n_{ij}^{SE} implies that the estimated expected number of subjects with sex i and education j is equal to the observed number of subjects with sex i and education j. Similarly, the equation $\hat{m}_{ik}^{SR} = n_{ik}^{SR}$ implies that the estimated expected number of subjects with sex i and response k is equal to the observed number of subjects with sex i and response k.

Other equalities between marginal totals are available which are redundant. For example, since the λ_i^{S} are not set to 0,

$$\hat{m}_i^{S} = \sum_j \sum_k \hat{m}_{ijk} = \sum_j \sum_k n_{ijk} = n_i^{S};$$

however, this equation is implied by the equation

$$\hat{m}_{ij}^{SE} = n_{ij}^{SE},$$

for

$$\hat{m}_i^{S} = \sum_j \hat{m}_{ij}^{SE} = \sum_j n_{ij}^{SE} = n_i^{S}.$$

In general, it suffices in a hierarchical model to consider only those equations $\hat{\mathbf{m}}^T = \mathbf{n}^T$ of highest order, i.e., those equations are considered for which the following conditions hold:

(a) The λ-parameters with superscript T are not set to 0.

(b) If T is contained in a distinct superscript U (each letter of T is in U and some letter of U is not in T), the λ-parameters with superscript U are set to 0.

The collection $\mathscr{E}$ of superscripts T which satisfy a and b is called a generating class by Haberman (1974a, page 159). The marginal tables $\mathbf{n}^T$ for T in $\mathscr{E}$ are said by Bishop (1971) to be a sufficient configuration. In the model of no three-factor interaction, the generating class consists of SE, SR, and ER, and the sufficient configuration contains $\mathbf{n}^{SE}$, $\mathbf{n}^{SR}$, and $\mathbf{n}^{ER}$. The maximum-likelihood equations

$$\hat{m}_{ij}^{SE} = n_{ij}^{SE},$$
$$\hat{m}_{ik}^{SR} = n_{ik}^{SR},$$

and

$$\hat{m}_{jk}^{ER} = n_{jk}^{ER}$$

must be considered.

The iterative proportional fitting algorithm generates approximations m_{ijkv} to $\hat{m}_{ijk}$ starting from an initial approximation m_{ijk0}. Each cycle contains one step for each superscript T in E. In this step, a proportional adjustment of m_{ijkv} is made such that the new approximation $m_{ijk(v+1)}$ has the same T-marginal totals $\mathbf{m}_{v+1}^T$ as does the observed table $\mathbf{n}$. In the model of no three-factor interaction, the algorithm may be reduced to the following steps. Calculations are illustrated in Table 3.6. Since the model of no three-factor interaction holds if each $\log m_{ijk} = 0$, an acceptable initial value of each m_{ijk0} is $1 = e^0$. The approximation m_{ijk1} may then be defined as

$$m_{ijk1} = m_{ijk0} n_{ij}^{SE} / m_{ij0}^{SE}$$
$$= \tfrac{1}{2} n_{ij}^{SE}.$$

Note that

$$m_{ij1}^{SE} = \sum_k m_{ijk1} = n_{ij}^{SE}.$$

The approximation m_{ijk2} may then be defined as

$$m_{ijk2} = m_{ijk1} n_{ik}^{SR} / m_{ik1}^{SR}$$
$$= (\tfrac{1}{2} n_{ij}^{SE}) n_{ik}^{SR} / (\tfrac{1}{2} n_i^S)$$
$$= n_{ij}^{SE} n_{ik}^{SR} / n_i^S,$$

which is also the maximum-likelihood estimate for the model of conditional independence of education and response given sex. Note that

$$m_{ik2}^{SR} = \sum_j m_{ijk2} = n_{ik}^{SE}.$$

The first cycle of the algorithm is completed by letting

$$m_{ijk3} = m_{ijk2} n_{jk}^{ER} / m_{jk2}^{ER},$$

so that

$$m_{jk3}^{ER} = n_{jk}^{ER}.$$

The next cycle begins by setting $m_{ij4}^{SE} = n_{ij}^{SE}$ by the equation

$$m_{ijk4} = m_{ijk3} n_{ij}^{SE} / m_{ij3}^{SE}.$$

In general, if v is a multiple of 3, then

$$m_{ijk(v+1)} = m_{ijkv} n_{ij}^{SE} / m_{ijv}^{SE},$$
$$m_{ijk(v+2)} = m_{ijk(v+1)} n_{ik}^{SR} / m_{ik(v+1)}^{SR},$$
$$m_{ijk(v+3)} = m_{ijk(v+2)} n_{jk}^{ER} / m_{jk(v+2)}^{ER}.$$

The algorithm has the property that

$$\mathbf{m}_{v+1}^{SE} = \mathbf{n}^{SE},$$
$$\mathbf{m}_{v+2}^{SR} = \mathbf{n}^{SR},$$

Table 3.6

Computation of Maximum-Likelihood Estimates for Table 3.4 for the Model of No Three-Factor Interaction

Sex	i	Education	j	Response	k	n_{ijk}	m_{ijk0}	n_{ij}^{SE}	m_{ij0}^{SE}	m_{ijk1}	n_{ik}^{SR}	m_{ik1}^{SR}	m_{ijk2}	n_{jk}^{ER}	m_{jk2}^{ER}	m_{ijk3}	m_{ijk6}	m_{ijk9}	$m_{ijk(12)}$
Male	1	≤8	1	Agree	1	72	1	119	2	55.5	226	324.0	41.50	158	86.26	76.01	77.31	77.08	77.05
				Disagree	2	47	1	119	2	59.5	422	324.0	77.50	85	156.74	42.03	42.17	41.98	41.95
		9–12	2	Agree	1	110	1	306	2	153.0	226	324.0	106.72	283	271.34	111.31	112.51	112.63	112.64
				Disagree	2	196	1	306	2	153.0	422	324.0	199.28	479	490.66	194.54	193.56	193.38	193.36
		≥13	3	Agree	1	44	1	223	2	111.5	226	324.0	77.77	72	155.39	36.04	36.16	36.29	36.30
				Disagree	2	179	1	223	2	111.5	422	324.0	145.23	366	282.61	188.08	186.29	186.65	186.69
Female	2	≤8	1	Agree	1	86	1	124	2	52.0	287	397.5	44.76	158	86.26	81.99	80.69	80.92	80.95
				Disagree	2	38	1	124	2	52.0	508	397.5	79.24	85	156.74	42.97	42.83	43.02	43.05
		9–12	2	Agree	1	173	1	456	2	228.0	287	397.5	164.62	283	271.34	171.69	170.49	170.37	170.36
				Disagree	2	283	1	456	2	228.0	508	397.5	291.38	479	490.66	284.46	285.44	285.62	285.64
		≥13	3	Agree	1	28	1	215	2	107.5	287	397.5	77.62	72	155.39	35.96	35.84	35.71	35.70
				Disagree	2	187	1	215	2	107.5	508	397.5	137.38	366	282.61	177.92	179.71	179.35	179.31

and

$$m_{v+3}^{ER} = n^{ER}.$$

As is evident from Table 3.6, m_{ijk3} is fairly close to the final approximation $m_{ijk(12)}$; however convergence is somewhat slower than is generally the case if the Newton–Raphson algorithm is used. The principal attraction of the algorithm is simplicity rather than speed. Computer programs for iterative proportional fitting are readily developed, as previously noted in this chapter and in Section 2.5.

The chi-square statistics X_{SER}^2 and L_{SER}^2 for the model of no three-factor interaction may be calculated using $m_{ijk(12)}$ for $\hat{m}_{ijk}$. One finds that $X_{SER}^2 = 5.95$ and $L_{SER}^2 = 5.98$. Earlier values of m_{ijk5} could be used without any harm. For example, based on m_{ijk6}, one would have $X_{SER}^2 = 5.95$ and $L_{SER}^2 = 5.99$. There are $(2 - 1)(3 - 1)(2 - 1) = 2$ degrees of freedom associated with these statistics, so X_{SER}^2 and L_{SER}^2 have approximate significance levels of 5 percent. Thus there is evidence that the three-factor interactions λ_{ijk}^{SER} are not all 0. The chi-square statistic

$$L_{ER \cdot S}^2 - L_{SER}^2 = 172.5 - 5.98 = 166.6$$

with $4 - 2 = 2$ degrees of freedom provides added confirmation of the partial association of education and response. On the other hand,

$$L_{SR \cdot E}^2 - L_{SER}^2 = 6.02 - 5.98 = 0.04$$

is a chi-square statistic, with $3 - 2 = 1$ degree of freedom, which provides no additional indication of partial association of sex and response beyond that provided by L_{SER}^2. Nonetheless, since L_{SER}^2 has an approximate 5 percent significance level, evidence does exist that sex has some effect on response.

This section has explored models for conditional and marginal independence and that of no three-factor interaction in the case of a $2 \times 3 \times 2$ table. Analyses of cross-product ratios and adjusted residuals have exploited the fact that two of the variables under study are dichotomous; however, the iterative proportional fitting procedures considered here for computation of maximum-likelihood estimates of cell means are applicable to $r \times s \times t$ tables. The comparison of the likelihood-ratio chi-square statistics $L_{ER \cdot S}^2$, $L_{SR \cdot E}^2$, and L_{SER}^2 considered in this section also illustrates a procedure of wider applicability. Thus some procedures of general importance have been introduced; however, a comprehensive study of hierarchical models and related methodology has not been provided in the first two sections of this chapter. In the next three sections, a general review of hierarchical models is undertaken.

3.3   Possible Hierarchical Models for Three-Way Tables

The examples in Sections 3.1 and 3.2 provide illustrations of only a few of the possible hierarchical models for three-way tables. This section provides an enumeration of all such models, together with maximum-likelihood equations, interpretations, and when available explicit expressions for maximum-likelihood estimates. In this section, an $r \times s \times t$ table with elements n_{ijk} and expected counts m_{ijk}, $1 \leq i \leq r$, $1 \leq j \leq s$, $1 \leq k \leq t$, is given. The interpretations used here are based on the assumption that n_{ijk} is the number of observations such that $A_h = i$, $B_h = j$, and $C_h = k$, where (A_h, B_h, C_h), $1 \leq h \leq N$, are independent and identically distributed triples with $p_{ijk} > 0$ equal to the probability that $A_h = i$, $B_h = j$, and $C_h = k$. It is assumed that A_h has integral values from 1 to r, B_h has integral values from 1 to s, and C_h has integral values from 1 to t. For example, in Section 3.1, A_h is race of victim, B_h is sex of victim, and C_h is type of assault. One has $r = s = t = 2$. In Section 3.2, the variables are S_h (sex of respondent), E_h (education of respondent), and R_h (response). Here $r = t = 2$ and $s = 3$.

Interpretations in this section assume N is fixed, so that the table $\mathbf{n}$ has a multinomial distribution with sample size N and probabilities p_{ijk}. One may also consider the case in which N has a Poisson distribution with mean M. Here each n_{ijk} has an independent Poisson distribution with mean $M p_{ijk}$. As an example, consider the analysis in Section 3.1.

Other sampling procedures can also be considered. Inferences may be conditional on the observed numbers $n_i{}^A$ of subjects with $A_h = i$, $1 \leq i \leq r$, provided that the $\lambda_i{}^A$ are not set to 0. For example, in Section 3.2, inferences are conditional on the observed numbers $n_1{}^S$ and $n_2{}^S$ of males and females. No model considered assumed that the $\lambda_i{}^S$ were 0. If inferences are conditional on the $n_i{}^A$, then the tables of n_{ijk}, $1 \leq j \leq s$, $1 \leq k \leq t$, are independent and have multinomial distributions. Inferences may also be conditional on the observed numbers n_{ij}^{AB} of $A_h = i$ and $B_h = j$, provided the λ_{ij}^{AB} are not assumed 0.

Repeated use is made of some simple equations relating marginal totals. Since the number of subjects $n_i{}^A$ with $A_h = i$ is the sum of the number of subjects n_{ij}^{AB} with $A_h = i$ and $B_h = j$ for $1 \leq j \leq s$,

$$n_i{}^A = \sum_j n_{ij}^{AB}.$$

Similarly,

$$n_i{}^A = \sum_k n_{ik}^{AC} = \sum_j \sum_k n_{ijk},$$

$$n_{ij}^{AB} = \sum_k n_{ijk},$$

$$N = \sum_i n_i{}^A = \sum_i \sum_j n_{ij}^{AB} = \sum_i \sum_j \sum_k n_{ijk}.$$

The enumeration used here follows Goodman (1970). It is based on a division of hierarchical models into nine classes, of which five are of primary interest.

Class 1. The saturated model   Here no restrictions are imposed on $\log m_{ijk}$, so

$$\log m_{ijk} = \lambda + \lambda_i^A + \lambda_j^B + \lambda_k^C + \lambda_{ij}^{AB} + \lambda_{ik}^{AC} + \lambda_{jk}^{BC} + \lambda_{ijk}^{ABC},$$

where

$$\sum_i \lambda_i^A = \sum_j \lambda_j^B = \sum_k \lambda_k^C = \cdots = \sum_i \lambda_{ijk}^{ABC} = \sum_j \lambda_{ijk}^{ABC} = \sum_k \lambda_{ijk}^{ABC} = 0.$$

The generating class $\mathscr{E}$ consists of ABC, so that

$$\hat{m}_{ijk} = \hat{m}_{ijk}^{ABC} = n_{ijk}.$$

Class 2. No three-factor interaction   Here

$$\log m_{ijk} = \lambda + \lambda_i^A + \lambda_j^B + \lambda_k^C + \lambda_{ij}^{AB} + \lambda_{ik}^{AC} + \lambda_{jk}^{BC}.$$

As already noted, $\mathscr{E}$ consists of AB, AC, and BC. One has

$$\hat{m}_{ij}^{AB} = n_{ij}^{AB},$$
$$\hat{m}_{ik}^{AC} = n_{ik}^{AC},$$
$$\hat{m}_{jk}^{BC} = n_{jk}^{BC},$$

and

$$\log \hat{m}_{ijk} = \hat{\lambda} + \hat{\lambda}_i^A + \hat{\lambda}_j^B + \hat{\lambda}_k^C + \hat{\lambda}_{ij}^{AB} + \hat{\lambda}_{ik}^{AC} + \hat{\lambda}_{jk}^{BC},$$

where the $\hat{\lambda}$-parameters satisfy the same constraints as the λ-parameters. Solution of these equations requires iterative methods such as the Newton–Raphson or the iterative proportional fitting algorithm.

The model holds if and only if the conditional cross-product ratios

$$q_{(ii')(jj')\cdot k}^{AB\cdot C} = \frac{m_{ijk}m_{i'j'k}}{m_{ij'k}m_{i'jk}}$$

measuring the interaction of A_h and B_h given C_h are independent of k, so that

$$\tau_{(ii')(jj')(kk')}^{ABC} = \log\left(\frac{m_{ijk}m_{i'j'k}m_{ij'k'}m_{i'jk'}}{m_{ij'k}m_{i'jk}m_{ijk'}m_{i'j'k'}}\right)$$

$$= \tau_{(ii')(jj')\cdot k}^{AB\cdot C} - \tau_{(ii')(jj')\cdot k'}^{AB\cdot C}$$

$$= 0.$$

This model has been used in both Section 3.1 and Section 3.2. It was introduced in the $2 \times 2 \times 2$ and $2 \times 2 \times 3$ cases by Bartlett (1935). Norton

(1945) considers the $2 \times 2 \times t$ case for $t \geq 2$. General discussions of this model are found in Roy and Kastenbaum (1956).

Class 3. Conditional independence This class contains three models. The three generating classes consist of AB and AC, AB and BC, and AC and BC. For illustrative purposes, consider the case in which $\mathscr{E}$ consists of AB and AC and

$$\log m_{ijk} = \lambda + \lambda_i^A + \lambda_j^B + \lambda_k^C + \lambda_{ij}^{AB} + \lambda_{ik}^{AC},$$

so that the AC-interactions λ_{ik}^{AC} and the ABC-interactions λ_{ijk}^{ABC} are 0.

As Birch (1963) notes, this model is equivalent to the hypothesis that given A_h, B_h and C_h are conditionally independent. Here conditional independence holds if

$$p_{jk \cdot i}^{BC \cdot A} = p_{j \cdot i}^{B \cdot A} p_{k \cdot i}^{C \cdot A}.$$

In this notation, p_i^A is the marginal probability that $A_h = i$, p_{ij}^{AB} is the marginal probability that $A_h = i$ and $B_h = j$,

$$p_{jk \cdot i}^{BC \cdot A} = p_{ijk}/p_i^A$$

is the conditional probability that $B_h = j$ and $C_h = k$ given that $A_h = i$,

$$p_{j \cdot i}^{B \cdot A} = p_{ij}^{Ab}/p_i^A$$

is the conditional probability that $B_h = j$ given that $A_h = i$, and

$$p_{k \cdot i}^{C \cdot A} = p_{ik}^{AC}/p_i^A$$

is the conditional probability that $C_h = k$ given that $A_h = i$.

To show the relationship between the hierarchical model under study and conditional independence, consider the log cross-product ratios

$$
\begin{aligned}
\tau_{(jj')(kk') \cdot i}^{BC \cdot A} &= \log\left(\frac{m_{ijk} m_{ij'k'}}{m_{ijk'} m_{ij'k}}\right) \\
&= \log\left(\frac{p_{ijk} p_{ij'k'}}{p_{ijk'} p_{ij'k}}\right) \\
&= \log\left[\frac{(p_{ijk}/p_i^A)(p_{ij'k'}/p_i^A)}{(p_{ijk'}/p_i^A)(p_{ij'k}/p_i^A)}\right] \\
&= \log\left(\frac{p_{jk \cdot i}^{BC \cdot A} p_{j'k' \cdot i}^{BC \cdot A}}{p_{jk' \cdot i}^{BC \cdot A} p_{j'k \cdot i}^{BC \cdot A}}\right).
\end{aligned}
$$

The same argument used in Section 2.6 may be employed here to show that all $\tau_{(jj')(kk') \cdot i}^{BC \cdot A}$ are 0 if and only if

$$p_{j'k' \cdot i}^{BC \cdot A} = p_{j' \cdot i}^{B \cdot A} p_{k' \cdot i}^{C \cdot A}.$$

Since

$$\tau_{(jj')(kk')\cdot i}^{BC\cdot A} = \log m_{ijk} - \log m_{ijk'} - \log m_{ij'k} + \log m_{ij'k'}$$
$$= (\lambda_{jk}^{BC} - \lambda_{jk'}^{BC} - \lambda_{j'k}^{BC} + \lambda_{j'k'}^{BC}) + (\lambda_{ijk}^{ABC} - \lambda_{ijk'}^{ABC} - \lambda_{ij'k}^{ABC} + \lambda_{ij'k'}^{ABC}),$$

if the BC-interactions $\lambda_{j'k'}^{BC}$ and the ABC-interactions $\lambda_{ij'k'}^{ABC}$ are all 0, then the $\tau_{(jj')(kk')\cdot i}^{BC\cdot A}$ are 0 and B_h and C_h are conditionally independent given A_h. On the other hand, if all the $\tau_{(jj')(kk')\cdot i}^{BC\cdot A}$ are 0, as is the case under conditional independence, then the restraints on the λ-parameters imply that

$$\frac{1}{rst}\sum_j \sum_k \sum_i \tau_{(jj')(kk')\cdot i}^{BC\cdot A} = \lambda_{j'k'}^{BC} = 0$$

and

$$\frac{1}{st}\sum_j \sum_k \tau_{(jj')(kk')\cdot i}^{BC\cdot A} = \lambda_{j'k'}^{BC} + \lambda_{ij'k'}^{ABC} = 0.$$

Thus the λ_{jk}^{BC} and λ_{ijk}^{ABC} are 0 and

$$\log m_{ijk} = \lambda + \lambda_i^A + \lambda_j^B + \lambda_k^C + \lambda_{ij}^{AB} + \lambda_{ik}^{AC}.$$

The maximum-likelihood estimate $\hat{m}_{ijk}$ of m_{ijk} satisfies the equation

$$\hat{m}_{ijk} = n_{ij}^{AB} n_{ik}^{AC} / n_i^A = N\hat{p}_i^A \hat{p}_{j\cdot i}^{B\cdot A} \hat{p}_{k\cdot i}^{C\cdot A},$$

where

$$\hat{p}_i^A = n_i^A / N,$$
$$\hat{p}_{j\cdot i}^{B\cdot A} = n_{ij}^{AB} / n_i^B,$$

and

$$\hat{p}_{k\cdot i}^{C\cdot A} = n_{ij}^{AC} / n_i^A.$$

Note that

$$m_{ijk} = Np_{ijk} = Np_i p_{jk\cdot i}^{BC\cdot A} = Np_i p_{j\cdot i}^{B\cdot A} p_{k\cdot i}^{C\cdot A}.$$

Note also that $\hat{p}_i^A$ is the fraction of observations with $A_h = i$, $\hat{p}_{j\cdot i}^{B\cdot A}$ is the fraction of the observations with $A_h = i$ such that $B_h = j$, and $\hat{p}_{k\cdot i}^{C\cdot A}$ is the fraction of the observations with $A_h = i$ such that $C_h = k$.

To verify that $\hat{m}_{ijk}$ is indeed the maximum-likelihood estimate, note that $\hat{m}_{ijk}$ must satisfy the constraints

$$\hat{m}_{ij}^{AB} = n_{ij}^{AB},$$
$$\hat{m}_{ik}^{AC} = n_{ik}^{AC},$$
$$\log \hat{m}_{ijk} = \hat{\lambda} + \hat{\lambda}_i^A + \hat{\lambda}_j^B + \hat{\lambda}_k^C + \hat{\lambda}_{ij}^{AC} + \hat{\lambda}_{ik}^{AC}.$$

The last constraint holds since

$$\hat{\tau}^{BC \cdot A}_{(jj')(kk') \cdot i} = \log\left(\frac{\hat{m}_{ijk}\hat{m}_{ij'k'}}{\hat{m}_{ijk'}\hat{m}_{ij'k}}\right) = 0.$$

To verify the first constraint, note that

$$\sum_k n^{AB}_{ij} n^{AC}_{ik}/n^A_i = (n^{AB}_{ij}/n^A_i)\sum_k n^{AC}_{ik} = (n^{AB}_{ij}/n^A_i)n^A_i = n^{AB}_{ij}.$$

Verification of the second constraint follows in a similar manner.

It is sometimes helpful to note that under the conditional independence model, the partial association coefficient

$$\tau^{AB \cdot C}_{(ii')(jj') \cdot k} = \log\left(\frac{m_{ijk}m_{i'j'k}}{m_{ij'k}m_{i'jk}}\right)$$

for A_h and B_h given C_h is equal to the marginal association coefficient

$$\tau^{AB}_{(ii')(jj')} = \log\left(\frac{m^{AB}_{ij}m^{AB}_{i'j'}}{m^{AB}_{ij'}m^{AB}_{i'j}}\right)$$

of A_h and B_h. Note that

$$m_{ijk} = N p^{AB}_{ij} p^{AC}_{ik}/p^A_i$$

and

$$m^{AB}_{ij} = N p^{AB}_{ij}.$$

It follows that

$$\tau^{AB \cdot C}_{(ii')(jj') \cdot k} = \log\left(\frac{p^{AB}_{ij}p^{AB}_{i'j'}}{p^{AB}_{ij'}p^{AB}_{i'j}}\right) = \tau^{AB}_{(ii')(jj')}.$$

The conditional independence model, as well as the models in Classes 4 and 5, are quite old. For example, see Pearson (1916) and Roy and Mitra (1956). Note that the conditional independence model appears in Section 3.2.

Class 4. Two variables independent of the third As in Class 3, three models are available. The generating class may consist of AB and C, A and BC, or AC and B. To illustrate results, let $\mathscr{E}$ consist of AB and C, so that

$$\log m_{ijk} = \lambda + \lambda^A_i + \lambda^B_j + \lambda^C_k + \lambda^{AB}_{ij}.$$

This model holds if and only if C_h is independent of the pair (A_h, B_h). Note that C_h is independent of (A_h, B_h) if

$$p_{ijk} = p^{AB}_{ij}p^C_k.$$

Equivalently, for each k and k',

$$\tau_{kk'\cdot ij}^{C\cdot AB} = \log\left(\frac{p_{k\cdot ij}^{C\cdot AB}}{p_{k'\cdot ij}^{C\cdot AB}}\right) = \log\left(\frac{m_{ijk}}{m_{ijk'}}\right)$$

is independent of i and j, where

$$p_{k\cdot ij}^{C\cdot AB} = p_{ijk}/p_{ij}^{AB}.$$

If

$$\log m_{ijk} = \lambda + \lambda_i^A + \lambda_j^B + \lambda_k^C + \lambda_{ij}^{AB},$$

then

$$\tau_{kk'\cdot ij}^{C\cdot AB} = \log m_{ijk} - \log m_{ijk'} = \lambda_k^C - \lambda_{k'}^C$$

is independent of i and j, so that independence holds. On the other hand, if the $\tau_{kk'\cdot ij}^{C\cdot AB}$ are independent of i and j, then

$$\tau_{(jj')(kk')\cdot i}^{BC\cdot A} = \tau_{kk'\cdot ij}^{C\cdot AB} - \tau_{kk'\cdot ij'}^{C\cdot AB} = \log\left(\frac{m_{ijk}m_{ij'k'}}{m_{ijk'}m_{ij'k}}\right) = 0.$$

As noted in the case of Class 3, it follows that the λ_{jk}^{BB} and λ_{ijk}^{ABC} are 0. Similarly,

$$\tau_{(ii')(kk')\cdot j}^{AC\cdot B} = 0$$

and the λ_{ik}^{AC} and λ_{ijk}^{ABC} are 0. Thus

$$\log m_{ijk} = \lambda + \lambda_i^A + \lambda_j^B + \lambda_k^C + \lambda_{ij}^{AB}.$$

The maximum-likelihood equations are

$$\hat{m}_{ij}^{AB} = n_{ij}^{AB},$$
$$\hat{m}_k^C = n_k^C,$$
$$\log \hat{m}_{ijk} = \hat{\lambda} + \hat{\lambda}_i^A + \hat{\lambda}_j^B + \hat{\lambda}_k^C + \hat{\lambda}_{ij}^{AB}.$$

The equations are satisfied by

$$\hat{m}_{ijk} = n_{ij}^{AB}n_k^C/N = N\hat{p}_{ij}^{AB}\hat{p}_k^C,$$

where

$$\hat{p}_{ij}^{AB} = n_{ij}^{AB}/N$$

is the fraction of observations such that $A_h = i$ and $B_h = j$ and

$$\hat{p}_k^C = n_k^C/N$$

is the fraction of observations such that $C_h = k$. Note that

$$\hat{\tau}_{kk' \cdot ij}^{C \cdot AB} = \log\left(\frac{N\hat{p}_{ij}^{AB}\hat{p}_k^C}{N\hat{p}_{ij}^{AB}\hat{p}_{k'}^C}\right) = \log\left(\frac{\hat{p}_k^C}{\hat{p}_{k'}^C}\right)$$

is independent of i and j,

$$\hat{m}_{ij}^{AB} = \sum_k \hat{m}_{ijk} = (n_{ij}^{AB}/N) \sum_k n_k^C = (n_{ij}^{AB}/N)N = n_{ij}^{AB},$$

and

$$\hat{m}_k^C = \sum_i \sum_j \hat{m}_{ijk} = (n_k^C/N) \sum_i \sum_j n_{ij}^{AB} = (n_k^C/N)N = n_k^C.$$

As in the example from Class 3, the partial association measure

$$\tau_{(ii')(jj') \cdot k}^{AB \cdot C} = \log\left(\frac{m_{ijk}m_{i'j'k}}{m_{ij'k}m_{i'jk}}\right)$$

for A_h and B_h given C_h is the same as the marginal association measure

$$\tau_{(ii')(jj')}^{AB} = \log\left(\frac{m_{ij}^{AB}m_{i'j'}^{AB}}{m_{ij'}^{AB}m_{i'j}^{AB}}\right)$$

for A_h and B_h. The claim is readily verified given that

$$m_{ijk} = Np_{ij}^{AB}p_k^C$$

and

$$m_{ij}^{AB} = Np_{ij}^{AB}.$$

Note that both $\tau_{(ii')(jj') \cdot k}^{AB \cdot C}$ and $\tau_{(ii')(jj')}^{AB}$ are equal to

$$\log\left(\frac{p_{ij}^{AB}p_{i'j'}^{AB}}{p_{ij'}^{AB}p_{i'j}^{AB}}\right).$$

Class 5. All variables are mutually independent   Here the generating class consists of A, B, and C and

$$\log m_{ijk} = \lambda + \lambda_i^A + \lambda_j^B + \lambda_k^C,$$

so that all two-factor and three-factor interactions are assumed 0. This model holds if and only if A_h, B_h, and C_h are mutually independent, so that

$$p_{ijk} = p_i^A p_j^B p_k^C.$$

Note that under independence,

$$\log m_{ijk} = \log N + \log p_i^A + \log p_j^B + \log p_k^C$$
$$= \lambda + \lambda_i^A + \lambda_j^B + \lambda_k^C,$$

where

$$\lambda = \log N + \frac{1}{r} \sum_f \log p_f{}^A + \frac{1}{s} \sum_g \log p_g{}^B + \frac{1}{t} \sum_h \log p_h{}^C,$$

$$\lambda_i{}^A = \log p_i{}^A - \frac{1}{r} \sum_f \log p_f{}^A,$$

$$\lambda_j{}^B = \log p_j{}^B - \frac{1}{s} \sum_g \log p_g{}^B,$$

and

$$\lambda_k{}^C = \log p_k{}^C - \frac{1}{t} \sum_h \log p_h{}^C.$$

On the other hand, if

$$\log m_{ijk} = \lambda + \lambda_1{}^A + \lambda_j{}^B + \lambda_k{}^C,$$

then it is surely true that

$$\log m_{ijk} = \lambda + \lambda_i{}^A + \lambda_j{}^B + \lambda_k{}^C + \lambda_{ij}^{AB}$$

and

$$\log m_{ijk} = \lambda + \lambda_i{}^A + \lambda_j{}^B + \lambda_k{}^C + \lambda_{ik}^{AC}.$$

Therefore, it follows from results for Class 4 that

$$p_{ijk} = p_{ij}^{AB} p_{jk}^C = p_{ik}^{AC} p_j{}^B.$$

Since

$$p_{ij}^{AB} = \sum_k p_{ijk} = p_j{}^B \sum_k p_{ik}^{AC} = p_i{}^A p_j{}^B,$$

it follows that

$$p_{ijk} = p_i{}^A p_j{}^B p_k{}^C.$$

The maximum-likelihood estimate

$$\hat{m}_{ijk} = n_i{}^A n_j{}^B n_k{}^C / N^2 = N \hat{p}_i{}^A \hat{p}_j{}^B \hat{p}_k{}^C,$$

where

$$\hat{p}_i{}^A = n_i{}^A / N,$$
$$\hat{p}_j{}^B = n_j{}^B / N,$$

and

$$\hat{p}_k{}^C = n_k{}^C / N.$$

The equations

$$\hat{m}_i{}^A = n_i{}^A,$$
$$\hat{m}_j{}^B = n_j{}^B,$$
$$\hat{m}_k{}^C = n_k{}^C,$$

and

$$\log \hat{m}_{ijk} = \lambda + \lambda_i{}^A + \lambda_j{}^B + \lambda_k{}^C$$

are satisfied. For details, see Exercise 3.6.

The remaining cases are less common, so they will receive only cursory attention.

Class 6. All categories of one variable equiprobable given the other two In the three models in this class, the generating class consists of the single pair AB, AC, or BC. The case of a generating class consisting of AC is considered here. In this case,

$$\log m_{ijk} = \lambda + \lambda_i{}^A + \lambda_k{}^C + \lambda_{ik}^{AC}.$$

The model holds if and only if given A_h and C_h, each category of B_h is equally probable, so that

$$p_{j \cdot ik}^{B \cdot AC} = 1/s.$$

The maximum-likelihood estimate

$$\hat{m}_{ijk} = \frac{1}{s} n_{ik}^{AC}$$

satisfies the equations

$$\hat{m}_{ik}^{AC} = n_{ik}^{AC}$$

and

$$\log \hat{m}_{ijk} = \hat{\lambda} + \hat{\lambda}_i{}^A + \hat{\lambda}_k{}^C + \hat{\lambda}_{ik}^{AC}.$$

Class 7. Categories of one variable equiprobable given the other two variables, and the other two variables independent Three models are in this class. The generating class may consist of A and B, A and C, or B and C. For illustrative purposes, let $\mathscr{E}$ consists of A and C, so that

$$\log m_{ijk} = \lambda + \lambda_i{}^A + \lambda_k{}^C.$$

The model holds if and only if A_h and C_h are independent and, given A_h and C_h, each category of B_h is equally probable. Thus

$$p_{ik}^{AC} = p_i{}^A p_k{}^C.$$

and

$$p_{j\cdot ik}^{B\cdot AC} = \frac{1}{s}.$$

The maximum-likelihood estimate

$$\hat{m}_{ijk} = s^{-1}n_i^A n_j^B/N = \frac{1}{s}\,N\hat{p}_i^A \hat{p}_k^C,$$

where

$$\hat{p}_i^A = n_i^A/N$$

and

$$\hat{p}_k^C = n_k^C/N.$$

The following equations hold.

$$\hat{m}_i^A = n_i^A,$$
$$\hat{m}_k^C = n_k^C,$$

and

$$\log \hat{m}_{ijk} = \hat{\lambda} + \hat{\lambda}_i^A + \hat{\lambda}_k^C.$$

Class 8. Given one variable, all combinations of categories of the other two variables are equally probable   Three models are in this class. The generating class contains the single element A, B, or C. If $\mathscr{E}$ contains A, then

$$\log m_{ijk} = \lambda + \lambda_i^A.$$

Given A_h, the model holds whenever each combination of categories of B_h and C_h is equally likely. Thus

$$p_{ik\cdot i}^{BC\cdot A} = \frac{1}{st}.$$

The maximum-likelihood estimate

$$\hat{m}_{ijk} = \left(\frac{1}{st}\right) n_i^A/N$$

satisfies the equations

$$\hat{m}_i^A = n_i^A,$$

and

$$\log \hat{m}_i = \hat{\lambda} + \hat{\lambda}_i^A.$$

Class 9. All combinations of the three variables are equally likely Here the generating class contains the null subscript and

$$\log m_{ijk} = \lambda.$$

The models hold when all combinations of categories of A_h, B_h, and C_h are equally likely, so that

$$p_{ijk} = \frac{1}{rst}.$$

The maximum-likelihood estimate

$$\hat{m}_{ijk} = \frac{N}{rst}$$

satisfies the constraint

$$\hat{M} = \sum_i \sum_j \sum_k \hat{m}_{ijk} = N.$$

3.4 Computational Procedures

Except in the model of no three-factor interaction, computation of maximum-likelihood estimate $\hat{m}_{ijk}$ can be accomplished through explicit expressions such as those presented in the preceding section. In the model of no three-factor interaction, the Newton–Raphson algorithm and the Deming–Stephan algorithm are available. The Deming–Stephan algorithm is also helpful in development of a computer package that does not require iterative computations when explicit expressions are available for maximum-likelihood estimates. As noted in Section 2.3, such expressions are available for three-way tables except in the model of the three-factor interaction.

The Newton–Raphson Algorithm

Versions of this algorithm have been used for three-way tables by Norton (1945), Dyke and Patterson (1952), Yates (1955), Bock (1970), and Haberman (1974a, page 50–54). A related but different approach is found in Kastenbaum and Lamphiear (1959). This algorithm's attractions include very rapid convergence and assistance in computation of asymptotic variances. However, it is often more difficult to implement than the iterative proportional fitting algorithm discussed later in this section, especially if the number of cells in the table is very large, say several hundred.

One example of use of the Newton–Raphson algorithm appears in Section 3.1. A similar case is also considered in Exercise 3.3. These techniques are primarily useful in $2 \times 2 \times 2$ or $2 \times 2 \times t$ tables since they do not involve inversion of matrices or solution of simultaneous equations.

More general versions of the Newton–Raphson algorithm are used in available programs such as Bock and Yates (1973). To illustrate possibilities, Table 3.4 will be used.

The simplest procedure is based on the general representation

$$
\log m_{ijk} = \lambda + \sum_{i'=1}^{r-1} \lambda_{i'}^A x_{ii'}^A + \sum_{j'=1}^{s-1} \lambda_{j'}^B x_{jj'}^B + \sum_{k'=1}^{t-1} \lambda_{k'}^C x_{kk'}^C
$$
$$
+ \sum_{i'=1}^{r-1} \sum_{j'=1}^{s-1} \lambda_{i'j'}^{AB} x_{ii'}^A x_{jj'}^B + \sum_{i'=1}^{r-1} \sum_{k'=1}^{t-1} \lambda_{i'k'}^{AC} x_{ii'}^A x_{kk'}^C
$$
$$
+ \sum_{j'=1}^{s-1} \sum_{k'=1}^{t-1} \lambda_{j'k'}^{BC} x_{jj'}^B x_{kk'}^C
$$

for the model of no three-factor-interaction in which

$$
\begin{aligned}
x_{ii'}^A &= 1, & i &= i', \\
&= 0, & i &\neq i', \quad i \neq r, \\
&= -1, & i &= r, \\
x_{jj'}^B &= 1, & j &= j', \\
&= 0, & j &\neq j', \quad j \neq s, \\
&= -1, & j &= s, \\
x_{kk'}^C &= 1, & k &= k', \\
&= 0, & k &\neq k', \quad k \neq t, \\
&= -1, & k &= t.
\end{aligned}
$$

This representation follows from the constraints on the λ-parameters. The algebra involved is somewhat tedious.

As an example, note that in Table 3.4,

$$
\log m_{ijk} = \lambda + \lambda_1^S x_{i1}^S + \lambda_1^E x_{j1}^E + \lambda_2^E x_{j2}^E + \lambda_1^R x_{k1}^R
$$
$$
+ \lambda_{11}^{SE} x_{i1}^S x_{j1}^E + \lambda_{12}^{SE} x_{i1}^S x_{j2}^E + \lambda_{11}^{SR} x_{i1}^S x_{k1}^R
$$
$$
+ \lambda_{11}^{ER} x_{j1}^E x_{k1}^R + \lambda_{21}^{ER} x_{j2}^E x_{k1}^R.
$$

Here

$$x_{i1}^S = 1, \qquad i = 1,$$
$$= -1 \qquad i = 2,$$
$$x_{j1}^E = 1, \qquad j = 1,$$
$$= 0, \qquad j = 2,$$
$$= -1, \qquad j = 3,$$
$$x_{j2}^E = 0, \qquad j = 1,$$
$$= 1, \qquad j = 2,$$
$$= -1, \qquad j = 3,$$
$$x_{k1}^R = 1, \qquad k = 1,$$
$$= -1, \qquad k = 2.$$

Note that

$$\log m_{111} = \lambda + \lambda_1^S + \lambda_1^E + \lambda_1^R + \lambda_{11}^{SE} + \lambda_{11}^{SR} + \lambda_{11}^{ER},$$

while

$$\log m_{231} = \lambda - \lambda_1^S - \lambda_1^E - \lambda_2^E + \lambda_1^R + \lambda_{11}^{SE} + \lambda_{12}^{SE} - \lambda_{11}^{SR} - \lambda_{11}^{ER} - \lambda_{22}^{ER}.$$

Since $\lambda_2^S = -\lambda_1^S, \lambda_3^E = -\lambda_1^E - \lambda_2^E, \lambda_{23}^{SE} = -\lambda_{13}^{SE} = \lambda_{11}^{SE} + \lambda_{12}^{SE}, \lambda_{21}^{SR} = -\lambda_{11}^{SR},$ and $\lambda_{31}^{ER} = -\lambda_{11}^{ER} - \lambda_{21}^{ER},$ this equation for $\log m_{231}$ is equivalent to the equation

$$\log m_{231} = \lambda + \lambda_2^S + \lambda_3^E + \lambda_1^R + \lambda_{23}^{SE} + \lambda_{21}^{SR} + \lambda_{31}^{ER}.$$

To convert the model of no three-factor interaction into a form similar to that considered in Chapter 1, note that this model implies that

$$\log m_{ijk} = \alpha + \sum_{c=1}^{9} \beta_c x_{ijkc},$$

where $\alpha = \lambda$ and the β_c and x_{ijkc} are defined as in Table 3.7. As in Chapter 1, one may let

$$m_{ijk0} = n_{ijk} + \tfrac{1}{2}$$

and

$$y_{ijk0} = \log m_{ijk0}.$$

Approximations β_{c0}, $1 \leqslant c \leqslant 9$, for $\hat{\beta}_c$, $1 \leqslant c \leqslant 9$, are then found as in the weighted-least-squares analysis of the model

$$y_{ijk0} = \alpha + \sum_c \beta_c x_{ijkc} + \varepsilon_{ijk},$$

Table 3.7

Parameters for the Newton–Raphson Algorithm for Table 3.4

c	1	2	3	4	5	6	7	8	9
β_c	λ_1^S	λ_1^E	λ_2^E	λ_1^R	λ_{11}^{SE}	λ_{12}^{SE}	λ_{21}^{SR}	λ_{11}^{ER}	λ_{21}^{ER}
x_{111c}	1	1	0	1	1	0	1	1	0
x_{112c}	1	1	0	-1	1	0	-1	-1	0
x_{121c}	1	0	1	1	0	1	1	0	1
x_{122c}	1	0	1	-1	0	1	-1	0	-1
x_{131c}	1	-1	-1	1	-1	-1	1	-1	-1
x_{132c}	1	-1	-1	-1	-1	-1	-1	1	1
x_{211c}	-1	1	0	1	-1	0	-1	1	0
x_{212c}	-1	1	0	-1	-1	0	1	-1	0
x_{221c}	-1	0	1	1	0	-1	-1	0	1
x_{222c}	-1	0	1	-1	0	-1	1	0	-1
x_{231c}	-1	-1	-1	1	1	1	-1	-1	-1
x_{232c}	-1	-1	-1	-1	1	1	1	1	1

where each ε_{ijk} is independently distributed with mean 0 and variance m_{ijk0}^{-1}. Thus the β_{c0}, $1 \leqslant c \leqslant 9$, are found using the equations

$$\theta_{c0} = \sum_i \sum_j \sum_k x_{ijkc} m_{ijk0} \Big/ \sum_i \sum_j \sum_k m_{ijk0},$$

$$w_c^0 = \sum_i \sum_j \sum_k (x_{ijkc} - \theta_{c0}) y_{ijk0} m_{ijk0},$$

$$S_{cd0} = \sum_i \sum_j \sum_k (x_{ijkc} - \theta_{c0})(x_{ijkd} - \theta_{d0}) m_{ijk0},$$

and

$$\boldsymbol{\beta}_0 = S_0^{-1} \mathbf{w}_0.$$

A new approximation m_{ijk1} for $\hat{m}_{ijk}$ is found using the equations

$$v_{ijk0} = \sum_c \beta_c x_{ijkc},$$

$$g_0 = \exp(\alpha_0) = N \Big/ \left(\sum_i \sum_j \sum_k \exp v_{ijk0} \right),$$

and

$$m_{ijk1} = g_0 \exp v_{ijk0}.$$

Table 3.8
Summary of Computations for the Newton–Raphson Algorithm for Table 3.4

c	β_{c0}	θ_{c0}	w_{i0}	S_{c10}	S_{c20}	S_{c30}	S_{c40}	S_{c50}	S_{c60}	S_{c70}	S_{c80}	S_{c90}	β_{c0}
1	λ_1^S	-0.10145	-211.12	1434.09	-32.78	-125.13	-17.30	-196.32	307.97	-414.46	-9.77	9.94	-0.06878
2	λ_1^E	-0.13458	-181.27	-32.78	658.76	483.60	310.88	1.25	-13.26	-43.64	-171.61	-280.81	-0.48981
3	λ_2^E	0.22360	254.50	-125.13	483.60	1131.55	191.24	10.91	-106.67	-5.59	-376.06	-511.91	0.63584
4	λ_1^R	-0.28778	-413.64	-17.30	310.88	191.24	1328.99	-50.74	-45.47	-139.81	-89.38	352.20	-0.25175
5	λ_{11}^{SE}	-0.00897	8.91	-196.32	1.25	10.91	-50.74	684.88	438.58	367.22	4.29	24.88	0.05012
6	λ_{12}^{SE}	-0.10904	-171.13	307.97	-13.26	-106.67	-45.47	438.58	1186.77	100.73	64.02	58.69	-0.13173
7	λ_{11}^{SR}	0.01725	68.59	-414.46	-43.64	-5.59	-139.81	367.22	100.73	1448.57	-19.33	-159.69	-0.00636
8	λ_{11}^{ER}	0.25328	184.56	-9.77	-171.61	-376.06	-89.38	4.29	64.02	-19.33	592.05	415.18	0.55648
9	λ_{21}^{ER}	0.06763	-65.93	9.94	-280.81	-511.91	352.20	24.88	58.69	-159.69	415.18	1197.37	-0.01186

c	θ_{c1}	a_{c1}	S_{c11}	S_{c21}	S_{c31}	S_{c41}	S_{c51}	S_{c61}	S_{c71}	S_{c81}	S_{c91}	δ_{c1}	β_{c1}
1	-0.10159	-0.4030	1428.11	-32.30	-124.02	-17.51	-196.14	304.19	-410.82	40.51	50.18	0.00030	-0.06848
2	-0.13505	-0.1271	-32.30	657.33	482.49	307.27	0.77	-13.66	6.92	-169.50	-277.84	-0.00130	-0.50011
3	0.22182	3.9099	-124.02	482.49	1127.61	187.73	10.25	-106.85	34.99	-371.36	-506.89	0.00909	0.64493
4	-0.28642	-3.6935	-17.51	307.27	187.73	1324.62	0.04	-4.42	-139.58	-90.88	347.60	-0.00396	-0.25571
5	-0.00866	-0.5008	-196.14	0.77	10.25	0.04	683.54	437.90	363.30	-6.28	-5.69	-0.00045	-0.05967
6	-0.10848	-1.4625	304.19	-13.66	-106.85	-4.42	437.90	1181.63	98.70	32.87	37.80	-0.00075	-0.13248
7	0.01696	0.5219	-410.82	6.92	34.99	-139.58	363.30	98.70	1442.58	-18.66	-158.17	0.00049	-0.00587
8	0.25162	3.9095	40.51	-169.50	-371.36	-90.88	-6.28	32.87	-18.66	592.29	415.09	0.00905	0.56553
9	0.06656	1.9545	50.18	-277.84	-506.89	347.60	-5.69	32.80	-158.17	415.09	1192.22	0.00328	-0.00858

i	j	k	n_{ijk}	m_{ijk0}	m_{ijk1}	m_{ijk2}	n_{ijk}	m_{ijk0}	m_{ijk1}	m_{ijk2}
2	1	1	72	72.5	77.186	77.048	86	86.5	81.146	80.947
		2	47	47.5	42.498	41.952	38	38.5	43.557	43.052
	2	1	110	110.5	112.36	112.64	173	173.5	169.94	170.35
		2	196	196.5	192.79	193.35	283	283.5	284.27	285.64
	3	1	44	44.5	37.352	36.320	28	28.5	36.872	35.711
		2	179	179.5	186.02	186.69	187	187.5	179.02	179.30

Table 3.9

Estimated Asymptotic Covariance Matrix for $\hat{\beta}$ in the Model
of No Three-Factor Interaction for Table 3.4[a]

c	$\hat{S}^{c1}$	$\hat{S}^{c2}$	$\hat{S}^{c3}$	$\hat{S}^{c4}$	$\hat{S}^{c5}$	$\hat{S}^{c6}$	$\hat{S}^{c7}$	$\hat{S}^{c8}$	$\hat{S}^{c9}$
1	9.0146	−0.1247	0.6239	0.1765	4.1583	−3.9876	1.8350	−0.0648	0.2157
2	−0.1247	26.3501	−8.6877	−6.5295	0.6873	−0.8177	−0.1540	−2.5282	5.2887
3	0.6239	−8.6877	16.1415	−0.7577	−0.9849	1.2699	0.3426	5.2911	3.2463
4	0.1765	−6.5295	−0.7577	11.2445	−0.2765	0.0050	0.6010	3.8442	−6.4398
5	4.1583	0.6873	−0.9849	−0.2765	24.6270	−9.8654	−4.4569	0.3614	−0.6395
6	−3.9876	−0.8177	1.2699	0.0050	−9.8654	13.2015	−0.4308	−0.0782	0.1323
7	1.8350	−0.1540	0.3426	0.6010	−4.4569	0.4308	8.7347	−0.4521	1.1438
8	−0.0648	−2.5282	5.2911	3.8442	0.3614	−0.0782	−0.4521	26.3049	−8.6191
9	0.2157	5.2887	3.2463	−6.4398	−0.6395	0.1323	1.1438	−8.6191	16.0376

[a] The elements $\hat{S}^{cd}$ of the inverse $\hat{S}^{-1}$ of $\hat{S}$ have been multiplied by 10,000 in the table to save space. Computations are based on the m_{ijk2} of Table 3.8.

The next iteration reduces to the equations

$$a_{c1} = \sum_i \sum_j \sum_k x_{ijkc} n_{ijk} - \sum_i \sum_j \sum_k x_{ijkc} m_{ijk1},$$

$$\theta_{c1} = N^{-1} \sum_i \sum_j \sum_k x_{ijkc} m_{ijk1},$$

$$S_{cd1} = \sum_i \sum_j \sum_k (x_{ijkc} - \theta_{c1})(x_{ijkd} - \theta_{d1}) m_{ijk1},$$

$$\delta_1 = S_1^{-1} a_1,$$

$$\beta_{c1} = \beta_{c0} + \delta_{c1},$$

$$v_{ijk1} = \sum_c \beta_c x_{ijk0},$$

$$g_1 = \exp(\alpha_i) = N \Big/ \sum_i \sum_j \sum_k \exp v_{ijk1},$$

$$m_{ijk2} = g_1 \exp v_{ijk1}.$$

Subsequent iterations are similar.

The detailed summaries of calculations used in Chapter 1 cannot be presented here due to space limitations; however, an abbreviated summary in Table 3.8 and Table 3.9 may assist the reader interested in working through the algorithm. Although the value of asymptotic variances is doubtful given that the model of no three-factor interaction is not fully satisfactory for Table 3.4, computation of estimated asymptotic standard deviations may still be instructive. For details, see Table 3.10. The EASD's for the $\hat{\lambda}_i^S$ are

Table 3.10

Parameter Estimates and Estimated Asymptotic Standard Deviations
for the Model of No Three-Factor Interaction for Table 3.4[a]

Parameter	Estimate	Formula (if not in Table 3.8)	EASD	Formula
λ_1^S	-0.068		0.030	
λ_2^S	0.068	$-\hat{\lambda}_1^S$	0.030	
λ_1^E	-0.500		0.051	$(\hat{S}^{22})^{1/2}$
λ_2^E	0.645		0.040	$(\hat{S}^{22})^{1/2}$
λ_3^E	-0.145	$-\hat{\lambda}_1^E - \hat{\lambda}_2^E$	0.050	$(\hat{S}^{22} + 2\hat{S}^{23} + \hat{S}^{33})^{1/2}$
λ_1^R	-0.256		0.034	$(\hat{S}^{44})^{1/2}$
λ_2^R	0.256	$-\hat{\lambda}_1^R$	0.034	$(\hat{S}^{44})^{1/2}$
λ_{11}^{SE}	0.050		0.050	$(\hat{S}^{55})^{1/2}$
λ_{21}^{SE}	-0.050	$-\hat{\lambda}_{11}^{SE}$	0.050	$(\hat{S}^{55})^{1/2}$
λ_{12}^{SE}	-0.132		0.036	$(\hat{S}^{66})^{1/2}$
λ_{22}^{SE}	0.132	$-\hat{\lambda}_{12}^{SE}$	0.036	$(\hat{S}^{66})^{1/2}$
λ_{13}^{SE}	0.083	$-\hat{\lambda}_{11}^{SE} - \hat{\lambda}_{12}^{SE}$	0.043	$(\hat{S}^{55} + 2\hat{S}^{56} + \hat{S}^{66})^{1/2}$
λ_{23}^{SE}	-0.083	$\hat{\lambda}_{11}^{SE} + \hat{\lambda}_{12}^{SE}$	0.043	$(\hat{S}^{55} + 2\hat{S}^{56} + \hat{S}^{66})^{1/2}$
λ_{11}^{SR}	-0.006		0.030	$(\hat{S}^{77})^{1/2}$
λ_{21}^{SR}	0.006	$-\hat{\lambda}_{11}^{SR}$	0.030	$(\hat{S}^{77})^{1/2}$
λ_{12}^{SR}	0.006	$-\hat{\lambda}_{11}^{SR}$	0.030	$(\hat{S}^{77})^{1/2}$
λ_{22}^{SR}	-0.006	$\hat{\lambda}_{11}^{SR}$	0.030	$(\hat{S}^{77})^{1/2}$
λ_{11}^{ER}	0.566		0.051	$(\hat{S}^{88})^{1/2}$
λ_{21}^{ER}	-0.009		0.040	$(\hat{S}^{99})^{1/2}$
λ_{31}^{ER}	-0.557	$-\hat{\lambda}_{11}^{ER} - \hat{\lambda}_{21}^{ER}$	0.050	$(\hat{S}^{88} + 2\hat{S}^{89} + \hat{S}^{99})^{1/2}$
λ_{12}^{ER}	-0.566	$-\hat{\lambda}_{11}^{ER}$	0.051	$(\hat{S}^{88})^{1/2}$
λ_{22}^{ER}	0.009	$-\hat{\lambda}_{21}^{ER}$	0.040	$(\hat{S}^{99})^{1/2}$
λ_{32}^{ER}	0.557	$\hat{\lambda}_{11}^{ER} + \hat{\lambda}_{21}^{ER}$	0.050	$(\hat{S}^{88} + 2\hat{S}^{89} + \hat{S}^{99})^{1/2}$
$\tau_{(12)(12)\cdot j}^{SR \cdot E}$	-0.023	$4\hat{\lambda}_{11}^{SR}$	0.118	$4(\hat{S}^{77})^{1/2}$
$\tau_{(12)(12)\cdot i}^{ER \cdot S}$	1.148	$2(\hat{\lambda}_{11}^{ER} - \hat{\lambda}_{21}^{ER})$	0.309	$4(\hat{S}^{88} - 2\hat{S}^{89} + \hat{S}^{99})^{1/2}$
$\tau_{(13)(12)\cdot i}^{ER \cdot S}$	2.245	$2(2\hat{\lambda}_{11}^{ER} + \hat{\lambda}_{21}^{ER})$	0.373	$4(4\hat{S}^{88} + 4\hat{S}^{89} + \hat{S}^{99})^{1/2}$
$\tau_{(23)(12)\cdot i}^{ER \cdot S}$	1.097	$2(\hat{\lambda}_{11}^{ER} + 2\hat{\lambda}_{21}^{ER})$	0.299	$4(\hat{S}^{88} + 4\hat{S}^{89} + 4\hat{S}^{99})^{1/2}$

[a] Note that $s(\tilde{\lambda}_1^S)$ and $s(\tilde{\lambda}_2^S)$ are omitted since the n_i^S are fixed. The maximum-likelihood estimates of parameters are computed by use of the β_{c1} in Table 3.8.

omitted since the n_i^S are fixed. In general, the estimated asymptotic standard deviation for a λ-parameter estimate with superscript T should not be used if $\mathbf{n}^T$ is fixed by the sampling procedure. The reader should note that the m_{ijk1} and β_{c0} are quite adequate approximations to the corresponding maximum-likelihood estimates.

If only a few λ-parameters are of interest, then computational labor can be reduced somewhat, at least in terms of the number of simultaneous equations to be solved. For example, assume that only λ_k^R, λ_{ik}^{SR}, and λ_{jk}^{ER} are of interest. Then one may write

$$\log m_{ijk} = \alpha_{ij}^{SE} + \sum_{c=1}^{4} \gamma_c u_{ijkc},$$

where

$$\alpha_{ij}^{SE} = \lambda + \lambda_i^S + \lambda_j^E + \lambda_{ij}^{SE},$$
$$\gamma_1 = \beta_4 = \lambda_1^R,$$
$$\gamma_2 = \beta_7 = \lambda_{11}^{SR},$$
$$\gamma_3 = \beta_8 = \lambda_{11}^{ER},$$
$$\gamma_4 = \beta_9 = \lambda_{21}^{ER},$$
$$u_{ijk1} = x_{ijk4},$$
$$u_{ijk2} = x_{ijk7},$$
$$u_{ijk3} = x_{ijk8},$$

and

$$u_{ijk4} = x_{ijk9}.$$

Let

$$n_{ijk0} = n_{ijk} + \tfrac{1}{2}$$

be an initial approximation for $\hat{m}_{ijk}$, and let

$$y_{ijk0} = \log m_{ijk0}.$$

Consider the regression model

$$y_{ijk0} = \alpha_{ij}^{SE} + \sum_{c=1}^{4} \gamma_c u_{ijkc} + \varepsilon_{ijk},$$

where the ε_{ijk} are independent random variables with respective means 0 and variances m_{ijk0}^{-1}. The weighted-least-squares estimates γ_{c0} of the γ_c satisfy the equations

$$\sum_d S_{cd0} \gamma_{d0} = S_{c10} \gamma_{10} + S_{c20} \gamma_{20} + S_{c30} \gamma_{30} + S_{c40} \gamma_{40} = w_{c0},$$

where

$$w_{c0} = \sum_i \sum_j \sum_k (u_{ijkc} - \theta_{ijc0}^{SE}) y_{ijk0} m_{ijk0},$$
$$S_{cd0} = \sum_i \sum_j \sum_k (u_{ijkc} - \theta_{ijc0}^{SE})(u_{ijkd} - \theta_{ijd0}^{SE}) m_{ijk0},$$

and

$$\theta_{ijc0}^{SE} = \left(\sum_k u_{ijkc} m_{ijk0} \right) \Big/ \sum_k m_{ijk0}.$$

Thus

$$\gamma_0 = S_0^{-1} w_0.$$

Similar formulas are encountered in the study of analysis of covariance. These weighted-least-squares estimates are also used in the version of the Newton–Raphson algorithm considered here.

As is evident from Table 3.11, the pairs of estimates γ_{10} and β_{40}, γ_{20} and β_{70}, γ_{30} and β_{80}, and γ_{40} and β_{90} coincide except for rounding errors. Given exact arithmetic, these pairs would be identical.

Table 3.11

Summary of Computations for the Alternate Newton–Raphson Algorithm for Table 3.4

c	γ_c	w_{c0}	S_{c10}	S_{c20}	S_{c30}	S_{c40}	γ_{c0}
1	λ_1^R	-312.41	1176.39	-95.71	-20.24	471.83	-0.25175
2	λ_{11}^{SR}	-0.84	-95.71	1176.39	-35.46	-101.29	-0.00636
3	λ_{11}^{SR}	260.08	-20.24	-35.46	462.96	241.60	0.55649
4	λ_{21}^{ER}	5.55	471.83	-191.29	241.60	955.03	-0.01186

c	a_{c1}	S_{c11}	S_{c21}	S_{c31}	S_{c41}	δ_{c1}	γ_{c1}
1	-3.4386	1179.59	-143.67	-24.27	465.83	-0.00397	-0.25572
2	0.6206	-143.67	1179.59	-6.15	-144.79	0.00049	-0.00586
3	5.1518	-24.27	-6.15	467.75	246.01	0.00909	0.56557
4	3.4637	465.83	-144.79	246.01	957.85	0.00329	-0.00857

c	$10,000\hat{S}^{c1}$	$10,000\hat{S}^{c2}$	$10,000\hat{S}^{c3}$	$10,000\hat{S}^{c4}$
1	11.246	0.6010	3.8428	-6.4416
2	0.6010	8.7352	-0.4520	1.1439
3	3.8428	-0.4520	26.3069	-8.6176
4	-6.4416	1.1439	-8.7176	16.0391

i	j	k	n_{ijk}	m_{ijk0}	m_{ijk1}	m_{ijk2}
1	1	1	72	72.5	76.745	77.050
		2	47	47.5	42.255	41.950
	2	1	110	110.5	112.67	112.64
		2	196	196.5	193.33	193.36
	3	1	44	44.5	37.290	36.316
		2	179	179.5	185.71	186.68
2	1	1	86	86.5	80.688	80.949
		2	38	38.5	43.312	43.051
	2	1	173	173.5	170.61	170.36
		2	283	283.5	285.39	285.64
	3	1	28	28.5	36.719	35.706
		2	187	187.5	178.28	179.29

The remainder of this Newton–Raphson algorithm now differs slightly from the previous version. Instead of ensuring that

$$\sum_i \sum_j \sum_k m_{ijk1} = N,$$

one now ensures that

$$m_{ij1}^{SE} = n_{ij}^{SE}$$

for each i and j. To do so, let

$$v_{ijk0} = \sum_c \gamma_c u_{ijkc} = \gamma_1 u_{ijk1} + \gamma_2 u_{ijk2} + \gamma_3 u_{ijk3} + \gamma_4 u_{ijk4},$$

$$g_{ij0}^{SE} = \exp(\alpha_{ij0}^{SE}) = n_{ij}^{SE} \Big/ \left(\sum_k \exp v_{ijk0} \right),$$

and

$$m_{ijk1} = g_{ij0}^{SE} \exp v_{ijk0}.$$

Further steps are similar to the initial one. For example,

$$y_{ijk1} = \log m_{ijk1} + (n_{ijk1} + (n_{ijk} - m_{ijk1})/m_{ijk1},$$

$$w_{c1} = \sum_i \sum_j \sum_k (u_{ijkc} - \theta_{ijc1}^{SE}) y_{ijk1} m_{ijk1}$$

$$S_{cd1} = \sum_i \sum_j \sum_k (u_{ijkc} - \theta_{ijc1}^{SE})(u_{ijkd} - \theta_{ijd1}^{SE}) m_{ijk1},$$

$$\theta_{ijc1}^{SE} = \left(\sum_k u_{ijkc} m_{ijk1} \right) \Big/ \sum_k m_{ijk1} = \frac{1}{n_{ij}^{SE}} \sum_k u_{ijkc} m_{ijk1},$$

and

$$\sum_d S_{cd1} \gamma_{d1} = w_{c1}.$$

As usual, computation of parameters at this point can be accomplished without explicit use of the working logarithms y_{ijk1}. Let

$$a_{c1} = \sum_i \sum_j \sum_k u_{ijkc} n_{ijk} - \sum_i \sum_j \sum_k u_{ijkc} m_{ijk1}$$

and

$$\sum_d S_{cd1} \delta_{d1} = a_{c1},$$

so that

$$\delta_1 = S_1^{-1} \mathbf{a}_1.$$

Then

$$\gamma_{c1} = \gamma_{c0} + \delta_{c1}.$$

To complete this cycle, let

$$v_{ijk1} = \sum_c \gamma_{c1} u_{ijkc}$$

$$g_{ij1}^{SE} = \exp(\alpha_{ij1}^{SE}) = n_{ij}^{SE} \left/ \left(\sum_k \exp v_{ijk1} \right) \right.,$$

and

$$m_{ijk2} = g_{ij1}^{SE} \exp v_{ijk1}.$$

For a summary of computations, see Table 3.11. Note that the program in the Appendix of Volume 2 can perform these calculations.

The estimated asymptotic covariance matrix $\hat{S}^{-1}$ of the $\hat{\gamma}$ is also shown in Table 3.11. Except for rounding errors and errors from use of slightly different approximations for $\hat{m}_{ijk}$, the estimates in Table 3.11 are the same as those those in Table 3.9. For example compare $\hat{S}^{44}$ in Table 3.9 and $\hat{S}^{11}$ in Table 3.11. The $\hat{S}^{cd}$ in Table 3.11 are the elements of the inverse $\hat{S}^{-1}$ of the matrix $\hat{S}$, where

$$\hat{S}_{cd} = \sum_i \sum_j \sum_k (u_{ijkc} - \hat{\theta}_{ijc}^{SE})(u_{ijkd} - \hat{\theta}_{ijd}^{SE})\hat{m}_{ijk}$$

and

$$\hat{\theta}_{ijkc}^{SE} = \left(\sum_k u_{ijkc}\hat{m}_{ijk} \right) \left/ \sum_k \hat{m}_{ijk} \right. = \frac{1}{n_{ij}^{SE}} \sum_k u_{ijkc}\hat{m}_{ijk}.$$

Note that it is *not* the case that $\hat{\theta}_{ijkc}^{SE}$ is equal to

$$\left(\sum_k u_{ijkc}n_{ijk} \right) \left/ n_{ij}^{SE} \right..$$

The principal gain obtained with the version of the Newton–Raphson algorithm just considered is use of a 4×4 matrix rather than the 9×9 matrix. Some further simplifications are discussed in Chapter 5.

Iterative Proportional Fitting

The iterative proportional fitting algorithm described in Section 3.2 may be applied to any hierarchical model for three-way tables rather than just to the model of no three-factor interaction. In those cases in which $\hat{m}_{ijk}$ has an explicit expression, the iterative proportional fitting algorithm produces the maximum-likelihood estimate at the end of a single cycle, as Bishop (1969) notes. In general, this feature of iterative proportional fitting is not shared by the Newton–Raphson algorithm. In addition, iterative proportional fitting never involves problems of matrix inversion or solutions of simultaneous equations.

On ther other hand, iterative proportional fitting is a slower algorithm than the Newton–Raphson algorithm in the model of no three-factor interaction, and iterative proportional fitting does not provide parameter estimates or estimated asymptotic variances as by-products of computation. Computer packages such as those by Haberman (1973c) or Fay and Goodman (1975) provide separate routines for such calculations, and Fay and Goodman (1975) only provide an upper bound for estimated asymptotic variances if the model is not saturated. Different users are likely to have different preferences for one algorithm over another.

To use an iterative proportional fitting algorithm, it is necessary to specify a generating class consisting of superscripts $T(g)$, $1 \leqslant g \leqslant h$. For example, in the model of no three-factor interaction for Table 3.4, $T(1) = SE$, $T(2) = SR$, and $T(3) = ER$. In the model

$$\log m_{ijk} = \lambda + \lambda_i{}^S + \lambda_j{}^E + \lambda_k{}^R + \lambda_{ij}^{SE} + \lambda_{jk}^{ER}$$

of conditional independence of sex and response given education, $T(1) = SE$ and $T(2) = ER$.

To begin the algorithm, initial values m_{ijk0} are required such that $\log m_{ijk0}$ satisfies the hierarchical model. For instance, in the model of no three-factor interaction, it is necessary that

$$\log m_{ijk0} = \lambda_0 + \lambda_{i0}^S + \lambda_{j0}^E + \lambda_{k0}^E + \lambda_{ij0}^{SE} + \lambda_{ik0}^{SR} + \lambda_{jk0}^{ER},$$

where

$$\sum_i \lambda_{i0}^S = \sum_j \lambda_{j0}^E = \sum_k \lambda_{k0}^R = \cdots = \sum_j \lambda_{jk0}^{ER} = \sum_k \lambda_{jk0}^{ER} = 0.$$

In the model of conditional independence of sex and response given education,

$$\log m_{ijk0} = \lambda_0 + \lambda_{i0}^S + \lambda_{j0}^E + \lambda_{k0}^R + \lambda_{ij0}^{SE} + \lambda_{jk0}^{ER}.$$

If each m_{ijk0} is 1, then this requirement is always met, for all the λ_0-parameters can be set to 0. Standard computer programs set all m_{ijk0} to 1 unless otherwise directed.

If v is divisible by h, then steps $v + 1$ to $v + h$ within a cycle of the algorithm involve successive adjustments which ensure that

$$\mathbf{m}_{v+g}^{T(g)} = \mathbf{n}^{T(g)}.$$

At step $v + g$, $m_{ijk(v+g)}$ is obtained from $m_{ijk(v+g-1)}$ by multiplication by the corresponding $T(g)$-marginal total of $\mathbf{n}$ and division by the corresponding

$T(g)$-marginal total of $\mathbf{m}_{v+g-1}$. In the model of no three-factor interaction, one has

$$m_{ijk1} = m_{ijk0}n_{ij}^{SE}/m_{ij0}^{SE},$$
$$m_{ijk2} = m_{ijk1}n_{ik}^{SR}/m_{ik0}^{SR},$$
$$m_{ijk3} = m_{ijk2}n_{jk}^{ER}/m_{jk0}^{ER},$$
$$m_{ijk4} = m_{ijk3}n_{ij}^{SE}/m_{ij3}^{SE}, \quad \text{etc.}$$

In the model of conditional independence of sex and response given education,

$$m_{ijk1} = m_{ijk0}n_{ij}^{SE}/m_{ij0}^{SE},$$
$$m_{ijk2} = m_{ijk1}n_{jk}^{ER}/m_{jk0}^{ER}.$$

Further iterations are unnecessary, for $m_{ijk2} = \hat{m}_{ijk}$. In general, if $\hat{m}_{ijk}$ has an expression in closed form, then $m_{ijkh} = \hat{m}_{ijk}$.

Parameter Estimation with Iterative Proportional Fitting

Given the $\hat{m}_{ijk}$, computation of specified ratios such as

$$\log\left(\frac{\hat{m}_{111}\hat{m}_{221}}{\hat{m}_{121}\hat{m}_{211}}\right)$$

is straightforward. Estimation of λ-parameters is a bit tedious but not difficult.

The maximum-likelihood estimates of the λ-parameters satisfy formulas such as

$$\hat{\lambda} = \hat{\mu},$$
$$\hat{\lambda}_i^A = \hat{\mu}_i^A - \hat{\lambda},$$
$$\hat{\lambda}_{ij}^{AB} = \hat{\mu}_{ij}^{AB} - \hat{\lambda} - \hat{\lambda}_i^A - \hat{\lambda}_j^B,$$
$$\hat{\lambda}_{ijk}^{ABC} = \hat{\mu}_{ijk} - \hat{\lambda} - \hat{\lambda}_i^A - \hat{\lambda}_j^B - \hat{\lambda}_k^C - \hat{\lambda}_{ij}^{AB} - \hat{\lambda}_{ik}^{AC} - \hat{\lambda}_{jk}^{BC},$$

where

$$\hat{u}_{ijk} = \log \hat{m}_{ijk},$$

$$\hat{\mu} = \frac{1}{rst}\sum_i\sum_j\sum_k \hat{\mu}_{ijk},$$

$$\hat{\mu}_i^A = \frac{1}{st}\sum_j\sum_k \hat{\mu}_{ijk},$$

$$\hat{\mu}_{ij}^{AB} = \frac{1}{t}\sum_k \hat{\mu}_{ijk},$$

and

$$\hat{\mu}_{ijk}^{ABC} = \hat{\mu}_{ijk} = \log \hat{m}_{ijk}.$$

Various systematic procedures exist for performing the computations. In the case of $2 \times 2 \times 2$ tables, the Yates (1937) algorithm is often helpful. A generalization due to Good (1958) is used in Fay and Goodman (1973). An alternate scheme described in Haberman (1974a, pages 200–205) is used in Haberman (1973c).

Given the $\hat{m}_{ijk}$, estimates of asymptotic variances can be obtained without matrix inversions or further iterative calculations if $\hat{m}_{ijk}$ can be expressed in closed form. Haberman (1974a, pages 205–212) provides some general formulas. It is also possible to obtain estimates of asymptotic variances in the model of no three-factor interaction by iterative calculations which involve no matrix inversions. An algorithm described in Fienberg (1972) is used in Haberman (1973c) for this purpose. This algorithm, which is based on a procedure by Ivanov (1962) for iterative solution of least-squares problems, can be used to find $s(\hat{\tau})$ whenever

$$\hat{\tau} = \sum_i \sum_j \sum_k c_{ijk} \log \hat{m}_{ijk}.$$

The reader should note that any λ-parameter can be expressed in the form

$$\sum_i \sum_j \sum_k c_{ijk} \log m_{ijk}.$$

For example,

$$\hat{\lambda}_{i'}^{A} = \frac{1}{rst} \sum_i \sum_j \sum_k (r\delta_{ii'} - 1) \log \hat{m}_{ijk}$$

$$\hat{\lambda}_{i'j'}^{AB} = \frac{1}{rst} \sum_i \sum_j \sum_k (r\delta_{ii'} - 1)(s\delta_{jj'} - 1) \log \hat{m}_{ijk},$$

and

$$\hat{\lambda}_{i'j'k'} = \frac{1}{rst} \sum_i \sum_j \sum_k (r\delta_{ii'} - 1)(s\delta_{jj'} - 1)(t\delta_{kk'} - 1) \log \hat{m}_{ijk},$$

where for any x and y,

$$\delta_{xy} = 1, \qquad x = y,$$
$$= 0, \qquad x \neq y.$$

Note that in the case of $\hat{\lambda}_{i'}^{A}$,

$$c_{ijk} = \frac{1}{rst}(r\delta_{ii'} - 1),$$

while for $\hat{\lambda}_{i'j'}^{AB}$, $c_{ijk} = (r\delta_{ii'} - 1)(s\delta_{jj'} - 1)$. Thus the method just considered can be employed to obtain asymptotic variances of λ-parameters.

As an example, consider the model of conditional independence of sex and response given education in Table 3.4. Let

$$\hat{\tau} = \hat{\tau}_{(12)(12) \cdot 1}^{ER \cdot S} = \hat{\tau}_{(12)(12) \cdot 2}^{ER \cdot S} = \log\left(\frac{\hat{m}_{111}\hat{m}_{222}}{\hat{m}_{112}\hat{m}_{121}}\right),$$

so that

$$\hat{\tau} = \sum_i \sum_j \sum_k c_{ijk} \log \hat{m}_{ijk},$$

where $c_{2jk} = 0, 1 \leqslant j \leqslant 2, 1 \leqslant k \leqslant 2, c_{111} = c_{122} = 1$, and $c_{121} = c_{112} = -1$. Note that the generating class includes $T(1) = SE$ and $T(2) = ER$.

To begin computations, let

$$y_{ijk0} = 0.$$

Let

$$z_{ijk0} = \hat{m}_{ijk} y_{ijk0} = 0.$$

Let

$$y_{ijk1} = y_{ijk0} + (c_{ij}^{SE} - z_{ij0}^{SE})/n_{ij}^{SE},$$

where

$$c_{ij}^{SE} = \sum_k c_{ijk} = c_{ij1} + c_{ij2}$$

and

$$z_{ij0}^{SE} = \sum_k z_{ijk0} = z_{ij10} + z_{ij20}.$$

Note that each c_{ij}^{SE} and each z_{ij0}^{SE} is 0. Thus

$$y_{ijk1} = y_{ijk0} = 0.$$

Next let

$$z_{ijk1} = \hat{m}_{ijk} y_{ijk1} = 0$$

and

$$y_{ijk2} = y_{ijk1} + (c_{jk}^{ER} - z_{jk1}^{ER})/n_{jk}^{ER}.$$

Here

$$z_{jk1}^{ER} = \sum_i z_{ijk1} = z_{ijk1} + z_{2jk1} = 0$$

and

$$c_{jk}^{ER} = \sum_i c_{ijk} = c_{1jk} + c_{2jk} = 1, \qquad j = k,$$
$$= -1, \qquad j \neq k.$$

Thus

$$y_{ijk2} = 1/n_{jk}^{ER}, \qquad j = k$$
$$= -1/n_{jk}^{ER}, \qquad j \neq k.$$

Since $\hat{m}_{ijk}$ can be expressed in closed form, further iterations are unnecessary. One may write

$$s^2(\hat{\tau}) = \sum_i \sum_j \sum_k \hat{m}_{ijk} y_{ijk2}^2$$
$$= \sum_j \sum_k \hat{m}_{jk}^{ER} (1/n_{jk}^{ER})^2$$
$$= \sum_j \sum_k \frac{1}{n_{jk}^{ER}}.$$

This result is not really surprising. Note that

$$\hat{\tau} = \log\left[\frac{(n_{11}^{SE} n_{11}^{ER}/n_1{}^E)(n_{12}^{SE} n_{22}^{ER}/n_2{}^E)}{(n_{11}^{SE} n_{12}^{ER}/n_1{}^E)(n_{12}^{SE} n_{11}^{ER}/n_2{}^E)}\right]$$
$$= \log\left(\frac{n_{11}^{ER} n_{22}^{ER}}{n_{12}^{ER} n_{21}^{ER}}\right).$$

Thus the estimate $s^2(\hat{\tau})$ is quite predictable given formulas in Chapter 2 for asymptotic variances of cross-product ratios.

If the c_{ijk} remain the same as in the preceding example but the model of no three-factor interaction is used, then calculations are iterative. One now has

$$y_{ijk0} = 0$$
$$z_{ijk0} = \hat{m}_{ijk} y_{ijk0},$$
$$y_{ijk1} = y_{ijk0} + (c_{ij}^{SE} - z_{ij0}^{SE})/n_{ij}^{SE},$$
$$z_{ijk1} = \hat{m}_{ijk} y_{ijk1},$$
$$y_{ijk2} = y_{ijk1} + (c_{ik}^{SR} - z_{ik1}^{SR})/n_{ik}^{SR},$$
$$z_{ijk2} = \hat{m}_{ijk} y_{ijk2},$$
$$y_{ijk3} = y_{ijk2} + (c_{jk}^{ER} - z_{jk2}^{ER})/n_{jk}^{ER},$$
$$z_{ijk3} = \hat{m}_{ijk} y_{ijk3},$$
$$y_{ijk4} = y_{ijk3} + (c_{ij}^{SE} - z_{ij3}^{SE})/n_{ij}^{SE},$$

etc. As v becomes large,

$$\sum_i \sum_j \sum_k \hat{m}_{ijk} y_{ijkv}^2$$

approaches $s^2(\hat{\tau})$. The reader who wishes to check calculations should consult Exercise 3.9.

3.5   Selection of Models

If a specific hierarchical model is of interest, then testing the model is a straightforward procedure. Chi-square statistics are readily obtained and formulas for adjusted residuals are straightforward, except in the model of no three-factor interaction. If data are to be analyzed with no specific hierarchical model selected in advance, then R^2 statistics, F statistics, standardized values of parameter estimates, and partitions of the likelihood-ratio chi-square all become important.

Chi-Square Statistics

No special problems arise in use of X^2 or L^2 with hierarchical models. As usual, little is known about the accuracy of the chi-square approximation. For some results on $2 \times 2 \times 2$ and $2 \times 2 \times 3$ tables, for the model of no three-factor interaction, see Odoroff (1970). The degrees of freedom for the chi-square statistics is the sum of all terms $b(T)$ for which the λ-parameters are assumed zero.

In an $r \times s \times t$ table formed by cross-classifying observations on variables A_h, B_h, and C_h, $1 \leqslant h \leqslant N$,

$$b(A) = r - 1,$$
$$b(B) = s - 1,$$
$$b(C) = t - 1,$$
$$b(AB) = (r - 1)(s - 1),$$
$$b(AC) = (r - 1)(t - 1),$$
$$b(BC) = (s - 1)(t - 1),$$

and

$$b(ABC) = (r - 1)(s - 1)(t - 1).$$

In the model of no three-factor interaction, only the λ_{ijk}^{ABC} are assumed zero, so there are

$$b(ABC) = (r - 1)(s - 1)(t - 1)$$

degrees of freedom. In the model

$$\log m_{ijk} = \lambda + \lambda_i^A + \lambda_j^B + \lambda_k^C + \lambda_{ij}^{AB} + \lambda_{ik}^{AC}$$

the λ_{jk}^{BC} and the λ_{ijk}^{ABC} are assumed 0. Thus there are

$$b(BC) + b(ABC) = (s - 1)(t - 1) + (r - 1)(s - 1)(t - 1) = r(s - 1)(t - 1)$$

degrees of freedom. Table 3.12, which can be obtained from Goodman (1970), provides a summary of formulas for degrees of freedom.

Table 3.12

Degrees of Freedom for Chi-Square Tests
for Hierarchical Log-Linear Models for Three-Way Tables

Generating class	Degrees of freedom
ABC	0
AB, AC, BC	$(r - 1)(s - 1)(t - 1)$
AB, AC	$r(s - 1)(t - 1)$
AB, C	$(rs - 1)(t - 1)$
A, B, C	$rst - r - s - t + 2$
AB	$rs(t - 1)$
A, B	$rst - r - s + 1$
A	$r(st - 1)$
Null	$rst - 1$

Partitions of Chi-Square Statistics

Previous analysis of Table 3.1 has already indicated that the model of no three-factor interaction and the saturated model are the only hierarchical models consistent with the data; however, examination of some additional chi-square statistics provides information concerning the relative importance of race and sex in determining the relative proportion of homicides involving firearms and explosives rather than cutting and piercing instruments. This examination also illustrates Goodman's (1969) partitioning method for analysis of three-way tables by means of likelihood-ratio chi-square statistics.

To obtain an appropriate partition of chi-square, define Models 1 through 4 as in Table 3.13. Note that

$$L_1^2 = 2 \sum_i \sum_j \sum_k n_{ijk} \log\left(\frac{n_{ijk}}{n_{ij}^{AB} n_k^C / N}\right),$$

$$L_2^2 = 2 \sum_i \sum_j \sum_k n_{ijk} \log\left(\frac{n_{ijk}}{n_{ij}^{AB} n_{ik}^{AC} / n_i^A}\right),$$

$$L_3^2 = 2 \sum_i \sum_j \sum_k n_{ijk} \log\left(\frac{n_{ijk}}{n_{ij}^{AB} n_{jk}^{BC} / n_j^B}\right),$$

Table 3.13

Likelihood-Ratio Chi-Square Statistics for Hierarchical Models for Table 3.1

Model number	Generating class	Likelihood- ratio chi- square	Degrees of freedom	Reduction in likelihood- ratio chi-square	Degrees of freedom for reduction	F	$R^{2\,d}$
1	AB, C	44.4	3			14.8	
2	AB, AC	7.4	2	37.0^a	1	3.7	0.83
3	AB, BC	40.2	2	4.2^b	1	3.7	0.83
4	AB, AC, BC	1.1	1	$6.3, 40.1^c$	1	1.1	0.98

$^a L_1{}^2 - L_2{}^2;$ $\quad^b L_2{}^2 - L_3{}^2;$
$^c L_2{}^2 - L_4{}^2; L_3{}^2 - L_4{}^2.$ $\quad^d$ Based on $L_1{}^2.$

and

$$L_4{}^2 = 2 \sum_i \sum_j \sum_k n_{ijk} \log(n_{ijk}/\hat{m}_{ijk4}).$$

Here $\hat{m}_4$ is the maximum-likelihood estimate of **m** under Model 4.

Interpretations based on Section 3.3 should be noted. Model 1 assumes that the type of assault is independent of the race and the sex of the victim. Model 2 assumes that given race, sex and type of assault are independent. Model 3 assumes that given sex, race and type of assault are independent. Model 4 is the model of no three-factor interaction used in Section 3.1. Since

$$\tau^{C \cdot AB}_{12 \cdot ij} = \lambda + \lambda_i^A + \lambda_j^B + \lambda_1^C + \lambda_{ij}^{AB} + \lambda_{i1}^{AC} + \lambda_{j1}^{BC}$$
$$- \lambda - \lambda_i^A - \lambda_j^B - \lambda_2^C - \lambda_{ij}^{AB} - \lambda_{i2}^{AC} - \lambda_{j2}^{BC}$$
$$= \lambda_1^C - \lambda_2^C + \lambda_{i1}^{AC} - \lambda_{i2}^{AC} + \lambda_{j1}^{BC} - \lambda_{j2}^{BC}$$
$$= 2(\lambda_1^C + \lambda_{i1}^{AC} + \lambda_{j1}^{BC}),$$

sex and race may be said to have additive effects on the log odds of assault by firearms and explosives rather than by cutting and piercing instruments.

An analysis of Table 3.1 may be based on the third and fourth columns of Table 3.13. The statistics $L_1{}^2$ and $L_3{}^2$ are so large that it is extremely unlikely that Model 1 or Model 3 holds. No evidence against Model 4 is present. The significance probability of 0.025 for $L_3{}^2$ casts doubt on Model 2. Thus both sex of victim and race of victim appear to influence the type of assault. The evidence concerning race is very strong. The influences of sex and race may be additive. The R^2 and F statistics also point out the importance of race and the relatively modest importance of sex.

An alternate analysis used the fifth and sixth columns of Table 3.13. In this example, the basic conclusions are unchanged. There exists no evidence that Model 4 fails and $L_3{}^2 - L_4{}^2$ is 40.1. Therefore, Model 3 is untenable. Since Model 3 and Model 4 differ only to the extent that λ^{AC} need not be zero in Model 4, it follows that very strong evidence exists that λ^{AC} is not zero. Since $L_2{}^2 - L_4{}^2$ is 6.3, evidence also exists that λ^{BC} is not zero. Note that $L_2{}^2 - L_4{}^2$ has approximate significance level 0.012, so that the evidence against Model 2 is stronger in this analysis than in the preceding one. Since Model 3 is untenable and Model 1 implies Model 3, Model 1 is even less tenable.

In this example, it is not really necessary to consider $L_1{}^2 - L_2{}^2$ or $L_1{}^2 - L_3{}^2$; however, the behavior of the statistics is of some general interest. Note that

$$L_1{}^2 - L_2{}^2 = 2 \sum_i \sum_k n_{ik}^{AC} \log\left(\frac{n_{ik}^{AC}}{n_i{}^A n_k{}^C / N}\right)$$

always provides a test of marginal independence of race of victim and type of assault. If given race, sex of victim and type of assault are conditionally independently, then $L_1{}^2 - L_2{}^2$ is also a test of the hypothesis that the race by type of assault interactions λ^{AC} are zero. Similarly,

$$L_1{}^2 - L_3{}^2 = 2 \sum_j \sum_k n_{jk}^{BC} \log\left(\frac{n_{jk}^{BC}}{n_j{}^B n_k{}^C / N}\right)$$

always provides a test of marginal independence of sex of victim and type of assault. If race of victim and type of assault are conditionally independent given sex of victim, then $L_1{}^2 - L_3{}^2$ is also a test that the sex by type of assault interactions λ^{BC} are zero. The conditional independence requirement can be relaxed to some extent without much effect on the test of zero interaction of sex of victim and type of assault; however, large departures from conditional independence reduce the sensitivity of the text. For example, $L_1{}^2 - L_3{}^2$ and $L_2{}^2 - L_4{}^2$ both test the hypothesis that λ^{BC} is zero. The former statistic is 4.2, while the latter is 6.3. Thus $L_2{}^2 - L_4{}^2$ provides a somewhat more sensitive test in this example. This result reflects the fact that Model 4 is consistent with the data but Model 3 is not.

Further discussion of the partition of chi-square statistics is found in Goodman (1969) and Haberman (1974a, pages 115–119). The Goodman article provides historical background on the development of this partition. As noted in Plackett (1962), earlier attempts by Lancaster (1951) and others to develop such partitions, led to test statistics with unsatisfactory properties. For other relevant references, see the Goodman article.

Residual Analysis

To illustrate residual analysis in three-way tables, consider the data in Table 3.14. In principal, sex of respondent should be unrelated to the variables to be classified. Any such relationship raises doubts about the accuracy of information provided by respondents.

Table 3.14

Married Subjects in 1974 General Social Survey, Cross-Classified by Sex of Respondent, Highest Degree Attained by Husband, and Highest Degree Attained by Wife[a]

Sex of respondent	Husband's highest degree[b]	Wife's highest degree[b]			
		Less than high school diploma	High school diploma or junior college degree	At least bachelor's degree	Total
Male	Less than high school diploma	135	60	1	196
	High school diploma or junior college degree	43	151	19	213
	Bachelor's degree	4	35	12	51
	Graduate degree	2	24	23	49
	Total	184	270	55	509
Female	Less than high school diploma	124	63	1	188
	High school diploma or junior college degree	39	219	18	276
	Bachelor's degree	1	24	26	51
	Graduate degree	0	17	14	31
	Total	164	323	59	546

[a] From National Opinion Research Center, 1974 General Social Survey, University of Chicago.

[b] Data on 10 married subjects are incomplete. Women with graduate degrees are not tabulated separately since there are only 19 in the sample.

Let n_{ijk} denote the number of respondents of sex i from couples with a husband with highest degree j and a wife with highest degree k. The basic model under study assumes that A, sex of respondent, is assumed independent of variable B, the highest degree of the husband, and of variable C, the highest degree of the wife. Thus the generating class consists of A and BC.

To begin analysis, note that for the model under study,

$$X^2 = \sum_{i=1}^{2} \sum_{j=1}^{4} \sum_{k=1}^{3} (n_{ijk} - n_i^A n_{jk}^{BC}/N)^2/(n_i^A n_{jk}^{BC}/N)$$

has value 26.4 and

$$L^2 = 2 \sum_{i=1}^{2} \sum_{j=1}^{4} \sum_{k=1}^{3} n_{ijk} \log\left(\frac{n_{ijk}N}{n^{iA}n_{jk}^{BC}}\right)$$

has value 27.5. Since the corresponding degrees of freedom are $(2-1)$ $(4.3-1) = 11$, rather strong evidence exists that sex of respondent is related to the responses under study. Note that the upper 0.5 percent point of the χ_{11}^2 distribution is 26.8.

To ascertain how the model fails, consider the adjusted residuals in Table 3.15. Note that

$$\hat{m}_{ijk} = n_i{}^A n_{jk}^{BC}/N.$$

The adjusted residual has the same form as in the independence case for an $r \times s$ table. One has

$$r_{ijk} = \frac{n_{ijk} - (n_i{}^A n_{jk}^{BC}/N)}{\{n_i{}^A n_{jk}^{BC}[1 - (n_i{}^A/N)][1 - (n_{jk}^{BC}/N)]/N\}^{1/2}}$$

The adjusted residuals for row totals for each sex of respondent are given by

$$\frac{n_{ik}^{AC} - (n_i{}^A n_k{}^C/N)}{\{n_i{}^A n_k{}^C[1 - (n_i{}^A/N)][1 - (n_k{}^C/N)]/N\}^{1/2}}$$

and the corresponding adjusted residuals for column totals are

$$\frac{n_{ij}^{AB} - (n_i{}^A n_j{}^B/N)}{\{n_i{}^A n_j{}^B[1 - (n_i{}^A/N)][1 - (n_j{}^B/N)]/N\}^{1/2}}.$$

It is sufficient to examine adjusted residuals for $k = 1$ since

$$r_{ij1} = -r_{ij2}.$$

The adjusted residual r_{221} of -3.6 is by far the largest r_{ij1} in magnitude. The only other adjusted residual r_{ij1} greater in magnitude than 2 is r_{331}, which is -2.1. Note also that all other r_{ij1} are positive. The size of r_{221} suggests that a large part of the failure of the model involves households in which both spouses have high school diplomas or junior college degrees. To appreciate the importance of these households, consider only those respondents such that either respondent's or spouse's highest attained degree is not a high school diploma or junior college degree. Construct an 11×2 contingency table in which the row categories are the 11 remaining possible combinations of highest attained degrees of husband and wife and the column categories are the two possible sexes of the respondent. The test for independence for this table yields an X^2 of 13.8 and an L^2 of 14.8 on 10 degrees of freedom. Neither chi-square statistic is significant at the 10 percent level.

Table 3.15

Observations, Fitted Expected Values, and Adjusted Residuals for Table 3.14
under the Model that Sex of Respondent Is Independent
of Highest Attained Degrees of Husband and Wife

Sex of respondent	Husband's highest degree	Less than high school diploma	High school diploma or junior college degree	Bachelor's or graduate degree	Total
Male	Less than high school diploma	135	60	1	196
		125.0	59.3	1.0	185.3
		1.4	0.1	0.1	1.4
	High school diploma or junior college degree	43	151	19	213
		39.6	178.5	17.8	235.9
		0.8	−3.6	0.4	−2.8
	Bachelor's degree	4	35	12	51
		2.4	28.5	18.3	49.2
		1.4	1.8	−2.1	0.4
	Graduate degree	2	24	23	49
		1.0	19.8	17.8	38.6
		1.5	1.3	1.7	2.4
	Total	184	270	55	509
		167.9	286.1	55.0	509
		2.1	−2.0	0.0	—
Female	Less than high school diploma	124	63	1	188
		134.0	63.7	1.0	198.7
		−1.4	−0.1	−0.1	−1.4
	High school diploma or junior college degree	39	219	18	276
		42.4	191.5	19.2	253.1
		−0.8	3.6	−0.4	2.8
	Bachelor's degree	1	24	26	51
		2.6	30.5	19.7	52.8
		−1.4	−1.8	2.1	−0.4
	Graduate degree	0	17	14	31
		1.0	21.2	19.2	41.4
		−1.5	−1.3	−1.7	−2.4
	Total	164	323	59	546
		180.1	306.9	59.0	546
		−2.1	2.1	0.0	—

A possible explanation of the effect of sex of respondent may be the quota sampling procedure used to select respondents within a block or segment. Given a household, the member interviewed is not randomly selected. Selection depends on the availability of the respondent for the interview. Although quotas are used to reduce bias and although interviews are conducted on weekends or holidays or after 4 p.m., the availability of respondents is still likely to have some relationship to the eduction of the spouses in the households.

If subjects answer questions accurately, the two-way marginal table of counts n_{jk}^{BC} may well provide an adequate comparison of the educations of husbands and wives. The sampling procedure for households may be better than the procedure for members of households. On the other hand, this example shows that care must be exercised in use of even the most carefully conducted surveys. The marginal table receives further attention in Chapters 7 and 8 of Volume 2.

As an aid in residual analysis, it is helpful to consider the formulas for r_{ijk} presented in Table 3.16. They may be verified by use of formulas in Haberman (1974a, page 212). The saturated model is omitted since $\hat{m}_{ijk} = n_{ijk}$. The model of no three-factor interaction is omitted since computation of r_{ijk} generally requires inversion of the matrix $\hat{S}$ corresponding to the version of the Newton–Raphson algorithm employed in finding maximum-likelihood estimates.

For rough results with this model, it is sometimes helpful to use approximate residuals such as the standardized residual

$$s_{ijk} = (n_{ijk} - \hat{m}_{ijk})/\hat{m}_{ijk}^{1/2}.$$

Under the model of no three-factor interaction, s_{ijk} has an approximate normal distribution with asymptotic mean 0 and asymptotic variance less than 1. If r, s, and t are large and all m_{ijk} are comparable in size, then s_{ijk} is a fairly good approximation for r_{ijk}. One may also use improved standardized residuals such as

$$s_{ijk}^* = \frac{(n_{ijk} - \hat{m}_{ijk})}{\left[\hat{m}_{ijk}\left(1 - \frac{n_{ij}^{AB}}{n_i^A}\right)\left(1 - \frac{n_{ik}^{AC}}{n_i^A}\right)\right]^{1/2}}$$

The asymptotic variance of s_{ijk}^* is less than 1 but is larger than that of s_{ijk}, especially if s or t is 2. Comparison of standardized and adjusted residuals can be found in Haberman (1973b).

The primary value of residual analysis is in detection of deviations from a model which only involves a limited number of cells. Table 3.15 illustrates this feature. Residual analysis can detect limited deviations efficiently without recourse to new computation of maximum-likelihood estimates.

Table 3.16

Adjusted Residuals for Hierarchical Models for Three-Way Tables

Generating class	r_{ijk}
AB, AC	$$\dfrac{n_{ijk} - (n_{ij}^{AB} n_{ik}^{AC}/n_i^A)}{\left[\dfrac{n_{ij}^{AB} n_{ik}^{AC}}{n_i^A}\left(1 - \dfrac{n_{ij}^{AB}}{n_i^A}\right)\left(1 - \dfrac{n_{ik}^{AC}}{n_i^A}\right)\right]^{1/2}}$$
AB, C	$$\dfrac{n_{ijk} - (n_{ij}^{AB} n_k^C/N)}{\left[\dfrac{n_{ij}^{AB} n_k^C}{N}\left(1 - \dfrac{n_{ij}^{AB}}{N}\right)\left(1 - \dfrac{n_k^C}{N}\right)\right]^{1/2}}$$
A, B, C	$$\dfrac{n_{ijk} - (n_i^A n_j^B n_k^C/N^2)}{\left[\dfrac{n_i^A n_j^B n_k^C}{N^2}\left(1 - \dfrac{n_i^A}{N}\dfrac{n_j^B}{N} - \dfrac{n_i^A}{N}\dfrac{n_k^C}{N} - \dfrac{n_j^B}{N}\dfrac{n_k^C}{N} + 2\dfrac{n_i^A}{N}\dfrac{n_j^B}{N}\dfrac{n_k^C}{N}\right)\right]^{1/2}}$$
AB	$$\dfrac{n_{ijk} - \left(\dfrac{1}{t}\right)n_{ij}^{AB}}{\left[\dfrac{1}{t}\left(1 - \dfrac{1}{t}\right)n_{ij}^{AB}\right]^{1/2}}$$
A, B	$$\dfrac{n_{ijk} - \left(\dfrac{1}{t}n_i^A n_j^B/N\right)}{\left\{\dfrac{1}{t}\dfrac{n_i^A n_j^B}{N}\left[1 - \dfrac{1}{t} + \dfrac{1}{t}\left(1 - \dfrac{n_i^A}{N}\right)\left(1 - \dfrac{n_j^B}{N}\right)\right]\right\}^{1/2}}$$
A	$$\dfrac{n_{ijk} - \left(\dfrac{1}{st}\right)n_i^A}{\left[\dfrac{1}{st}\left(1 - \dfrac{1}{st}\right)n_i^A\right]^{1/2}}$$
Empty	$$\dfrac{n_{ijk} - \left(\dfrac{1}{rst}\right)N}{\left[\dfrac{1}{rst}\left(1 - \dfrac{1}{rst}\right)N\right]^{1/2}}$$

More general deviations from a model may be suggested by residual analysis; however, their exploration often requires comparison of the model under study with more general models by tools such as the partition of likelihood-ratio chi-squares and by comparison of parameter estimates to their estimated asymptotic standard deviations.

Standardized Parameter Estimates

Comparison of parameter estimates and corresponding estimated standard deviations provides a valuable tool in the determination of which hierarchical models can conceivably fit the data. This tool has been extensively exploited by Goodman. For example, see Goodman (1970). In practice, such comparisons are commonly based on parameter estimates from the saturated model. Computational simplicity and the generality of the saturated model both favor this approach.

As an example, consider Table 3.1. Under the saturated model, parameter estimates λ, $\lambda_1{}^A$, $\lambda_1{}^B$, $\lambda_1{}^C$, λ_{11}^{AB}, λ_{11}^{AC}, λ_{11}^{BC}, and λ_{111}^{ABC} are readily obtained. For example, if

$$\hat{\mu}_{ijk} = \log \hat{m}_{ijk} = \log n_{ijk},$$

then

$$\hat{\lambda}_{111}^{ABC} = \frac{1}{8}(\hat{\mu}_{111} - \hat{\mu}_{211} - \hat{\mu}_{121} + \hat{\mu}_{221} - \hat{\mu}_{112} + \hat{\mu}_{212} + \hat{\mu}_{122} - \hat{\mu}_{222})$$

The estimated asymptotic standard deviation of $\hat{\lambda}_{111}^{ABC}$ is

$$s(\hat{\lambda}_{111}^{ABC}) = \frac{1}{8}\left(\sum_i \sum_j \sum_k \frac{1}{n_{ijk}}\right)^{1/2}.$$

Thus the standardized value

$$z_{ABC} = \frac{\hat{\lambda}_{111}^{ABC}}{s(\hat{\lambda}_{111}^{ABC})}$$

$$= \frac{\hat{\mu}_{111} - \hat{\mu}_{211} - \hat{\mu}_{121} + \hat{\mu}_{221} - \hat{\mu}_{112} + \hat{\mu}_{212} + \hat{\mu}_{221} - \hat{\mu}_{222}}{\left(\dfrac{1}{n_{111}} + \dfrac{1}{n_{211}} + \dfrac{1}{n_{121}} + \dfrac{1}{n_{221}} + \dfrac{1}{n_{112}} + \dfrac{1}{n_{212}} + \dfrac{1}{n_{221}} + \dfrac{1}{n_{222}}\right)^{1/2}}$$

provides a test of the hypothesis that λ_{111}^{ABC} is 0. Since all λ_{ijk}^{ABC} equal λ_{111}^{ABC} or $-\lambda_{111}^{ABC}$, one also obtains a test of whether the λ_{ijk}^{ABC} are zero. This test has already been used in Section 3.1. Similar tests that other λ-parameters are 0 can also be based on the saturated model.

Since Table 3.1 is a $2 \times 2 \times 2$ table, computations can be greatly simplified. Let

$$\hat{\lambda}_{1(A)}^A = \hat{\lambda}_1{}^A,$$
$$\hat{\lambda}_{1(A)B}^{AB} = \hat{\lambda}_{11}^{AB},$$
$$\hat{\lambda}_{1(ABC)}^{ABC} = \hat{\lambda}_{111}^{ABC},$$

etc. Then for any superscript T,

$$s(\hat{\lambda}_{1(T)}^T) = \frac{1}{8}\left(\sum_i \sum_j \sum_k \frac{1}{n_{ijk}}\right).$$

Thus little work is required to obtain estimated asymptotic standard deviations. To obtain estimates $\hat{\lambda}_{1(T)}^T$, the Yates (1937) algorithm may be used.

To use the algorithm, let

$$y_{ijk0} = \log n_{ijk}$$

and for $0 \leqslant t \leqslant 2, 1 \leqslant i \leqslant 2, 1 \leqslant j \leqslant 2$, let

$$y_{ij1(t+1)} = y_{1ijt} + y_{2ijt}$$

and

$$y_{ij2(t+1)} = y_{1ijt} - y_{2ijt}.$$

Then each y_{ijk3} corresponds to a $\hat{\lambda}^T_{1(T)}$. The correspondence has A in T if $i = 2$, A not in T if $i = 1$, B in T if $j = 2$, B not in T if $j = 1$, C in T if $k = 2$, and C not in T if $k = 1$.

Computations are summarized in Table 3.17. Extra digits have been used in presenting the y_{ijkt} to reduce rounding error during intermediate calculations. Except for z_{BC} and z_{ABC}, the standardized values z_T are so large that the significance probabilities $2[1 - \Phi(|z_T|)]$ are negligible. Thus considerable evidence exists that λ, λ_1^A, λ_1^B, λ_{11}^{AB}, λ_1^C, and λ_{11}^{AC} are not zero. The approximate significance probability $2[1 - (2.41)] = 0.016$ for z_{BC} is also low. Thus substantial evidence exists that λ_{11}^{BC} is not zero. On the other hand, Table 3.17 provides no indication that the three-factor interaction λ_{111}^{ABC} is not zero. Thus the only hierarchical models that are likely to be appropriate for these data are the saturated model and the model of no three-factor interaction. Thus a simple algorithm has succeeded in greatly reducing the number of models that need be considered.

Table 3.17

Estimated Interactions in Table 3.1 Under the Saturated Model

T	i	j	k	y_{ijk0}	y_{ijk1}	y_{ijk2}	y_{ijk3}	$\hat{\lambda}^T_{1(T)}$	z_T
Null	1	1	1	8.2713	16.8312	30.6218	55.7023	6.96	—[a]
A	2	1	1	8.5599	13.7906	25.0805	−0.9469	−0.12	−8.62
B	1	2	1	6.9565	13.9281	−0.1662	5.8163	0.73	52.95
AB	2	2	1	6.8341	11.1524	−0.7807	−0.7081	−0.09	6.45
C	1	1	2	6.6946	−0.2886	3.0406	5.5413	0.69	50.45
AC	2	1	2	7.2335	0.1224	2.7757	0.6145	0.08	5.59
BC	1	2	2	5.4553	−0.5389	−0.4110	0.2649	0.03	2.41
ABC	2	2	2	5.6971	−0.2418	−0.2971	−0.1139	−0.01	−1.04

$$\left[\sum_{i=1}^{2} \sum_{j=1}^{2} \sum_{k=1}^{2} 1/n_{ijk}\right]^{1/2} = 0.1098.$$

[a] The hypothesis that $\lambda = 0$ is of no interest, so z is omitted.

EXERCISES

3.1 Woolf (1955) proposes an estimate of $\tau_{(12)(12)}^{AC(B)}$ under the model of no three-factor interaction. His estimate is

$$\frac{(d_1/\hat{v}_1) + (d_2/\hat{v}_2)}{(1/\hat{v}_1) + (1/\hat{v}_2)},$$

where

$$d_j = \log\left(\frac{n_{1j1}n_{2j2}}{n_{1j2}n_{2j1}}\right)$$

and

$$\hat{v}_j = \frac{1}{n_{1j1}} + \frac{1}{n_{1j2}} + \frac{1}{n_{2j1}} + \frac{1}{n_{2j2}}.$$

His estimated asymptotic standard deviation for his estimate is

$$1\bigg/\left(\frac{1}{\hat{v}_1} + \frac{1}{\hat{v}_2}\right)^{1/2}.$$

Compare Woolf's (1955) estimate and estimated asymptotic standard deviation to $\hat{\tau}_{(12)(12)}^{AC}$ and $s(\hat{\tau}_{(12)(12)}^{AC})$. Use the data in Table 3.1.

Solution

The Woolf estimate of $\tau_{(12)(12)}^{AC(B)}$ is 0.273 and the *EASD* is 0.044. Thus the Woolf and maximum-likelihood estimates and the corresponding estimated asymptotic standard deviations agree to three decimal places. These results are not unusual in large samples.

3.2 Under the model for Table 3.1 of no three-factor interaction, find the estimates $\hat{\tau}_{12 \cdot jk}^{A \cdot BC}$, $s(\hat{\tau}_{12 \cdot jk}^{A \cdot BC})$, $\hat{\tau}_{12 \cdot ik}^{B \cdot AC}$, $s(\hat{\tau}_{12 \cdot ik}^{B \cdot AC})$, $\hat{\tau}_{12 \cdot ij}^{C \cdot AB}$, and $s(\hat{\tau}_{12 \cdot ij}^{C \cdot AB})$. Compare these estimates with corresponding estimates based on the saturated model in which no λ-parameters are assumed zero. Use the general result of Haberman (1974a, page 79) that under a model of no three-factor interaction for a $2 \times 2 \times 2$ table, if

$$\hat{\tau} = \sum_i \sum_j \sum_k a_{ijk} \log \hat{m}_{ijk},$$

then

$$s^2(\hat{\tau}) = \sum_i \sum_j \sum_k \left[a_{ijk}^2/\hat{m}_{ijk}\right] - h^{-1}\left(\sum_i \sum_j \sum_k \left[a_{ijk}c_{ijk}/\hat{m}_{ijk}\right]\right)^2,$$

where as in Section 3.1,

$$h = \sum_i \sum_j \sum_k \left(1/\hat{m}_{ijk}\right)$$

and

$$c_{ijk} = 1, \quad i + j + k \text{ odd,}$$
$$= -1, \quad i + j + k \text{ even.}$$

Solution

Results are summarized in Table 3.18. As an example of the formulas used, consider $\hat{\tau}_{12 \cdot jk}^{A \cdot BC}$. Under the saturated model,

$$\hat{\tau}_{12 \cdot jk}^{A \cdot BC} = \log\left(\frac{n_{1jk}}{n_{2jk}}\right)$$

and

$$s^2(\hat{\tau}_{12 \cdot jk}^{A \cdot BC}) = \frac{1}{n_{ijk}} + \frac{1}{n_{2jk}}.$$

Under the model of no three-factor interaction,

$$\hat{\tau}_{12 \cdot jk}^{A \cdot BC} = \log\left(\frac{\hat{m}_{ijk}}{\hat{m}_{2jk}}\right)$$

and

$$s^2(\hat{\tau}_{12 \cdot jk}^{A \cdot BC}) = \frac{1}{\hat{m}_{ijk}} + \frac{1}{\hat{m}_{2jk}} - h^{-1}\left(\frac{1}{\hat{m}_{1jk}} - \frac{1}{\hat{m}_{2jk}}\right)^2.$$

Table 3.18

Estimated Log Odds Ratios for Table 3.1 for the Model of No Three-Factor Interactions

	Saturated model		Unsaturated model	
Parameter	Estimate	*EASD*	Estimate	*EASD*
$\tau_{12 \cdot 11}^{A \cdot BC}$	-0.289	0.021	-0.284	0.021
$\tau_{12 \cdot 12}^{A \cdot BC}$	-0.539	0.044	-0.557	0.044
$\tau_{12 \cdot 21}^{A \cdot BC}$	0.122	0.045	0.103	0.045
$\tau_{12 \cdot 22}^{A \cdot BC}$	-0.242	0.087	-0.170	0.087
$\tau_{12 \cdot 11}^{B \cdot AC}$	1.315	1.320	0.035	0.035
$\tau_{12 \cdot 12}^{B \cdot AC}$	1.239	1.188	0.074	0.073
$\tau_{12 \cdot 21}^{B \cdot AC}$	1.726	1.714	0.036	0.035
$\tau_{12 \cdot 22}^{B \cdot AC}$	1.536	1.575	0.064	0.065
$\tau_{12 \cdot 11}^{C \cdot AB}$	1.577	0.038	1.590	0.038
$\tau_{12 \cdot 22}^{C \cdot AB}$	1.501	0.072	1.453	0.065
$\tau_{12 \cdot 21}^{C \cdot AB}$	1.326	0.030	1.328	0.030
$\tau_{12 \cdot 22}^{C \cdot AB}$	1.137	0.067	1.179	0.064

The estimated and corresponding $EASD$'s are quite similar for both models. The relatively large values of the $\hat{\tau}_{12 \cdot ik}^{B \cdot AC}$ corresponds to the predominance of males among victims. The figures for $\hat{\tau}_{12 \cdot jk}^{A \cdot BC}$ are more striking than might at first be apparent. Note that they imply that in all but one group, there are significantly more black than white victims. This fact is remarkable given that there are about 7.4 times as many whites as blacks in the general population.

3.3 A version of the Newton–Raphson algorithm similar to Norton's (1945) may be used to compute the maximum-likelihood estimates $\hat{m}_{ijk}$ in the model of no three-factor interaction for Table 3.4. In this algorithm, m_{ijk0} is an initial approximation to $\hat{m}_{ijk}$. The estimates $m_{ijk0} = n_{ijk} + \frac{1}{2}$ will be used in this example. Step $v \geqslant 0$ of the algorithm is defined by the equations

$$y_{ijkv} = \log m_{ijk0}, \qquad v = 0,$$

$$= \log m_{ijkv} + (n_{ijk} - m_{ijkv})/m_{ijkv}, \qquad v > 0,$$

$$\gamma_{jv} = y_{ij1v} - y_{1j2v} - y_{2j1v} + y_{2j2v},$$

$$h_{jv} = \frac{1}{m_{1j1v}} + \frac{1}{m_{1j2v}} + \frac{1}{m_{2j1v}} + \frac{1}{m_{2j2v}},$$

$$\bar{\gamma}_v = \sum_j (\gamma_{jv}/h_{jv}) \Big/ \sum_i (1/h_{jv}),$$

$$\mu_{ijkv} = y_{ijkv} - \frac{(-1)^{i+k}(\gamma_{jv} - \bar{\gamma}_v)}{h_{jv} m_{ijkv}},$$

$$g_v = N \Big/ \sum_i \sum_j \sum_k \exp \mu_{ijkv},$$

$$m_{ijk(v+1)} = g_v \exp \mu_{ijkv}.$$

Use this algorithm to compute the $\hat{m}_{ijk}$.

Solution

Computations are summarized in Table 3.19. Except for rounding errors, the m_{ijkv} produced here are the same as in Table 3.8.

3.4 Consider a $2 \times s \times 2$ contingency table with observed counts n_{ijk}, $1 \leqslant i \leqslant 2, 1 \leqslant j \leqslant s, 1 \leqslant k \leqslant 2$. Show that the adjusted residuals r_{ijk} under the model

$$\log m_{ijk} = \lambda + \lambda_i^A + \lambda_j^B + \lambda_k^C + \lambda_{ij}^{AB} + \lambda_{jk}^{BC}$$

satisfy the equation

$$r_{1j1} = -r_{2j1} = -r_{1j2} = r_{2j2} = \frac{n_{1j1}n_{2j2} - n_{1j2}n_{2j2}}{(n_{1j}^{AB} n_{2j}^{AB} n_{j1}^{BC} n_{j2}^{BC}/n_j^{B})^{1/2}}.$$

Table 3.19

Computation of Maximum-Likelihood Estimates in Table 3.4 for the Model
of No Three-Factor Interaction: The Method of Exercise 3.3

Education in years	j	Sex	i	Response	k	n_{ijk}	m_{ijk0}	y_{ijk0}
$\leqslant 8$	1	Male	1	Agree	1	72	72.5	4.2836
				Disagree	2	47	47.5	3.8607
		Female	2	Agree	1	86	86.5	4.4601
				Disagree	2	38	38.5	3.6507
9–12	2	Male	1	Agree	1	110	110.5	4.7050
				Disagree	2	196	196.5	5.2807
		Female	2	Agree	1	173	173.5	5.1562
				Disagree	2	283	283.5	5.6472
$\geqslant 13$	3	Male	1	Agree	1	44	44.5	3.7955
				Disagree	2	179	179.5	5.1902
		Female	2	Agree	1	28	28.5	3.3499
				Disagree	2	187	187.5	5.2338
Total							1443(N)	

j	i	k	$(-1)^{i+k}y_{ijk0}$	$1/m_{ijk0}$	μ_{ijk0}	$\exp\mu_{ijk0}$	m_{ijk1}	y_{ijk1}	$(-1)^{i+k}y_{ijk1}$
1	1	1	4.2836	0.013793	4.3524	77.666	77.186	4.2790	4.2790
		2	-3.8607	0.021053	3.7557	42.763	42.498	3.8554	-3.85539
	2	1	-4.4601	0.011561	4.4025	81.651	81.146	4.4561	-4.4561
		2	3.6507	0.025974	3.7803	43.828	43.557	3.6465	3.6465
Total			$-0.3866(\gamma_{10})$	$0.072380(h_{10})$					$-0.3859(\gamma_{11})$
2	1	1	4.7050	0.009050	4.7279	113.06	112.36	4.7007	4.7007
		2	-5.2807	0.005089	5.2678	193.99	192.79	5.2783	-5.2783
	2	1	-5.1562	0.005764	5.1416	170.99	169.94	5.1535	-5.1535
		2	5.6472	0.003527	5.6561	286.04	284.27	5.6455	5.6455
Total			$-0.0846(\gamma_{20})$	$0.023430(h_{20})$					$-0.0855(\gamma_{21})$
3	1	1	3.7955	0.022472	3.6266	37.584	37.352	3.7984	3.7984
		2	5.1902	0.005571	5.2321	187.18	186.02	5.1881	-5.1881
	2	1	3.3499	0.035088	3.6136	37.101	36.872	3.3668	-5.3668
		2	5.2338	0.005333	5.1937	180.13	179.02	5.2321	5.2321
Total			$0.4892(\gamma_{30})$	$0.068464(h_{30})$					$0.4755(\gamma_{31})$
Total						1452.0	1443.0		

$$\bar{\gamma}_0 = \frac{-0.3866/0.07238 - 0.0846/0.023430 + 0.4892/0.068464}{1/0.07238 + 1/0.023430 + 1/0.068464} = -0.0254;$$

$$g_0 = 1443/1452.0 = 0.99382$$

Table 3.19 (continued)

j	i	k	$1/m_{ijk1}$	μ_{ijk1}	$\exp \mu_{ijk1}$	m_{ijk1}
1	1	1	0.012956	4.3445	77.051	77.048
		2	0.023530	3.7365	41.953	41.952
	2	1	0.012323	4.3938	80.950	80.947
		2	0.022958	3.7624	43.054	43.052
Total			0.071768(h_{11})			
2	1	1	0.008900	4.7242	112.64	112.64
		2	0.005187	5.2645	193.36	193.35
	2	1	0.005885	5.1379	170.36	170.35
		2	0.003518	5.6548	285.65	285.64
Total			0.023490(h_{21})			
3	1	1	0.026773	3.5924	36.321	36.320
		2	0.005376	5.2295	186.69	186.69
	2	1	0.027121	3.5755	35.712	35.711
		2	0.005586	5.1891	179.31	179.30
Total			0.06486(h_{31})			
Total						1443.0

$$\bar{\gamma}_1 = \frac{-0.3859/0.071768 - 0.0855/0.023490 + 0.4755/0.06486}{1/0.071768 + 1/0.023490 + 1/0.06486} = -0.023;$$

$$g_1 = 1443/1443.0 = 0.99997.$$

Solution

Note that

$$r_{ijk} = \frac{n_{ijk} - n_{ij}^{AB} n_{jk}^{BC}/n_j^B}{\left[n_{ij}^{AB} n_{jk}^{BC} (1 - n_{ij}^{AB}/n_j^B)(1 - n_{jk}^{BC}/n_j^B)/n_j^B \right]^{1/2}}.$$

Observe that

$$n_j^B = n_{1j}^{AB} + n_{2j}^{AB} = n_{j1}^{BC} + n_{j2}^{BC},$$

so that

$$1 - n_{ij}^{AB}/n_j^B = \frac{n_j^B - n_{ij}^{AB}}{n_j^B} = \frac{n_{1j}^{AB} + n_{2j}^{AB} - n_{ij}^{AB}}{n_j^B}$$

and

$$1 - n_{jk}^{BC}/n_j^B = \frac{n_{j1}^{BC} + n_{j2}^{BC} - n_{jk}^{BC}}{n_j^B}.$$

Thus for each i, j and k,

$$n_{ij}^{AB} n_{jk}^{BC} (1 - n_{ij}^{AB}/n_j^{B})(1 - n_{jk}^{BC}/n_j^{B})/n_j^{B} = n_{1j}^{AB} n_{2j}^{AB} n_{j1}^{BC} n_{j2}^{BC}/(n_j^{B})^3.$$

Note also that

$$n_{ij}^{AB} = n_{ij1} + n_{ij2},$$
$$n_j^{B} = n_{1j1} + n_{1j2} + n_{2j1} + n_{2j2},$$
$$n_{jk}^{BC} = n_{1jk} + n_{2jk}.$$

Thus

$$n_{1j1} - n_{1j}^{AB} n_{j1}^{BC}/n_j^{B} = \frac{n_{1j1} n_j^{B} - n_{1j}^{AB} n_{j2}^{BC}}{n_j^{B}}$$

$$= \frac{n_{1j1}(n_{1j1} + n_{1j2} + n_{2j1} + n_{2j2}) - (n_{1j1} + n_{1j2})(n_{1j1} + n_{2j1})}{n_j^{B}}$$

$$= \frac{n_{1j1} n_{2j2} - n_{1j2} n_{2j1}}{n_j^{B}},$$

so that

$$r_{1j1} = \frac{n_{1j1} n_{2j2} - n_{1j2} n_{2j1}}{(n_{1j}^{AB} n_{2j}^{AB} n_{j1}^{BC} n_{j2}^{BC}/n_j^{B})^{1/2}}.$$

To show that $r_{1j1} = -r_{2j2} = -r_{2j1} = r_{2j2}$, observe that

$$(n_{ij1} - n_{ij}^{AB} n_{j2}^{BC}/n_j^{B}) + (n_{ij2} - n_{ij}^{AB} n_{j2}^{BC}/n_j^{B})$$
$$= n_{ij}^{AB} - (n_{ij}^{AB}/n_j^{B})(n_{j1}^{BC} + n_{j2}^{BC})$$
$$= n_{ij}^{AB} - (n_{ij}^{AB}/n_j^{B})n_j^{B}$$
$$= 0.$$

Thus $r_{1j1} = -r_{1j2}$ and $r_{2j1} = -r_{2j2}$. A similar argument shows that $r_{1j2} = -r_{2j2}$, so that $r_{1j1} = -r_{1j2} = -r_{2j1} = r_{2j2}$.

3.5 Table 3.20 contains a cross-classification from the 1974 General Social Survey comparable to the cross-classification in Table 3.4 based on the 1975 General Social Survey. Is the model for no three-factor interaction consistent with Table 3.20?

Solution

As in Table 3.4, maximum-likelihood estimates can be obtained by itera-tive proportional fitting or by various versions of the Newton–Raphson algorithms. Maximum-likelihood estimates for the means m_{ijk} are shown in Table 3.21. The chi-square statistics are $X^2 = 3.29$ and $L^2 = 3.31$. Since there are two degrees of freedom, the significance level of these statistics is about 20 percent. Thus little evidence exists in this table that the model of

Table 3.20

Subjects in the 1974 General Social Survey, Cross-Classified by Attitude Toward Women,
Staying at Home, Sex of Respondent, and Education of Respondent[a]

| | | Response | | | | |
| | | Agree | | Disagree | | |
Sex of respondent	Education of respondent in years[b]	No.	Percent.	No.	Percent.	Total number
Male	⩽ 8	89	67.4	43	32.6	132
	9–12	102	35.9	182	64.1	284
	⩾ 13	48	19.9	193	80.1	241
	Total	239	36.4	418	63.6	657
Female	⩽ 8	83	74.1	29	25.9	112
	9–12	152	34.9	284	65.1	436
	⩾ 13	33	14.8	190	85.2	223
	Total	268	34.8	503	65.2	771
Total	⩽ 8	172	70.5	72	29.5	244
	9–12	254	35.3	466	64.7	720
	⩾ 13	81	17.5	383	82.5	464
	Total	507	35.7	921	64.3	464

[a] From National Opinion Research Center, 1974 General Social Survey, University of Chicago.

[b] Three subjects did not state their educational level and 53 others were not sure whether they agreed with the statement asked. The statement is given in Table 3.4.

Table 3.21

Maximum-Likelihood Estimates of Cell Means
of Table 3.20 for the Model of No Three-Factor Interaction

Sex of respondent	Education of respondent in years	Response	
		Agree	Disagree
Male	⩽ 8	93.7	38.3
	9–12	102.3	181.7
	⩾ 13	43.0	198.0
Female	⩽ 8	78.3	33.7
	9–12	151.7	284.3
	⩾ 13	38.0	185.0

no three-factor interaction does not hold. The general pattern, however, is similar to that of Table 3.4. A comparison of the two tables is undertaken in the next chapter.

3.6 Show that in the model

$$\log m_{ijk} = \lambda + \lambda_i^A + \lambda_j^B + \lambda_k^C,$$
$$\hat{m}_{ijk} = n_i^A n_j^B n_k^C / N^2.$$

Solution

Note that

$$\hat{m}_i^A = \sum_j \sum_k n_i^A n_j^B n_k^C / N^2$$
$$= (n_i^A / N^2) \sum_j n_j^B \sum_k n_k^C$$
$$= (n_i^A / N) \sum_j n_j^B$$
$$= (n_i^A / N) N$$
$$= n_i^A.$$

Similarly, $\hat{m}_j^B = n_j^B$ and $\hat{m}_k^C = n_k^C$.

If

$$\hat{\lambda}_i^A = \log n_i^A - \frac{1}{r} \sum_{i'} \log n_{i'}^A,$$

$$\hat{\lambda}_j^B = \log n_j^B - \frac{1}{s} \sum_{j'} \log n_{j'}^B,$$

$$\hat{\lambda}_k^C = \log n_k^C - \frac{1}{t} \sum_{k'} \log n_{k'}^C,$$

and

$$\hat{\lambda} = \frac{1}{r} \sum_{i'} \log n_{i'}^A + \frac{1}{s} \sum_{j'} \log n_{j'}^B + \frac{1}{t} \sum_{k'} \log n_{k'}^C - 2 \log N,$$

then

$$\log \hat{m}_{ijk} = \log n_i^A + \log n_j^B + \log n_k^C - 2 \log N$$
$$= \hat{\lambda} + \hat{\lambda}_i^A + \hat{\lambda}_j^B + \hat{\lambda}_k^C.$$

3.7 Consider the model of complete independence

$$\log m_{ijk} = \lambda + \lambda_i^A + \lambda_j^B + \lambda_k^C$$

for an $r \times s \times t$ table. Write down an iterative proportional fitting algorithm for this model. Show that $m_{ijk3} = m_{ijk}$ if $m_{ijk0} = 1$.

Solution

Let

$$m_{ijk1} = m_{ijk0} n_i^A / m_{i0}^A = \left(\frac{1}{st}\right) n_i^A,$$

$$m_{ijk2} = m_{ijk1} n_j^B / m_{j1}^B = \left[\left(\frac{1}{st}\right) n_i^A n_j^B\right] \Big/ \left(\frac{1}{t} N\right) = \left[\left(\frac{1}{t}\right) n_i^A n_j^B\right] \Big/ N,$$

$$m_{ijk3} = m_{ijk2} n_k^C / m_{k1}^C = \frac{1}{t} (n_i^A n_j^B / N) n_k^C \Big/ \left(\frac{1}{t} N\right) = n_i^A n_j^B n_k^C / N^2 = \hat{m}_{ijk}.$$

Changing the order of the algorithm does not change the conclusion. For instance, the algorithm

$$m_{ijk1} = m_{ijk0} n_j^B / m_{j0}^B,$$
$$m_{ijk2} = m_{ijk1} n_i^A / m_{i0}^A,$$
$$m_{ijk3} = m_{ijk2} n_k^C / m_{k0}^C$$

also leads to $m_{ijk3} = \hat{m}_{ijk}$.

3.8   Consider the $2 \times s \times 2$ table of Exercise 3.4. Cochran (1954) considers several linear combinations of the r_{1j1} for use in testing the hypothesis of conditional independence of variables A_h and C_h given B_h. All tests utilize the fact that the differences

$$d_{1j1} = n_{1j1} - (n_{1j}^{AB} n_{j1}^{BC} / n_j^B)$$

have approximate independent $N(0, m_{1j}^{AB} m_{2j}^{AB} m_{j1}^{BC} m_{j2}^{BC}/(m_j^C)^3)$ distributions if conditional independence holds. Find the adjusted residual corresponding to

$$d = \sum_j \left[n_{1j1} - (n_{1j}^{AB} n_{j1}^{BC} / n_j^B) \right].$$

Find the approximate distribution of

$$\sum_j r_{1j1}$$

if the model holds.

Solution

The asymptotic variance of d is

$$\sum_j \left[\frac{m_{1j}^{AB} m_{2j}^{AB} m_{j1}^{BC} m_{j2}^{BC}}{(m_j^B)^3} \right].$$

Thus the adjusted residual

$$t = d \Big/ \left[\sum_j \frac{n_{1j}^{AB} n_{2j}^{AB} n_{j1}^{BC} n_{j2}^{BC}}{(n_j^B)^3} \right]^{1/2}.$$

Since the r_{1j1} are approximately independent $N(0, 1)$ observations, the sum

$$\sum_j r_{1j1}$$

has an approximate $N(0, s)$ distribution under the model.

The test statistic d is especially useful if the λ_{ijk}^{ABC} are 0. For a careful discussion, see Birch (1964). A very similar statistic to d is also found in Mantel and Haenzsel (1959). Related statistics also appear in Mantel (1963).

3.9 Use the iterative algorithm presented in Section 3.4 to estimate the asymptotic variance of $\hat{\tau}_{12 \cdot 11}^{R \cdot SE}$ in Table 3.4 under the model of no three-factor interaction.

Solution

Computations are summarized in Table 3.22. Note that convergence is rather rapid.

3.10 Consider dichotomous random variables A and B and a polytomous random variable C with t classes. Let possible values of A and B be 1 and 2, and let C assume integral values from 1 to t. Let $p_{ijk} > 0$ be the probability that $A = i$, $B = j$, and $C = k$, $1 \le i \le 2, i \le j \le 2, 1 \le k \le t$. Consider the log cross-product ratios

$$\tau_{(12)(12) \cdot k}^{AB \cdot C} = \log \left(\frac{p_{11k}p_{2k}}{p_{21k}p_{12k}} \right), \qquad 1 \le k \le t,$$

and

$$\tau_{(12)(12)}^{AB} = \log \left(\frac{p_{11}^{AB}p_{22}^{AB}}{p_{21}^{AB}p_{12}^{AB}} \right),$$

where for $1 \le i \le 2, 1 \le j \le 2$,

$$p_{ij}^{AB} = \sum_k p_{ijk}$$

is the marginal probability that $A = i$ and $B = j$. Show that $\tau_{(12)(12) \cdot k}^{AB \cdot C} = \tau_{(12)(12)}^{AB}, 1 \le k \le t$, if either A and C are conditionally independent given B or B and C are conditionally independent given A.

Solution

First assume conditional independence of A and C given B. Then

$$p_{ijk} = p_{ij}^{AB}p_{jk}^{BC}/p_j^{B},$$

where

$$p_{jk}^{BC} = \sum_i p_{ijk}$$

Table 3.22

Computation of $s(\hat{\tau}_{12 \cdot 11}^{R \cdot SE})$ Under the Model of No Three-Factor Interaction for Table 3.4

i	j	k	c_{ijk}	y_{ijk0}	$\hat{m}_{ijk}$	z_{ijk0}	c_{ij}^{SE}	z_{ij}^{SE}	n_{ij}^{SE}	y_{ijk1}
1	1	1	1	0	77.05	0	0	0	119	0
		2	−1	0	41.95	0	0	0	119	0
	2	1	0	0	112.64	0	0	0	306	0
		2	0	0	193.36	0	0	0	306	0
	3	1	0	0	36.30	0	0	0	223	0
		2	0	0	186.69	0	0	0	223	0
2	1	1	0	0	80.95	0	0	0	124	0
		2	0	0	43.05	0	0	0	124	0
	2	1	0	0	170.36	0	0	0	456	0
		2	0	0	285.64	0	0	0	456	0
	3	1	0	0	35.70	0	0	0	215	0
		2	0	0	179.31	0	0	0	215	0

i	j	k	z_{ijk1}	c_{ik}^{SR}	z_{ik}^{SR}	n_{ik}^{SR}	y_{ijk2}^{1000x}	z_{ijk2}	c_{jk}^{ER}	z_{jk2}^{ER}	n_{jk}^{ER}
1	1	1	0	1	0	226	4.4248	0.34093	1	0.34093	158
		2	0	−1	0	422	−2.3697	−0.09941	−1	−0.09941	85
	2	1	0	1	0	226	4.4248	0.49841	0	0.49841	283
		2	0	−1	0	422	−2.3697	−0.45820	0	−0.45820	479
	3	1	0	1	0	226	4.4248	0.16062	0	0.16062	72
		2	0	−1	0	422	−2.3697	−0.44239	0	−0.44239	366
2	1	1	0	0	0	287	0	0	1	0.34093	258
		2	0	0	0	508	0	0	−1	−0.09941	85
	2	1	0	0	0	287	0	0	0	0.49841	283
		2	0	0	0	508	0	0	0	−0.45820	49
	3	1	0	0	0	287	0	0	0	0.16062	72
		2	0	0	0	508	0	0	0	−0.44239	366

i	j	k	y_{ijk3}^{1000x}	z_{ijk3}	z_{ij3}^{SE}	y_{ijk4}^{1000x}	z_{ijk4}	z_{ik4}^{SR}	y_{ijk5}^{1000x}
1	1	1	8.5961	0.66233	0.11845	7.6007	0.58563	0.97776	7.6991
		2	−12.9649	−0.54388	0.11845	−13.9603	−0.58563	−0.97776	−14.0130
	2	1	2.6636	0.30003	0.02679	2.5761	0.29017	0.97776	2.6745
		2	−1.4131	−0.27324	0.02679	−1.5007	−0.29017	−0.97776	−1.5534
	3	1	2.1940	0.07964	−0.13710	2.8087	0.10196	0.97776	2.9072
		2	−1.1609	−0.21674	−0.13710	−0.5462	−0.10196	−0.97776	−0.5989
2	1	1	4.1713	0.33767	−0.11845	5.1266	0.41500	0.02257	5.0480
		2	−10.5952	−0.45612	−0.11845	−9.6399	−0.41500	−0.02258	−9.5955
	2	1	−1.76116	−0.30003	−0.00679	−1.7024	−0.29002	0.02257	−1.7810
		2	0.9566	0.27324	−0.02679	1.0153	0.29002	−0.02258	1.0598
	3	1	−2.2308	−0.07964	0.13710	−2.8685	−0.10240	0.02257	−2.9471
		2	1.2087	0.21674	0.13710	0.5711	0.10240	−0.02258	0.6155

i	j	k	z_{ijk5}	z_{jk5}^{ER}	y_{ijk5}^{1000x}	i	j	k	z_{ijk5}	z_{jk5}^{ER}	y_{ijk5}^{1000x}
1	1	1	0.59322	1.00185	7.6874	2	1	1	0.40863	1.00185	5.0363
		2	−0.58784	−1.00093	−1.4002			2	−0.41309	−1.00093	−9.5845
	2	1	0.30125	−0.00217	2.6821		2	1	−0.30342	−0.00217	−1.7734
		2	−0.30036	0.00236	−1.5583			2	0.30272	0.00236	1.0549
	3	1	0.10553	0.00032	2.9027		3	1	−0.10521	0.00032	−2.9515
		2	−0.11180	−0.00143	−0.5949			2	0.11037	−0.00143	0.6194

$$\left[\sum_i \sum_j \sum_k \hat{m}_{ijk} y_{ijk3}^2\right]^{1/2} = 0.1477; \qquad \left[\sum_i \sum_j \sum_k \hat{m}_{ijk} y_{ijk6}^2\right]^{1/2} = 0.1472.$$

and

$$p_j{}^B = \sum_i \sum_k p_{ijk}.$$

Thus

$$\tau^{AB \cdot C}_{(12)(12) \cdot k} = \log \left[\frac{(p^{AB}_{11} p^{BC}_{1k}/p_1{}^B)(p^{AB}_{22} p^{BC}_{2k}/p_2{}^B)}{(p^{AB}_{21} p^{BC}_{1k}/p_1{}^B)(p^{AB}_{12} p^{BC}_{2k}/p_2{}^B)} \right] = \log \left(\frac{p^{AB}_{11} p^{AB}_{22}}{p^{AB}_{22} p^{AB}_{12}} \right) = \tau^{AB}_{(12)(12)}.$$

A similar argument applies to conditional independence of B and C given A.

3.11 Let A, B, C, and p_{ijk} be defined as in Exercise 3.10. Show that for any g and h, the p_{ijk} can be chosen so that

$$\tau^{AB \cdot C}_{(12)(12) \cdot k} = g, \qquad 1 \leqslant k \leqslant t,$$

and

$$\tau^{AB}_{(12)(12)} = h.$$

Solution

Many solutions are possible. For example, let $x = e^g$ and let y and z be any positive numbers such that

$$x(y + y^{-1})^2/(z + z^{-1})^2 = e^h.$$

Let

$$u^{-1} = x^{1/2}(y + y^{-1}) + z + z^{-1},$$

and let

$$
\begin{aligned}
p_{ijk} &= \frac{ux^{1/2}}{2y(t-1)}, & i = j = 1,\ \ 1 \leqslant k \leqslant t-1, \\[2mm]
&= \frac{uz}{2(t-1)}, & i = 2,\ \ j = 1,\ \ 1 \leqslant k \leqslant t-1, \\[2mm]
&= \frac{u}{2z(t-1)}, & i = 1,\ \ j = 2,\ \ 1 \leqslant k \leqslant t-1, \\[2mm]
&= \frac{ux^{1/2}y}{2(t-1)}, & i = j = 2,\ \ 1 \leqslant k \leqslant t-1, \\[2mm]
&= \tfrac{1}{2}ux^{1/2}y, & i = j = 1,\ \ k = t, \\[1mm]
&= \tfrac{1}{2}u/z, & i = 2,\ \ j = 1,\ \ k = t, \\[1mm]
&= \tfrac{1}{2}uz, & i = 1,\ \ j = 2,\ \ k = t, \\[1mm]
&= \tfrac{1}{2}ux^{1/2}/y, & i = j = 2,\ \ k = t.
\end{aligned}
$$

Note that

$$p_{22}^{AB} = p_{11}^{AB} = \tfrac{1}{2}ux^{1/2}(y + y^{-1})$$

and

$$p_{21}^{AB} = p_{12}^{AB} = \tfrac{1}{2}u(z + z^{-1}),$$

so that $\tau_{(12)(12)}^{AB} = h$ and $\tau_{(12)(12) \cdot k}^{AB \cdot C} = g$, $1 \leqslant k \leqslant t$.

An important feature of this particular solution arises if $g = h$. Then one may let $y = z$, so

$$\tau_{(12)(1t) \cdot j}^{AC \cdot B} = -4\log y, \qquad 1 \leqslant j \leqslant 2,$$

and

$$\tau_{(12)(1t) \cdot i}^{BC \cdot A} = -2\log y, \qquad 1 \leqslant i \leqslant 2.$$

Thus neither of the conditional independence hypotheses of Exercise 3.10 need hold, even though $\tau_{(12)(12) \cdot k}^{AB \cdot C} = \tau_{(12)(12)}^{AB}$, $1 \leqslant k \leqslant t$.

4 Complete Higher-Way Tables

The basic principals for use with hierarchical models in three-way tables continue to apply when higher-way tables are examined in which four or more variables are cross-classified. Three principal changes occur as the number of variables increases. The number of possible hierarchical models increases very rapidly; a decreasing fraction of hierarchical models have explicit expressions for maximum-likelihood estimates; and interpretation of hierarchical models becomes increasingly complex.

This chapter consists of two examples illustrating application of hierarchical models to four-way tables. The first, which appears in Section 4.1, compares Tables 3.4 and 3.20 in Chapter 3 to determine any differences between the 1974 and 1975 General Social Surveys in responses to a question on women's roles. The second example, which is found in Section 4.2, is a cross-classification of white Christian subjects in the 1972 General Social Survey by year of survey, religion, education, and attitudes toward nontherapeutic abortions.

The analyses in this chapter emphasize use of log odds, conditional log odds, cross-product ratios, and conditional cross-product ratios in interpreting hierarchical models. The standardized parameter estimates of Section 3.5 are a major tool in the selection of models, as is the use of partitions of the likelihood-ratio chi-square statistic.

Maximum-likelihood estimations in this chapter are found using closed-form estimates and iterative proportional fitting. This choice proves convenient in this chapter for two reasons. First, many models are compared without regard to the estimated asymptotic standard deviations of parameters. Second, since the hierarchical models of greatest interest have closed-form maximum-likelihood estimates, the needed estimated asymptotic standard deviations may be obtained without recourse to the Newton–Raphson algorithm. It should not be supposed that the Newton–Raphson algorithm is inappropriate for analysis of higher-way tables.

This chapter considers cases in which the Newton–Raphson algorithm can be avoided. In Chapter 5 and in Chapter 6 of Volume 2, practical alternatives to the Newton–Raphson algorithm generally will not be available.

4.1   Comparison of Three-Way Tables: Year of Survey, Sex, Education, and Place

This section explores possible hierarchical models for a $2 \times 2 \times 3 \times 2$ table. Standardized parameter estimates are used to limit the class of potential hierarchical models. Then closed-form expressions are found for maximum-likelihood estimates of some of the hierarchical models of interest, while iterative proportional fitting is used for another model. Partitions of likelihood-ratio chi-square statistics are then used to determine which models are consistent with the data. The parameters of the most attractive model are then estimated and interpreted.

The analysis of this section is based on the observation that Table 3.4 and 3.20 in Chapter 3 are rather similar in terms of the corresponding percentages of subjects of various sexes and educations who agree with the statement that women should run their homes and let men run the country. Two related questions may be raised. Are there any differences in responses between the two tables? If not, how can data from the two tables be most efficiently used to describe relationships between sex and education of respondent and responses?

To begin analysis, let n_{ijkl} be the number of subjects h in year $Y_h = i$ ($i = 1$ for 1974 and $i = 2$ for 1975) of sex $S_h = j$ and education $E_h = k$, with response $R_h = l$. Thus the table of counts n_{ijkl} is a $2 \times 2 \times 3 \times 2$ table. Given the sampling procedure used, inferences are conditional on the observed marginal totals n_{ij}^{YS} for the number of subjects in year i of sex j. Given $Y_h = i$ and $S_h = j$, $1 \leqslant h \leqslant N$, it is assumed that the pairs (E_h, R_h) are independently distributed with conditional probability $p_{kl \cdot ij}^{ER \cdot YS} > 0$ that $E_h = k$ and $R_h = l$ given that $Y_h = i$ and $S_h = j$. The expected value m_{ijkl} of n_{ijkl} is thus $n_{ij}^{YS} p_{kl \cdot ij}^{ER}$.

Estimation in the Saturated Model

The possible hierarchical models for this table are based on the decomposition

$$\mu_{ijkl} = \log m_{ijkl} = \lambda + \lambda_i^{Y} + \lambda_j^{S} + \lambda_k^{E} + \lambda_l^{R} + \lambda_{ij}^{YS} + \lambda_{ik}^{YE} + \lambda_{il}^{YR} + \lambda_{jk}^{SE}$$
$$+ \lambda_{jl}^{SR} + \lambda_{kl}^{ER} + \lambda_{ijk}^{YSE} + \lambda_{ijl}^{YSR} + \lambda_{ikl}^{YER} + \lambda_{jkl}^{SER} + \lambda_{ijkl}^{YSER},$$

where

$$\sum_i \lambda_i^Y = \sum_j \lambda_j^S = \sum_k \lambda_k^E = \sum_l \lambda_l^R = \sum_i \lambda_{ij}^{YS} = \sum_j \lambda_{ij}^{YS} = \cdots = \sum_i \lambda_{ijk}^{YSE} = \sum_j \lambda_{ijk}^{YSE}$$

$$= \sum_k \lambda_{ijk}^{YSE} = \cdots = \sum_i \lambda_{ijkl}^{YSER} = \sum_j \lambda_{ijkl}^{YSER} = \sum_k \lambda_{ijkl}^{YSER} = \sum_l \lambda_{ijkl}^{YSER} = 0.$$

To limit the possible hierarchical models that need ever be considered, it is helpful to estimate the λ-parameters and their corresponding estimated asymptotic standard deviations under the staturated model in which no λ-parameters are assumed 0. In this model, m_{ijkl} has maximum-likelihood estimate n_{ijkl} and $\mu_{ijkl} = \log m_{ijkl}$ has maximum-likelihood estimate $\hat{\mu}_{ijkl} = \log n_{ijkl}$.

To aid in estimation of asymptotic standard deviations, it is helpful to note that

$$\hat{\lambda}_v^T = \frac{1}{2 \cdot 2 \cdot 3 \cdot 2} \sum_i \sum_j \sum_k \sum_l c_{ijkl}(\lambda_v^T) \hat{\mu}_{ijkl},$$

where the $c_{ijkl}(\lambda_v^T)$ are defined as in Table 4.2. Note that

$$\hat{\lambda} = \frac{1}{2 \cdot 2 \cdot 3 \cdot 2} \sum_i \sum_j \sum_k \sum_l \hat{\mu}_{ijkl},$$

$$\hat{\lambda}_{l'}^R = \frac{1}{2 \cdot 2 \cdot 3 \cdot 2} \sum_i \sum_j \sum_k \sum_l (2\delta_{ll'} - 1) \hat{\mu}_{ijkl},$$

$$\hat{\lambda}_{k'l'}^{ER} = \frac{1}{2 \cdot 2 \cdot 3 \cdot 2} \sum_i \sum_j \sum_k \sum_l (3\delta_{kk'} - 1)(2\delta_{ll'} - 1) \hat{\mu}_{ijkl},$$

$$\hat{\lambda}_{j'k'l'}^{SER} = \frac{1}{2 \cdot 2 \cdot 3 \cdot 2} \sum_i \sum_j \sum_k \sum_l (2\delta_{jj'} - 1)(3\delta_{kk'} - 1)(2\delta_{ll'} - 1) \hat{\mu}_{ijkl},$$

and

$$\hat{\lambda}_{i'j'k'l'}^{YSER} = \frac{1}{2 \cdot 2 \cdot 3 \cdot 2} \sum_i \sum_j \sum_k \sum_l (2\delta_{ii'} - 1)(2\delta_{jj'} - 1)(3\delta_{kk'} - 1)(2\delta_{ll'} - 1) \hat{\mu}_{ijkl},$$

where $\delta_{xy} = 1$ if $x = y$ and $\delta_{xy} = 0$ if $x \neq y$. Relevant references include Goodman (1970) or Haberman (1974a, pages 200–205; 1975).

To aid in interpreting these formulas, consider $\hat{\lambda}_{11}^{ER}$. Here $k' = l' = 1$, so that $\hat{\mu}_{ijkl}$ has coefficient $(3\delta_{k2} - 1)(2\delta_{l1} - 1)$. If $k = l = 1$, then $\delta_{k1} = \delta_{l1} = 1$ and the coefficient is $(3 \times 1 - 1)(2 \times 1 - 1) = 2$. If $k > 1$ but $l = 1$, then $\delta_{k1} = 0$ and $\delta_{l1} = 1$. The coefficient is then $(3 \times 0 - 1)(2 \times 1 - 1) = -1$. If $k = 1$ and $l = 2$, then $\delta_{k1} = 1$, $\delta_{l1} = 0$, and the coefficient is $(3 \times 1 - 1)(2 \times 0 - 1) = -2$. Finally, $k > 1$ and $l = 2$ yields $\delta_{k1} = 0$, $\delta_{l1} = 0$, and $(3 \times 0 - 1) (2 \times 0 - 1) = 1$.

The formulas for estimates of λ-parameters follow from a few basic rules. Observe that the denominator $2 \cdot 2 \cdot 3 \cdot 2 = 24$ is the number of combinations

TABLE 4.1   Coefficients of λ-Parameters in a $2 \times 2 \times 3 \times 2$ Table[a]

Table 4.1 (continued)

[a] Entries are coefficients $c_{ijkl}(\lambda_v{}^T)$. To obtain entries not tabulated, note that

$$c_{ijkl}(\lambda_2{}^Y) = -c_{ijkl}(\lambda_1{}^Y),$$

$$c_{ijkl}(\lambda_2{}^S) = -c_{ijkl}(\lambda_1{}^S),$$

$$c_{ijkl}(\lambda_2{}^R) = -c_{ijkl}(\lambda_1{}^R),$$

$$c_{ijkl}(\lambda_{22}^{YS}) = -c_{ijkl}(\lambda_{12}^{YS}) = -c_{ijkl}(\lambda_{21}^{YS}) = c_{ijkl}(\lambda_{11}^{YS}),$$

$$c_{ijkl}(\lambda_{2k'}^{YE}) = -c_{ijkl}(\lambda_{1k'}^{YE}),$$

$$c_{ijkl}(\lambda_{22}^{YR}) = -c_{ijkl}(\lambda_{12}^{YR}) = -c_{ijkl}(\lambda_{21}^{YR}) = c_{ijkl}(\lambda_{11}^{YR}),$$

$$c_{ijkl}(\lambda_{2k'}^{SE}) = -c_{ijkl}(\lambda_{1k'}^{SE}),$$

$$c_{ijkl}(\lambda_{22}^{SR}) = -c_{ijkl}(\lambda_{12}^{SR}) = -c_{ijkl}(\lambda_{21}^{SR}) = c_{ijkl}(\lambda_{11}^{SR}),$$

$$c_{ijkl}(\lambda_{k'1}^{ER}) = -c_{ijkl}(\lambda_{k'1}^{ER}),$$

$$c_{ijkl}(\lambda_{22k'}^{YSE}) = -c_{ijkl}(\lambda_{12k'}^{YSE}) = -c_{ijkl}(\lambda_{21k'}^{YSE}) = c_{ijkl}(\lambda_{11k'}^{YSE}),$$

$$c_{ijkl}(\lambda_{222}^{YSR}) = -c_{ijkl}(\lambda_{122}^{YSR}) = -c_{ijkl}(\lambda_{212}^{YSR}) = -c_{ijkl}(\lambda_{221}^{YSR})$$
$$= c_{ijkl}(\lambda_{112}^{YSR}) = c_{ijkl}(\lambda_{121}^{YSR}) = c_{ijkl}(\lambda_{211}^{YSR}) = -c_{ijkl}(\lambda_{111}^{YSR}),$$

$$c_{ijkl}(\lambda_{2k'2}^{YER}) = -c_{ijkl}(\lambda_{1k'2}^{YER}) = -c_{ijkl}(\lambda_{2k'1}^{YER}) = c_{ijkl}(\lambda_{1k'1}^{YER}),$$

$$c_{ijkl}(\lambda_{2k'2}^{SER}) = -c_{ijkl}(\lambda_{1k'2}^{SER}) = -c_{ijkl}(\lambda_{2k'1}^{SER}) = c_{ijkl}(\lambda_{1k'1}^{SER}),$$

$$c_{ijkl}(\lambda_{22k'1}^{YSER}) = -c_{ijkl}(\lambda_{12k'2}^{YSER}) = -c_{ijkl}(\lambda_{21k'2}^{YSER}) = -c_{ijkl}(\lambda_{22k'1}^{YSER})$$
$$= c_{ijkl}(\lambda_{11k'2}^{YSER}) = c_{ijkl}(\lambda_{12k'1}^{YSER}) = c_{ijkl}(\lambda_{22k'1}^{YSER}) = -c_{ijkl}(\lambda_{11k'1}^{YSER}).$$

of i, j, k, and l in Table 4.1. The coefficient $c_{ijkl}(\lambda_v{}^T)$ is a product in which each multiplicand corresponding to A has v categories and if subscript indices b of $\hat{\lambda}_v{}^T$ and a of $\hat{\mu}_{ijkl}$ correspond to A, then the multiplicand is $(v\delta_{ab} - 1)$. For example, E is in the superscript of $\hat{\lambda}_{k'l'}^{ER}$. Corresponding to E is a trichotomous variable. The subscript index k' of $\hat{\lambda}_{k'l'}^{ER}$ corresponds to E, as does the index k of $\hat{\mu}_{ijkl}$. Thus the term $(3\delta_{kk'} - 1)$ is a multiplicand of $c(\lambda_{k'l'}^{ER})$. The second multiplicand of $\hat{\lambda}_{k'l'}^{ER}$ corresponds to R. The letter R corresponds to a dichotomous variable and to the subscript indices l' of $\hat{\lambda}_{k'l'}^{ER}$ and l of $\hat{\mu}_{ijkl}$. Thus the multiplicand $(2\delta_{ll'} - 1)$ is present.

If the marginal totals $\mathbf{n}^T$ are not fixed by the sampling procedure, then

$$s^2(\hat{\lambda}_v{}^T) = \left(\frac{1}{2 \cdot 2 \cdot 3 \cdot 2}\right)^2 \sum_i \sum_j \sum_k \sum_l [c_{ijkl}(\lambda_v{}^T)]^2 / n_{ijkl}.$$

For example, the total $n_g{}^R$ of responses $R_h = g$ is not fixed, so that

$$s^2(\hat{\lambda}_{l'}^R) = \left(\frac{1}{2 \cdot 2 \cdot 3 \cdot 2}\right)^2 \sum_i \sum_j \sum_k \sum_l (2\delta_{ll'} - 1)^2 / n_{ijkl}$$

$$= \frac{1}{576} \sum_i \sum_j \sum_k \sum_l \frac{1}{n_{ijkl}}.$$

Here the fact that $(2\delta_{ll'} - 1)$ is 1 or -1 has been used. Similarly, the total $n_{k'l'}^{ER}$ of subjects with education $E_h = k'$ and response $R_h = l'$ is not fixed, so that

$$s^2(\hat{\lambda}_{k'l'}^{ER}) = \left(\frac{1}{2 \cdot 2 \cdot 3 \cdot 2}\right)^2 \sum_i \sum_j \sum_k \sum_l (3\delta_{kk'} - 1)^2 (2\delta_{ll'} - 1)^2 / n_{ijkl}$$

$$= \frac{1}{576} \sum_i \sum_j \sum_k \sum_l (3\delta_{kk'} - 1)^2 / n_{ijkl}.$$

These formulas do not apply to $\hat{\lambda}_{i'j'}^{YS}$ since the sampling procedure restricts the total $n_{i'j'}^{YS}$ of subjects of sex $S_h = j'$ interviewed in year $Y_h = i'$.

If the n_{ijkl} are not very large, then it is advisable to add $\frac{1}{2}$ to each n_{ijkl} before computing the $\hat{\mu}_{ijkl}$ and the estimated asymptotic standard deviations. Although this practice will not be used in this section, it will be employed in Section 4.2. Note that Goodman (1970) does recommend addition of $\frac{1}{2}$ to the n_{ijkl} in saturated models for multidimensional tables.

Table 4.2 lists estimated λ-parameters and corresponding estimated asymptotic standard deviations and standardized values. Calculations used the C-TAB program described in Haberman (1973c). The parameter λ has been ignored since it is not of real interest.

If a parameter λ_v^T is 0 and if the table $\mathbf{n}^T$ of marginal totals is not fixed, then the standardized value $\hat{\lambda}_v^T / s(\hat{\lambda}_v^T)$ has an approximate standard normal distribution. Given this observation, there is very strong evidence that λ_1^E, λ_2^E, λ_1^R, λ_{12}^{SE}, λ_{11}^{ER}, and λ_{31}^{ER}, are not 0. In each case, the standardized value is at least 6.67 in absolute value. The simplest hierarchical model that sets none of these parameters to 0 has the form

$$\log m_{ijkl} = \lambda + \lambda_i^Y + \lambda_j^S + \lambda_k^E + \lambda_l^R + \lambda_{ij}^{YS} + \lambda_{jk}^{SE} + \lambda_{kl}^{ER}. \tag{4.1}$$

The terms λ_i^T, λ_j^S, and λ_{ij}^{YS} have been added to reflect constraints on the marginals n_{ij}^{YS} resulting from the sampling procedure. The standardized values associated with $\hat{\lambda}_{111}^{SER}$ and $\hat{\lambda}_{131}^{SER}$ are both at least 2.5 in magnitude. The probability that a standard normal deviate has absolute value at least 2.51 is 0.012, while the probability of the same deviate exceeding 2.89 in magnitude is 0.004. Thus the hierarchical model

$$\log m_{ijk} = \lambda + \lambda_i^Y + \lambda_j^S + \lambda_k^E + \lambda_l^R + \lambda_{ij}^{YS} + \lambda_{jk}^{SE} + \lambda_{jl}^{SR} + \lambda_{kl}^{ER} + \lambda_{jkl}^{SER} \tag{4.2}$$

may be of interest.

No indication can be found from Table 4.2 that other λ-parameters may also be needed.

Table 4.2

Estimated Parameters for Comparison of Tables 3.4 and 3.20 by the Saturated Model[a]

Parameter	Estimate	EASD	Standardized value
λ_1^Y	-0.005	—	—
λ_1^S	-0.015	—	—
λ_1^Y	-0.508	0.037	-13.65
λ_2^E	0.620	0.029	21.47
λ_3^E	-0.113	0.036	3.18
λ_1^R	-0.237	0.024	-9.85
λ_{11}^{YS}	0.016	—	—
λ_{11}^{YE}	-0.018	0.037	-0.48
λ_{12}^{YE}	-0.030	0.029	-1.02
λ_{13}^{YE}	0.048	0.036	1.34
λ_{11}^{YR}	0.023	0.024	0.97
λ_{11}^{SE}	0.078	0.037	2.06
λ_{12}^{SE}	-0.193	0.029	-6.67
λ_{13}^{SE}	0.115	0.036	3.24
λ_{11}^{SR}	0.004	0.024	0.17
λ_{11}^{ER}	0.615	0.037	16.53
λ_{21}^{ER}	-0.047	0.029	-1.62
λ_{31}^{ER}	-0.568	0.036	-16.01
λ_{111}^{YSE}	0.038	0.037	1.01
λ_{121}^{YSE}	-0.019	0.029	-0.66
λ_{131}^{YSE}	-0.018	0.036	-0.52
λ_{111}^{YSR}	0.002	0.024	0.10
λ_{111}^{YER}	0.043	0.037	1.17
λ_{121}^{YER}	-0.040	0.029	-1.39
λ_{131}^{YER}	-0.003	0.036	-0.10
λ_{111}^{SER}	-0.094	0.037	-2.51
λ_{121}^{SER}	-0.009	0.029	-0.31
λ_{131}^{SER}	0.103	0.036	2.89
λ_{1111}^{YSER}	0.006	0.037	0.15
λ_{1211}^{YSER}	0.014	0.029	0.48
λ_{1311}^{YSER}	-0.020	0.036	-0.55

[a] Note that $s(\hat{\lambda}_1^Y)$, and $s(\hat{\lambda}_1^S)$, and $s(\hat{\lambda}_{11}^{YS})$ are not computed due to restrictions on the n_{ij}^{YS}. Note also that

$$\hat{\lambda}_2^Y = -\hat{\lambda}_1^Y, \quad \hat{\lambda}_2^S = -\hat{\lambda}_1^S, \quad \hat{\lambda}_2^R = -\hat{\lambda}_1^R,$$

$$\hat{\lambda}_{22}^{YS} = -\hat{\lambda}_{12}^{YS} = -\hat{\lambda}_{21}^{YS} = \hat{\lambda}_{22}^{YS}, \quad \hat{\lambda}_{2k'}^{YE} = -\hat{\lambda}_{1k'}^{YE},$$

$$\hat{\lambda}_{22}^{YR} = -\hat{\lambda}_{12}^{YR} = -\hat{\lambda}_{21}^{YR} = \hat{\lambda}_{21}^{YR}, \quad \hat{\lambda}_{2k'}^{SE} = -\hat{\lambda}_{1k'}^{SE},$$

$$\hat{\lambda}_{22}^{SR} = -\hat{\lambda}_{12}^{SR} = -\hat{\lambda}_{21}^{SR} = \hat{\lambda}_{11}^{SR}, \quad \hat{\lambda}_{k'2}^{ER} = -\hat{\lambda}_{k'1}^{ER},$$

$$\hat{\lambda}_{22k'}^{YSE} = -\hat{\lambda}_{12k'}^{YSE} = -\hat{\lambda}_{21k'}^{YSE} = \hat{\lambda}_{11k'}^{YSE},$$

$$\hat{\lambda}_{222}^{YSR} = -\hat{\lambda}_{122}^{YSR} = -\hat{\lambda}_{212}^{YSR} = -\hat{\lambda}_{221}^{YSR} = \hat{\lambda}_{112}^{YSR} = \hat{\lambda}_{121}^{YSR} = \hat{\lambda}_{211}^{YSR}$$
$$= -\hat{\lambda}_{111}^{YSR},$$

$$\hat{\lambda}_{2k'2}^{YER} = -\hat{\lambda}_{1k'2}^{YER} = -\hat{\lambda}_{2k'1}^{YER} = \hat{\lambda}_{1k'1}^{YER},$$

$$\hat{\lambda}_{2k'2}^{SER} = -\hat{\lambda}_{1k'2}^{SER} = -\hat{\lambda}_{2k'1}^{SER} = \hat{\lambda}_{1k'1}^{SER},$$

$$\hat{\lambda}_{22k'2}^{YSER} = -\hat{\lambda}_{12k'2}^{YSER} = -\hat{\lambda}_{21k'2}^{YSER} = -\hat{\lambda}_{22k'1}^{YSER} = \hat{\lambda}_{11k'2}^{YSER}$$
$$= \hat{\lambda}_{12k'1}^{YSER} = \hat{\lambda}_{21k'1}^{YSER} = -\hat{\lambda}_{11k'2}^{YSER}.$$

Closed-Form Expressions for Computation of Maximum-Likelihood Estimates for Unsaturated Models

Both (4.1) and (4.2) specify hierarchical models for which the maximum-likelihood estimate $\hat{m}_{ijkl}$ can be expressed in closed form. In (4.1),

$$\hat{m}_{ijkl} = \frac{n_{ij}^{YS} n_{jk}^{SE} n_{kl}^{ER}}{n_j^S n_k^E}. \tag{4.3}$$

In (4.2),

$$\hat{m}_{ijkl} = n_{ij}^{YS} n_{jkl}^{SER} / n_j^S. \tag{4.4}$$

These estimates are listed in Table 4.3.

To find these estimates, general procedures discussed in Bishop (1971), Bishop, Fienberg, and Holland (1975, pages 73–83), Goodman (1970, 1971b), and Haberman (1974a, pages 166–190) may be used. These procedures for determination of closed-form expressions for maximum-likelihood estimates actually result in less computational gain than might at first be apparent, for, as will be noted later in this section, if a closed-form expression is available for the maximum-likelihood estimates $\hat{m}_{ijkl}$, then the iterative proportional fitting algorithm normally yields those estimates after one cycle. Nonetheless, other gains are achieved if a closed-form expression is available for $\hat{m}_{ijkl}$. The model can be interpreted in terms of hypotheses of independence, conditional independence, and equiprobability. Estimated asymptotic standard deviations can be found without iterative computations or use of the Newton–Raphson algorithm. Adjusted residuals can be expressed in a simple form. Some comparisons of models can be performed using marginal tables rather than complete tables. Thus some attention to closed-form estimation seems warranted.

Closed-form estimates can involve many different subscripts. To simplify notation, an abbreviated expression such as

$$\hat{\mathbf{m}} = \frac{\mathbf{n}^{YS} \times \mathbf{n}^{SE} \times \mathbf{n}^{ER}}{\mathbf{n}^S \times \mathbf{n}^E}$$

will be used rather than (4.3). Similarly, (4.4) can be written as

$$\hat{\mathbf{m}} = \frac{\mathbf{n}^{YS} \times \mathbf{n}^{SER}}{\mathbf{n}^S}.$$

To determine formulas for explicit expressions for $\hat{m}_{ijkl}$, let the generating class consist of elements $T(a)$, $1 \leqslant a \leqslant b$. In the saturated model, b is 1 and $T(1) = YSER$. In (4.1), $b = 3$, $T(1) = YS$, $T(2) = SE$, and $T(3) = ER$. In (4.2), $b = 2$, $T(1) = YS$, and $T(2) = SER$.

Table 4.3

Estimated Cell Means for Tables 3.4 and 3.20[a]

Year	Sex	Education in years	Response	
			Agree	Disagree
1974	Male	≤ 8	89	43
			85.6	40.7
			81.1	45.3
		9–12	102	182
			107.6	189.4
			106.7	190.3
		≥ 13	48	193
			39.6	194.0
			46.3	187.3
	Female	≤ 8	83	29
			78.7	37.5
			83.2	33.0
		9–12	152	284
			159.1	280.0
			160.0	279.2
		≥ 13	33	190
			36.6	179.1
			30.0	185.6
1975	Male	≤ 8	72	47
			84.5	40.2
			79.9	44.7
		9–12	110	196
			106.2	186.8
			105.3	187.7
		≥ 13	44	179
			39.1	191.3
			45.7	184.7
	Female	≤ 8	86	38
			81.2	38.6
			85.8	34.0
		9–12	173	283
			164.1	288.8
			165.0	287.8
		≥ 13	28	187
			37.7	184.6
			31.0	191.4

[a] The first line is the observed count n_{ijkl}. The second line is the estimated cell mean m_{ijkl} for the hierarchical model with generating class YS, SE, and ER. The third line is the estimated cell mean $\hat{m}_{ijkl}$ for the hierarchical model with generating class YS and SER.

Let U consist of all letters in some $T(a)$, $1 \leqslant a \leqslant b$. In the saturated model, in (4.1), or in (4.2), U is $YSER$. If each variable corresponds to some letter in U, let $d = 1$. Thus in (4.1), (4.2), or the saturated model, $d = 1$.

Otherwise, let d be the product of the number of categories in each variable not represented by a letter in U. For example, if U were YD, then d would be $3 \times 2 = 6$, where 3 is the number of categories of E_h and 2 is the number of categories of R_h. If $a > 1$, let $V(a)$ consist of all letters in $T(a)$ that are also in some $T(a')$, $a' < a$, and let $c(a)$ be the smallest integer such that each letter of $V(a)$ is also in $T(c(a))$. In (4.1), $V(2) = S$, $c(2) = 1$, $V(3) = E$, and $c(3) = 2$. In (4.2), $V(2) = S$ and $c(2) = 1$.

If $b = 1$, then

$$\hat{\mathbf{m}} = \frac{1}{d} \mathbf{n}^{T(1)}. \tag{4.5}$$

For example, in the saturated model $n_{ijkl}^{YSER} = n_{ijkl}$, $d = 1$, and

$$\hat{m}_{ijkl} = n_{ijkl}.$$

If $b = 2$, then

$$\hat{\mathbf{m}} = \frac{1}{d} \mathbf{n}^{T(1)} \times \mathbf{n}^{T(2)} / \mathbf{n}^{V(2)}. \tag{4.6}$$

If $V(2)$ is empty, then $\mathbf{n}^{V(2)}$ is N. In (4.2), $d = 1$, $T(1) = YS$, $T(2) = SER$, and $V(2) = S$, so that

$$\hat{m}_{ijkl} = n_{ij}^{YS} n_{jkl}^{SER} / n_j^S.$$

If $b \geqslant 3$ and $c(a) < a$ for $2 \leqslant a \leqslant b$, then

$$\hat{\mathbf{m}} = \frac{1}{d} \mathbf{n}^{T(1)} \times \left(\frac{\mathbf{n}^{T(2)}}{\mathbf{n}^{V(2)}} \right) \times \cdots \times \left(\frac{\mathbf{n}^{T(b)}}{\mathbf{n}^{V(b)}} \right). \tag{4.7}$$

In (4.1), $d = 1$, $T(1) = YS$, $T(2) = SE$, $T(3) = ER$, $V(1) = S$, $V(2) = E$, $c(2) = 1$, and $c(3) = 2$, so that

$$\hat{\mathbf{m}} = \mathbf{n}^{YS} (\mathbf{n}^{SE} / \mathbf{n}^S)(\mathbf{n}^{ER} / \mathbf{n}^E),$$

or

$$\hat{m}_{ijkl} = \frac{n_{ij}^{YS} n_{jk}^{SE} n_{kl}^{ER}}{n_j^S n_k^E}.$$

Results are shown in Table 4.3. For further examples, see Exercises 4.1 and 4.2 or the references.

Test of Fit

Chi-square statistics

$$L^2 = 2 \sum_i \sum_j \sum_k \sum_l n_{ijkl} \log(n_{ijkl} / \hat{m}_{ijkl})$$

and

$$X^2 = \sum_i \sum_j \sum_k \sum_l (n_{ijkl} - \hat{m}_{ijkl})^2 / \hat{m}_{ijkl}$$

are easily computed for the models specified by (4.1) or (4.2). To find degrees of freedom, the same procedure used in Chapter 3 may be used here. In (4.1), the parameters λ_{ik}^{YE}, λ_{il}^{YR}, λ_{jl}^{SR}, λ_{ijk}^{YSE}, λ_{ijl}^{YSR}, λ_{ikl}^{YER}, λ_{jkl}^{SER}, and λ_{ijkl}^{YSER} are all set to zero. The degrees of freedom are $b(YE) + b(YR) + b(SR) + b(YSE) + b(YSR) + b(YER) + b(SER) + b(YSER)$. Since Y_h has 2 categories, S_h has 2 categories, E_h has 3 categories, and R_h has 2 categories,

$$b(YE) = (2 - 1)(3 - 1) = 2,$$
$$b(YR) = (2 - 1)(2 - 1) = 1,$$
$$b(SR) = (2 - 1)(2 - 1) = 1,$$
$$b(YSE) = (2 - 1)(2 - 1)(3 - 1) = 2,$$
$$b(YSR) = (2 - 1)(2 - 1)(2 - 1) = 1,$$
$$b(YER) = (2 - 1)(3 - 1)(2 - 1) = 2,$$
$$b(SER) = (2 - 1)(3 - 1)(2 - 1) = 2,$$

and

$$b(YSER) = (2 - 1)(2 - 1)(3 - 1)(2 - 1) = 2.$$

Thus there are 13 degrees of freedom. Since $L^2 = 14.89$ and $X^2 = 14.58$, the model specified by (4.1) appears to fit the data quite well.

Since (4.2) specifies a less restrictive model than (4.1), it is only natural to assume that the model specified by (4.2) also fits the data. Using (4.2), one finds that $L^2 = 6.13$ and $X^2 = 6.12$. Since λ_{ik}^{YE}, λ_{il}^{YR}, λ_{ijk}^{YSE}, λ_{ijl}^{YSR}, λ_{jkl}^{YER}, and λ_{ijkl}^{YSER} are assumed zero, there are $b(YE) + b(YR) + b(YSE) + b(YSR) + b(YER) + b(YSER) = 10$ degrees of freedom.

The difference of 8.76 between the likelihood-ratio chi-square statistics for the two models is rather large, for the corresponding number of degrees of freedom is only $13 - 10 = 3$. The significance level of the difference is about 3 percent.

Note that the difference in likelihood-ratio chi-square statistics is

$$2 \sum_i \sum_j \sum_k \sum_l n_{ijkl} \log\left(\frac{n_{ijkl} n_j^S n_k^E}{n_{ij}^{YS} n_{jk}^{SE} n_{kl}^{ER}}\right) - 2 \sum_i \sum_j \sum_k \sum_l n_{ijkl} \log\left(\frac{n_{ijkl} n_j^S}{n_{ij}^{YS} n_{jkl}^{SER}}\right)$$

$$= 2 \sum_i \sum_j \sum_k \sum_l n_{ijkl} \log\left(\frac{n_{jkl}^{SER} n_k^E}{n_{jk}^{SE} n_{kl}^{ER}}\right)$$

$$= 2 \sum_j \sum_k \sum_l n_{jkl}^{SER} \log\left(\frac{n_{jkl}^{SER} n_k^E}{n_{jk}^{SE} n_{kl}^{ER}}\right),$$

which is a test statistic for the marginal hypothesis of conditional independence of sex and response given education. Thus there is some evidence that sex and response are not conditionally independent given education. In terms of λ-parameters, the difference in fit between models specified by (4.1) and (4.2) suggests that λ_{ik}^{SR} or λ_{ijk}^{SER} is not 0. This suggestion is consistent with the indication in Table 4.2 that λ_{ijk}^{SER} may not be 0.

The Deming-Stephan Algorithm

To determine more clearly whether the λ_{ijk}^{SER} are 0, consider the model

$$\log m_{ijkl} = \lambda + \lambda_i^Y + \lambda_j^S + \lambda_k^E + \lambda_l^R + \lambda_{ij}^{YS} + \lambda_{jk}^{SE} + \lambda_{jl}^{SR} + \lambda_{kl}^{ER}. \quad (4.8)$$

This model differs from (4.1) since λ_{jl}^{SR} is included. It differs from (4.2) since λ_{jkl}^{SER} is excluded. Thus it can provide information concerning the importance of these parameters.

No closed-form expression for $\hat{m}_{ijkl}$ is available under (4.8). Note that the generating class consists of YS, SE, SR, and ER. If $T(1) = YS$, $T(2) = SE$, $T(3) = SR$, and $T(4) = ER$, then $V(2) = S$, $V(3) = S$, $V(4) = ER$, $c(2) = 1$, $c(3) = 2$, and $c(4) = 4$. Since $c(4) = 4$, this arrangement does not yield a closed-form estimate. Other arrangements also fail. For instance, let $T(1) = ER$, $T(2) = SR$, $T(3) = SE$, and $T(4) = YS$. Then $c(3) = 3$ and no closed-form estimate is produced.

Given that (4.8) is being used to help compare two models, asymptotic variances are not likely to be required. Therefore, the Deming–Stephan algorithm is attractive. To implement the algorithm, let

$$m_{ijkl0} = 1,$$
$$m_{ijkl1} = m_{ijkl0} n_{ij}^{YS}/m_{ij0}^{YS},$$
$$m_{ijkl2} = m_{ijkl1} n_{jk}^{SE}/m_{jk1}^{SE},$$
$$m_{ijkl3} = m_{ijkl2} n_{jl}^{SR}/m_{jl2}^{SR},$$
$$m_{ijkl4} = m_{ijkl3} n_{kl}^{ER}/m_{kl3}^{ER},$$
$$m_{ijkl5} = m_{ijkl4} n_{ij}^{YS}/m_{ij4}^{YS},$$

etc. Using this algorithm, the estimates shown in Table 4.4 may be obtained. The corresponding chi-square statistics are $L^2 = 14.84$ and $X^2 = 14.54$. There are 12 degrees of freedom, as may be readily verified (see Exercise 4.3). The difference between the likelihood-ratio chi-square statistics for the models specified by (4.8) and (4.2) is 8.72. The corresponding degrees of freedom is $12 - 10 = 2$, so the approximate significance level is 1 percent. Thus there is evidence that the λ_{jkl}^{SER} are not all zero.

Given this evidence that the λ^{SER}-parameters may not all be zero, the hypothesis that the λ^{SR}-parameters are all zero is of less interest. In any event, the difference is negligible between the likelihood-ratio chi-square statistics for the models specified by (4.1) and (4.8).

At this point, the model specified by (4.2) appears very satisfactory. The assumption in (4.1) that the λ_{jkl}^{SER} are zero appears questionable. Thus it is appropriate to turn to the problems of interpretation of (4.2) and of estimation of λ-parameters.

The iterative proportional fitting algorithm used to produce Table 4.4 may also be used efficiently to produce Table 4.3. If $\hat{m}_{ijkl0} = 1$ and if a hierarchical model with a generating class of b members has a closed-form expression for $\hat{m}_{ijkl}$, then $m_{ijklb} = \hat{m}_{ijkl}$, so that one cycle of the iterative proportional fitting algorithm yields the closed-form maximum-likelihood estimate. For some examples, see Exercise 4.4. For a general discussion, see Haberman (1974a, pages 190–198). The results noted here for four-way tables hold also for five-way and six-way tables. A few exceptions begin to appear when seven-way tables are examined.

Interpretation of Models

Goodman (1970) has noted that any hierarchical model with a closed-form maximum-likelihood estimate can be interpreted in terms of independence, conditional independence, and equiprobability. In the case of hierarchical models without closed-form maximum-likelihood estimates, interpretation is more complicated. For some work in this area, see Goodman (1972, 1973a, 1973b, 1975).

Table 4.4

Maximum-Likelihood Estimates of Cell Means for Tables 3.4 and 3.20 for the Hierarchical Model with Generating Class YS, SE, SR, and ER

Year	Sex	Education in years	Response	
			Agree	Disagree
1974	Male	$\leqslant 8$	85.9	40.5
		9–12	108.3	188.7
		$\geqslant 13$	39.9	193.7
	Female	$\leqslant 8$	78.5	37.7
		9–12	158.4	280.7
		$\geqslant 13$	36.3	179.3
1975	Male	$\leqslant 8$	84.7	40.0
		9–12	106.9	186.1
		$\geqslant 13$	39.4	191.0
	Female	$\leqslant 8$	80.9	38.9
		9–12	163.4	289.5
		$\geqslant 13$	37.4	184.9

The following general observation is often helpful. Let each letter representing a variable be in $U(g)$, $1 \leqslant g \leqslant h$, $h \geqslant 2$, or W. Let no letter be in both $U(g)$ and $U(f)$, $g \neq f$, or both $U(g)$ and W, $1 \leqslant g \leqslant h$. Let a λ^T-parameter be assumed zero if and only if some letter of T is in $U(g)$ and some letter of T is in $U(f)$ for some $g \neq f$.

Assume that no $U(g)$ is empty. If W is empty, then the hierarchical model is equivalent to the model that the variables represented by different $U(g)$ are mutually independent. Otherwise, the model is equivalent to a model in which given the variables represented by W, the variables corresponding to different $U(g)$ are conditionally independent.

For example, consider (4.2). Let $U(1) = Y$, $U(2) = ER$, and $W = S$. Note that each λ^T-parameter is zero if T includes Y and either E or R. Thus (4.2) is equivalent to the assertion that

(a)   given sex S_h of respondent, education E_h of respondent and response R_h are independent of year Y_h of survey.

In (4.1), the model holds if and only if (4.2) holds and

$$\log m_{ijkl} = \lambda + \lambda_i^Y + \lambda_j^S + \lambda_k^E + \lambda_l^R + \lambda_{ij}^{YS} + \lambda_{ik}^{YE} + \lambda_{kl}^{ER} + \lambda_{ijk}^{YSE}. \tag{4.9}$$

Note that (4.2) requires that λ_{ijk}^{YSE} and λ_{ik}^{YE} be zero. Given these restraints, (4.9) reduces to (4.1). In (4.9), let $U(1) = YS$, $U(2) = R$, and $W = E$. Then a λ^T-parameter is assumed zero if and only if E is in T and Y or S is in T. Thus (4.1) holds if and only if assertion (a) holds and

(b)   given education E_h of respondent, response B_h is independent of year Y_h of survey and sex S_h of respondent.

The analysis of likelihood-ratio chi-squares for models specified by (4.1) and (4.2) suggests that assertion (a) is consistent with the data but the combination of assertions (a) and (b) may not be. Thus there is some evidence against (b). The comparison of likelihood-ratio chi-squares suggests that given education, some association is present between sex and response.

The model defined by (4.8) can be interpreted by noting that it holds if (4.2) holds and $\lambda_{jkl}^{SER} = 0$. Since under (4.2),

$$\tau_{(kk')(12) \cdot ij}^{ER \cdot YS} = \log \left(\frac{m_{ijk1} m_{ijk'2}}{m_{ijk2} m_{ijk'1}} \right)$$

$$= (\lambda_{k1}^{ER} - \lambda_{k'1}^{ER} - \lambda_{k2}^{ER} + \lambda_{k'2}^{ER})$$

$$+ (\lambda_{jk1}^{SER} - \lambda_{jk'1}^{SER} - \lambda_{jk2}^{SER} + \lambda_{jk'2}^{SER}), \qquad 1 \leqslant k < k' \leqslant 3,$$

(4.8) holds whenever assertion (a) holds and the log cross-product ratios $\tau_{(kk')(12) \cdot ij}^{ER \cdot YS}$ are independent of year $Y_h = i$ of survey and sex $S_h = j$ of respondent. Thus (4.8) requires that the interaction between education and response be independent of sex of respondent and year of survey. Equivalently, one may state that (4.8) holds if and only if (a) holds and the log cross-product ratios

$$\tau^{SR \cdot YE}_{(12)(12) \cdot ik} = \log \left(\frac{m_{i1k1} m_{i2k2}}{m_{i1k2} m_{i2k1}} \right)$$

are independent of year $Y_h = i$ of survey and education $E_h = k$ of respondent. Some additional examples of interpretations are given in Exercise 4.5.

Parameter Estimation under (4.2)

For comparison with earlier results of Chapter 3, the principal parameters of interest are the log cross-product ratios $\tau^{SR \cdot YE}_{(12)(12) \cdot ik}$, $\tau^{ER \cdot YS}_{(12)(12) \cdot ij}$, $\tau^{ER \cdot YS}_{(13)(12) \cdot ij}$, and $\tau^{ER \cdot YS}_{(23)(12) \cdot ij}$. One has

$$\hat{\tau}^{SR \cdot YE}_{(12)(12) \cdot ik} = \log \left(\frac{\hat{m}_{i1k1} \hat{m}_{i2k2}}{\hat{m}_{i1k2} \hat{m}_{i2kl}} \right)$$

$$= \log \left[\frac{(n^{YS}_{il} n^{SER}_{ik1}/n_1{}^S)(n^{YS}_{i2} n^{SER}_{2k2}/n_2{}^S)}{(n^{YS}_{il} n^{SER}_{ik2}/n_1{}^S)(n^{YS}_{i2} n^{SER}_{2k1}/n_2{}^S)} \right]$$

$$= \log \left[\frac{n^{SER}_{1k1} n^{SER}_{2k2}}{n^{SER}_{1k2} n^{SER}_{2k1}} \right],$$

$$= \hat{\tau}^{SR \cdot E}_{(12)(12) \cdot k},$$

so that

$$s^2(\hat{\tau}^{SR \cdot YE}_{(12)(12) \cdot ik}) = s^2(\hat{\tau}^{SR \cdot E}_{(12)(12) \cdot k}) = \frac{1}{n^{SER}_{1k1}} + \frac{1}{n^{SER}_{1k2}} + \frac{1}{n^{SER}_{2k1}} + \frac{1}{n^{SER}_{2k2}}.$$

Similarly,

$$\hat{\tau}^{ER \cdot YS}_{(kk')(12) \cdot ij} = \log \left[\frac{n^{SER}_{jk1} n^{SER}_{jk'2}}{n^{SER}_{jk2} n^{SER}_{jk'1}} \right] = \hat{\tau}^{ER \cdot S}_{(kk')(12) \cdot j}$$

and

$$s^2(\hat{\tau}^{ER \cdot YS}_{(kk')(12) \cdot ij}) = s^2(\hat{\tau}^{ER \cdot S}_{(kk')(12) \cdot j}) = \frac{1}{n^{SER}_{jk1}} + \frac{1}{n^{SER}_{jk2}} + \frac{1}{n^{SER}_{jk'1}} + \frac{1}{n^{SER}_{jk'2}}.$$

Note that the estimates are independent of the year i. Results for both years have been combined and log cross-product ratios computed as if a saturated model were used for the table cross-classifying sex, education, and response.

Table 4.5

Estimated Partial Association Coefficients
for Tables 3.4 and 3.20 Combined

Parameter	Estimate	EASD
$\tau^{SR \cdot E}_{(1\,2)(1\,2) \cdot 1}$	-0.34	0.20
$\tau^{SR \cdot E}_{(1\,2)(1\,2) \cdot 2}$	-0.02	0.11
$\tau^{SR \cdot E}_{(1\,2)(1\,2) \cdot 3}$	0.42	0.18
$\tau^{ER \cdot S}_{(1\,2)(1\,2) \cdot 1}$	1.16	0.16
$\tau^{ER \cdot S}_{(1\,2)(1\,2) \cdot 2}$	1.48	0.16
$\tau^{ER \cdot S}_{(1\,3)(1\,2) \cdot 1}$	1.98	0.18
$\tau^{ER \cdot S}_{(1\,3)(1\,2) \cdot 2}$	2.75	0.20
$\tau^{ER \cdot S\,2}_{(2\,3)(1\,2) \cdot 1}$	0.82	0.14
$\tau^{ER \cdot S}_{(2\,3)(1\,2) \cdot 2}$	1.26	0.15

Results are summarized in Table 4.5. As an example, note that

$$\hat{\tau}^{SR \cdot E}_{(1\,2)(1\,2) \cdot 1} = \log\left(\frac{161 \times 67}{90 \times 169}\right) = -0.34$$

and

$$s(\hat{\tau}^{SR \cdot E}_{(1\,2)(1\,2) \cdot 1}) = \left(\frac{1}{161} + \frac{1}{90} + \frac{1}{169} + \frac{1}{67}\right)^{1/2} = 0.20.$$

In general, results are quite similar to those in Section 3.2, except that the estimated asymptotic standard deviations have been reduced. Thus the 1974 data have simply added further confirmation to results obtained from 1975 data.

This section has described most of the tools for analysis of hierarchical models. Topics have included standardized parameter estimates, closed-form estimates, iterative proportional fitting, chi-square tests, and estimation of interactions when a model has closed-form maximum-likelihood estimates. The next section considers these tools in connection with a more complex table.

4.2 Year of Survey, Religion, Education, and Attitudes toward Nontherapeutic Abortions

The basic procedures described in Section 4.1 are readily applied to four-way tables in which none of the variables under study is dichotomous. Examination of cross-product ratios becomes more complex, and residual analysis must be explored; however, the main principals are the same in analysis of a table such as Table 4.6. The data in Table 4.6 have been compiled from the 1972, 1973, and 1974 General Survey of the National Opinion Research Center. They provide information concerning factors associated with attitudes toward nontherapeutic abortions and they provide information concerning temporal changes in these attitudes. Subjects have been excluded from tabulation if they did not give definite answers to the abortion questions, if they are neither Protestant nor Catholic, if they are not white, or if they did not report their educational level. In all, 1420 of 4601 subjects have been excluded. The breakdown of Protestants into Southern and Northern categories has been used to obtain information on regional variations. A similar breakdown among Catholics has been avoided due to the smaller number of Catholic respondents available. Exclusions made have been based on the small sample sizes for the excluded groups.

Table 4.6 is an example of a $3 \times 3 \times 3 \times 3$ contingency table. In this table, n_{ijkl} is the number of subjects in year i, of religion j, education k, and attitudes l. The expected value m_{ijkl} or n_{ijkl} is positive for all possible combinations of years, races, religions, educations, and attitudes, so that $\mathbf{n} = \{n_{ijkl}\}$ is a complete factorial table.

To construct a suitable probability model, let A_h be the year of the survey in which respondent h was interviewed, let B_h be the religion of the respondent, let C_h be the educational level of the respondent, and let D_h be the response, so that n_{ijkl} is the number of subjects h such that $A_h = i$, $B_h = j$, $C_h = k$, and $D_h = l$.

Let $p_{ikl \cdot i}^{BCD \cdot A} > 0$ be the conditional probability that $B_h = j$, $C_h = k$, and $D_h = l$ given that $A_h = i$. Assume that given all A_h, $1 \leqslant h \leqslant N = 3181$, the triples (B_h, C_h, D_h) are independent, so that given that n_i^A subjects are cross-classified in Table 4.6 in year i, m_{ijkl} is $n_i^A p_{jkl \cdot i}^{BCD \cdot A}$ and each table of counts n_{ijkl}, $1 \leqslant j \leqslant 3$, $1 \leqslant k \leqslant 3$, $1 \leqslant l \leqslant 3$, has an independent multinomial distribution. This probability model is only approximate given the complex sampling procedure used; however, it is probably adequate for most purposes.

The Saturated Model

As in Section 4.1, it is helpful to begin analysis by examination of the saturated model

$$\log m_{ijkl} = \lambda + \lambda_i^A + \lambda_j^B + \lambda_k^C + \lambda_l^D + \lambda_{ij}^{AB} + \lambda_{ik}^{AC} + \lambda_{il}^{AD} + \lambda_{jk}^{BC} + \lambda_{jl}^{BD}$$
$$+ \lambda_{kl}^{CD} + \lambda_{ijk}^{ABC} + \lambda_{ijl}^{ABD} + \lambda_{ikl}^{ACD} + \lambda_{jkl}^{BCD} + \lambda_{ijkl}^{ABCD}$$

where the λ-parameters are subject to the usual constraints. Estimates of λ-parameters may be found as in Section 4.1. Results are shown in Table

Table 4.6

Attitudes toward Nontherapeutic Abortions Among White, Christian Subjects in the 1972–1974 General Social Surveys[a]

Year	Religion[b]	Education in years	Positive	Mixed	Negative	Total
1972	North. Prot.	$\leqslant 8$	9	16	41	66
		9–12	85	52	105	242
		$\geqslant 13$	77	30	38	145
	South. Prot.	$\leqslant 8$	8	8	46	62
		9–12	35	29	54	118
		$\geqslant 13$	37	15	22	74
	Catholic	$\leqslant 8$	11	14	38	63
		9–12	47	35	115	197
		$\geqslant 13$	25	21	42	88
1973	North. Prot.	$\leqslant 8$	17	17	42	76
		9–12	102	38	84	224
		$\geqslant 13$	88	15	31	134
	South. Prot.	$\leqslant 8$	14	11	34	59
		9–12	61	30	59	150
		$\geqslant 13$	49	11	19	79
	Catholic	$\leqslant 8$	6	16	26	48
		9–12	60	29	108	197
		$\geqslant 13$	31	18	50	99
1974	North. Prot.	$\leqslant 8$	23	13	32	68
		9–12	106	50	88	244
		$\geqslant 13$	79	21	31	131
	South. Prot.	$\leqslant 8$	5	15	37	57
		9–12	38	39	54	131
		$\geqslant 13$	52	12	32	96
	Catholic	$\leqslant 8$	8	10	24	42
		9–12	65	39	89	193
		$\geqslant 13$	37	18	43	98

Attitudes[c] (spanning Positive, Mixed, Negative)

[a] From National Opinion Research Center, 1972, 1973, and 1974 General Social Surveys, University of Chicago.

[b] Southern Protestants are those respondents living in Alabama, Arkansas, Delaware, Florida, Georgia, Kentucky, Louisiana, Maryland, Mississippi, North Carolina, Oklahoma, South Carolina, Tennessee, Texas, Virginia, Washington, D.C., or West Virginia.

[c] Responses to three questions are used to determine this variable. Subjects are asked, "Please tell me whether or not you think it should be possible for a pregnant woman to obtain a *legal* abortion if . . . (six conditions are given)." The following three are used in the table:

B.   If she is married and does not want any more children.

D.   If the family has a very low income and cannot afford any more children.

F.   If she is not married and does not want to marry the man.

Further details on these questions may be found in National Opinion Research Center (1974, page 53). The attitude toward abortion is said to be positive if the subject answers "yes" to all three questions, the attitude is judged negative if the subject answers "no" to all three questions, and the attitude is said to be mixed if the subject answers "yes" to at least one question and "no" to at least one question. Subjects are excluded from tabulation if they do not answer "yes" or "no" to all three questions.

4.7. The only change in this section involves use of $n_{ijkl} + \frac{1}{2}$ rather than n_{ijkl} due to the relatively small n_{ijkl} in some cells. Thus $\hat{\mu}_{ijkl} = \log(n_{ijkl} + \frac{1}{2})$. Since Table 4.6 is a $3 \times 3 \times 3 \times 3$ table,

$$\hat{\lambda} = \frac{1}{3 \cdot 3 \cdot 3 \cdot 3} \sum_i \sum_j \sum_k \sum_l \hat{\mu}_{ijkl},$$

$$\hat{\lambda}_{l'}^D = \frac{1}{3 \cdot 3 \cdot 3 \cdot 3} \sum_i \sum_j \sum_k \sum_l (3\delta_{ll'} - 1)\hat{\mu}_{ijkl},$$

$$\hat{\lambda}_{k'l'}^{CD} = \frac{1}{3 \cdot 3 \cdot 3 \cdot 3} \sum_i \sum_j \sum_k \sum_l (3\delta_{kk'} - 1)(3\delta_{ll'} - 1)\hat{\mu}_{ijkl},$$

$$\hat{\lambda}_{j'k'l'}^{BCD} = \frac{1}{3 \cdot 3 \cdot 3 \cdot 3} \sum_i \sum_j \sum_k \sum_l (3\delta_{jj'} - 1)(3\delta_{kk'} - 1)(3\delta_{ll'} - 1)\hat{\mu}_{ijkl},$$

A standardized value such as $\hat{\lambda}_{3111}^{ABCD}/s(\hat{\lambda}_{3111}^{ABCD})$ has an approximate normal distribution if the corresponding parameter value λ_{3111}^{ABCD} is zero. The probability that a given standardized value exceeds 2 in magnitude is therefore about

$$2[1 - \Phi(2)] = 0.045.$$

However, since 255 interaction estimates have been computed, care in interpretation is essential. Of the 252 interactions for which standardized values have been computed,

$$252(2)[1 - \Phi(2)] = 12.3$$

is the approximate expected number of standardized values that would exceed 2 in magnitude even if all λ^T-parameters were zero for T not equal to A. Note that the corresponding approximate expected numbers for standardized values exceeding 3 and 4 in magnitude are respectively

$$252(2)[1 - \Phi(3)] = 0.7$$

and

$$252(2)[1 - \Phi(4)] = 0.02.$$

Thus standardized values exceeding 4 in magnitude are relatively unlikely to occur by chance. It would not be surprising if some standardized value exceeded 3 in magnitude, and standardized values exceeding 2 in magnitude can be expected to occur fairly frequently.

Given these observations, Table 4.7 provides very strong evidence that $\lambda_1^B, \lambda_2^B, \lambda_1^C, \lambda_2^C, \lambda_2^D, \lambda_3^D, \lambda_{11}^{BD}, \lambda_{13}^{BD}, \lambda_{31}^{BD}, \lambda_{33}^{BD}, \lambda_{11}^{CD}, \lambda_{31}^{CD}$, and λ_{33}^{CD} are not zero. Each corresponding standardized value is at least 4 in magnitude. Except for the λ^{CD}-parameters, each standardized value is at least 5.9 in magnitude.

Table 4.7

Estimated Parameters for the Saturated Model for Table 4.6

Variable combination[a]	Category combination	Parameter	Parameter estimate[b]	Estimated asymptotic standard deviation	Standardized value
Year (A)	1972	λ_1^A	−0.002	−	−
	1973	λ_2^A	0.000	−	−
	1974	λ_3^A	0.002	−	−
Religion (B)	N. Prot.	λ_1^B	0.239	0.030	7.90
	S. Prot.	λ_2^B	−0.198	0.034	−5.90
	Cath.	λ_3^B	−0.041	0.032	−1.25
Education (C)	≤8 yr	λ_1^C	−0.597	0.037	−16.10
	9–12 yr	λ_2^C	0.614	0.028	22.34
	≤ 13 yr	λ_3^C	−0.173	0.031	0.55
Attitudes (D)	Positive	λ_1^D	0.023	0.033	0.68
	Mixed	λ_2^D	−0.406	0.034	−11.81
	Negative	λ_3^D	0.384	0.029	13.42
Year × Religion	1972, N. Prot.	λ_{11}^{AB}	0.019	0.043	0.44
(AB)	1972, S. Prot.	λ_{12}^{AB}	−0.067	0.048	−1.39
	1972, Cath.	λ_{13}^{AB}	0.048	0.045	1.05
	1973, N. Prot.	λ_{21}^{AB}	−0.028	0.043	−0.64
	1973, S. Prot.	λ_{22}^{AB}	0.047	0.047	1.00
	1973, Cath.	λ_{23}^{AB}	−0.020	0.046	−0.42
	1974, N. Prot.	λ_{31}^{AB}	0.008	0.043	0.20
	1974, S. Prot.	λ_{32}^{AB}	0.020	0.048	0.41
	1974, Cath.	λ_{33}^{AB}	−0.028	0.046	−0.60
Year × Education	1972, ≤8 yr	λ_{11}^{AC}	0.011	0.052	0.22
(AC)	1972, 9–12 yr	λ_{12}^{AC}	−0.030	0.039	−0.77
	1972, ≥ 13 yr	λ_{13}^{AC}	0.019	0.044	0.43
	1973, ≤8 yr	λ_{21}^{AC}	0.056	0.052	1.07
	1973, 9–12 yr	λ_{22}^{AC}	0.003	0.039	0.07
	1973, ≥ 13 yr	λ_{23}^{AC}	−0.058	0.045	−1.30
	1974, ≤8 yr	λ_{31}^{AC}	−0.067	0.053	−1.26
	1974, 9–12 yr	λ_{32}^{AC}	0.027	0.039	0.70
	1974, ≥ 13 yr	λ_{33}^{AC}	0.040	0.044	0.89
Year × Attitudes	1972, Positive	λ_{11}^{AD}	−0.135	0.047	−2.86
(AD)	1972, Mixed	λ_{12}^{AD}	0.053	0.049	1.09
	1972, Negative	λ_{13}^{AD}	0.082	0.040	2.05
	1973, Positive	λ_{21}^{AD}	0.104	0.046	2.24
	1973, Mixed	λ_{22}^{AD}	−0.071	0.049	−1.45
	1973, Negative	λ_{23}^{AD}	−0.033	0.041	−0.80
	1974, Positive	λ_{31}^{AD}	0.032	0.048	0.67
	1974, Mixed	λ_{32}^{AD}	0.018	0.049	0.37
	1974, Negative	λ_{33}^{AD}	−0.050	0.041	−1.22
Religion ×	N. Prot., ≤8 yr	λ_{11}^{BC}	−0.027	0.049	−0.55
Education (BC)	N. Prot., 9–12 yr	λ_{12}^{BC}	0.024	0.036	0.65
	N. Prot., ≥ 13 yr	λ_{13}^{BC}	0.003	0.042	0.08
	S. Prot., ≤8 yr	λ_{21}^{BC}	0.117	0.055	2.14
	S. Prot., 9–12 yr	λ_{22}^{BC}	−0.089	0.041	−2.17
	S. Prot., ≥ 13 yr	λ_{23}^{BC}	−0.028	0.047	−0.60
	Cath., ≤8 yr	λ_{31}^{BC}	−0.090	0.054	−1.68
	Cath., 9–12 yr	λ_{32}^{BC}	0.065	0.039	1.66
	Cath., ≥13 yr	λ_{33}^{BC}	0.025	0.044	0.56

Table 4.7 (continued)

Variable combination[a]	Category combination	Parameter	Parameter estimate[b]	Estimated asymptotic standard deviation	Standardized value
Religion ×	N. Prot., Positive	λ_{11}^{BD}	0.212	0.043	4.89
Attitude (BD)	N. Prot., Mixed	λ_{12}^{BD}	−0.048	0.046	−1.03
	N. Prot., Negative	λ_{13}^{BD}	−0.165	0.039	−4.26
	S. Prot., Positive	λ_{21}^{BD}	−0.003	0.049	−0.06
	S. Prot., Mixed	λ_{22}^{BD}	−0.000	0.051	−0.00
	S. Prot., Negative	λ_{23}^{BD}	0.003	0.042	0.07
	Cath., Positive	λ_{31}^{BD}	−0.209	0.048	−4.32
	Cath., Mixed	λ_{32}^{BD}	0.048	0.049	0.98
	Cath., Negative	λ_{33}^{BD}	0.162	0.040	4.00
Education ×	≤ 8 yr, Positive	λ_{11}^{CD}	−0.505	0.057	−8.86
Attitudes (CD)	≤ 8 yr, Mixed	λ_{12}^{CD}	0.166	0.055	3.04
	≤ 8 yr, Negative	λ_{13}^{CD}	0.339	0.045	7.51
	9–12 yr, Positive	λ_{21}^{CD}	0.060	0.039	1.52
	9–12 yr, Mixed	λ_{22}^{CD}	−0.021	0.042	−0.50
	9–12 yr, Negative	λ_{23}^{CD}	−0.039	0.035	−1.12
	≥ 13 yr, Positive	λ_{31}^{CD}	0.445	0.043	10.37
	≥ 13 yr, Mixed	λ_{32}^{CD}	−0.145	0.049	−2.98
	≥ 13 yr, Negative	λ_{33}^{CD}	−0.300	0.041	−7.33
Year × Religion	1972, N. Prot.,	λ_{111}^{ABC}	−0.156	0.071	−2.21
× Education	≤ 8 yr,				
(ABC)	1972, Cath., ≤ 8 yr,	λ_{131}^{ABC}	0.153	0.074	2.07
Year × Religion	1973, S. Prot.,	λ_{221}^{ABD}	0.154	0.067	2.30
× Attitudes	Positive				
(ABD)	1973, Cath.,	λ_{231}^{ABD}	−0.155	0.069	−2.24
	Positive				
	1974, S. Prot.,	λ_{321}^{ABD}	−0.208	0.072	−2.90
	Positive				
Religion ×	S. Prot., ≤ 8 yr,	λ_{213}^{BCD}	0.182	0.066	2.77
Education ×	Negative				
Attitude (BCD)	S. Prot., 9–12 yr,	λ_{222}^{BCD}	0.154	0.062	2.47
	Mixed				
	S. Prot., ≥ 13 yr,	λ_{231}^{BCD}	0.169	0.063	2.67
	Positive				
	Cath., ≤ 8 yr,	λ_{313}^{BCD}	−0.221	0.065	−3.40
	Negative				
	Cath., 9–12 yr,	λ_{322}^{BCD}	−0.155	0.059	−2.61
	Mixed				
	Cath., ≥ 13 yr,	λ_{331}^{BCD}	−0.227	0.063	−3.63
	Positive				
	Cath., ≥ 13 yr,	λ_{333}^{BCD}	0.162	0.056	2.87
	Negative				
Year × Religion	1974, N. Prot.,	λ_{3111}^{ABCD}	0.226	0.104	2.17
× Education	≤ 8 yr, Positive				
× Attitudes	1974, N. Prot.,	λ_{1131}^{ABCD}	−0.160	0.078	−2.04
(ABCD)	≥ 13 yr,				
	Positive				
	1974, S. Prot.,	λ_{3212}^{ABCD}	0.236	0.113	2.08
	≤ 8 yr, Mixed				

[a] Three-way and four-way interactions are only listed if the standardized value exceeds 2 in magnitude.
[b] Due to rounding, some estimates may not add to 0.

Thus no realistic hierarchical model can set λ, λ_i^A, λ_j^B, λ_k^C, λ_l^D, λ_{jl}^{BD}, or λ_{kl}^{CD} to zero.

Since seven of 27 standardized values $\hat{\lambda}_{jkl}^{BCD}/s(\hat{\lambda}_{jkl}^{BCD})$ exceed 2.47 in absolute value and the largest is 3.63 in absolute value, it is quite unlikely that λ_{jkl}^{BCD} can be excluded from any hierarchical model. If λ_{jkl}^{BCD} is not excluded, then λ_{jk}^{BC} must be added to the list of needed λ-parameters.

No large standardized values are associated with λ_{ij}^{AB}, λ_{ik}^{AC}, or λ_{ikl}^{ACD}. That three of 81 standardized values for λ_{ijkl}^{ABCD} exceed 2 in magnitude is not surprising, nor is it noteworthy that of 27 standardized values for λ^{ABC}-parameters, one is -2.21 and one is 2.17. Since three of nine standardized values for λ^{AD}-parameters have respective values -2.86, 2.05, and 2.24, there is some suggestion that the year-by-attitudes interactions λ_{il}^{AD} are not all 0. There is somewhat less evidence that some λ_{ijl}^{ABD} are not zero.

Some Hierarchical Models

The preliminary analysis of Table 4.7 suggests the following models may be of initial interest:

(1) $\log m_{ijkl} = \lambda + \lambda_i^A + \lambda_j^B + \lambda_k^C + \lambda_l^D + \lambda_{jl}^{BD} + \lambda_{kl}^{CD}$,

(2) $\log m_{ijkl} = \lambda + \lambda_i^A + \lambda_j^B + \lambda_k^C + \lambda_l^D + \lambda_{jk}^{BC} + \lambda_{jl}^{BD} + \lambda_{kl}^{CD}$,

(3) $\log m_{ijkl} = \lambda + \lambda_i^A + \lambda_j^B + \lambda_k^C + \lambda_l^D + \lambda_{jk}^{BC} + \lambda_{jl}^{BD} + \lambda_{kl}^{CD} + \lambda_{jkl}^{BCD}$,

(4) $\log m_{ijkl} = \lambda + \lambda_i^A + \lambda_j^B + \lambda_k^C + \lambda_l^D + \lambda_{il}^{AD} + \lambda_{jk}^{BC} + \lambda_{jl}^{BD} + \lambda_{kl}^{CD} + \lambda_{jkl}^{BCD}$,

(5) $\log m_{ijkl} = \lambda + \lambda_i^A + \lambda_j^B + \lambda_k^C + \lambda_l^D + \lambda_{ij}^{AB} + \lambda_{il}^{AD} + \lambda_{jk}^{BC} + \lambda_{jl}^{BD} + \lambda_{kl}^{CD}$
$\qquad + \lambda_{jkl}^{BCD}$,

(6) $\log m_{ijkl} = \lambda + \lambda_i^A + \lambda_j^B + \lambda_k^C + \lambda_l^D + \lambda_{ij}^{AB} + \lambda_{il}^{AD} + \lambda_{jk}^{BC} + \lambda_{jl}^{BD} + \lambda_{kl}^{CD}$
$\qquad + \lambda_{ijl}^{ABD} + \lambda_{jkl}^{BCD}$.

These models are chosen in order of increasing complexity. Table 4.7 suggests that all the parameters in Model 1 must be used in any hierarchical model for Table 4.6. There are indications in order of decreasing evidence that the λ_{jkl}^{BCD}, λ_{il}^{AD}, and λ_{ijl}^{ABD} are needed. Model 3 includes all parameters of Model 1, together with λ_{jkl}^{BCD} and λ_{jk}^{BC}. The last parameter is needed to preserve the hierarchy principle. To indicate the relative importance of λ_{jkl}^{BCD} and λ_{jk}^{BC}, Model 2 adds λ_{jk}^{BC} alone to Model 1. Model 4 adds the interaction λ_{il}^{AD} to Model 3, while Model 6 adds λ_{ijl}^{ABD} and λ_{ij}^{AB} to Model 4. The parameter λ_{ij}^{AB} is needed to preserve the hierarchical structure. Its importance is assessed through Model 5, which adds λ_{ij}^{AB} to Model 4 but omits λ_{ijl}^{ABD}.

The iterative proportional fitting algorithm is attractive here since Models 1, 3, 4, and 6 all have explicit expressions for maximum-likelihood

Table 4.8

Iterative Proportional Fitting for Models 1 to 6 for Table 4.6

Model	Generating class	Initial iterations[a]	m_{ijkl}
1	A, BD, CD	$m_{ijkl1} = m_{ijkl0} n_i^A / m_{i0}^A$ $m_{ijkl2} = m_{ijkl1} n_{jl}^{BD} / m_{jl1}^{BD}$ $m_{ijkl3} = m_{ijkl2} n_{kl}^{CD} / m_{kl2}^{CD}$	$\dfrac{n_i^A n_{jl}^{BD} n_{kl}^{CD}}{N n_l^D}$
2	A, BC, BD, CD	$m_{ijkl1} = m_{ijkl0} n_i^A / m_{i0}^A$ $m_{ijkl2} = m_{ijkl1} n_{jk}^{BC} / m_{jk1}^{BC}$ $m_{ijkl3} = m_{ijkl2} n_{jl}^{BD} / m_{jl2}^{BD}$ $m_{ijkl4} = m_{ijkl3} n_{kl}^{CD} / m_{kl3}^{CD}$	No explicit equation
3	A, BCD	$m_{ijkl1} = m_{ijkl0} n_i^A / m_{i0}^A$ $m_{ijkl2} = m_{ijkl1} n_{jkl}^{BCD} / m_{jkl1}^{BCD}$	$\dfrac{n_i^A n_{jkl}^{BCD}}{N}$
4	AD, BCD	$m_{ijkl1} = m_{ijkl0} n_{il}^{AD} / m_{il0}^{AD}$ $m_{ijkl2} = m_{ijkl1} n_{jkl}^{BCD} / m_{jkl0}^{BCD}$	$\dfrac{n_{il}^{AD} n_{jkl}^{BCD}}{n_l^D}$
5	AD, AB, BCD	$m_{ijkl1} = m_{ijkl0} n_{il}^{AD} / m_{il0}^{AD}$ $m_{ijkl2} = m_{ijkl1} n_{ij}^{AB} / m_{ij1}^{AB}$ $m_{ijkl3} = m_{ijkl2} n_{jkl}^{BCD} / m_{jkl2}^{BCD}$	No explicit equation
6	ABD, BCD	$m_{ijkl1} = m_{ijkl0} n_{ijl}^{ABD} / m_{ijl0}^{ABD}$ $m_{ijkl2} = m_{ijkl1} n_{jkl}^{BCD} / m_{jkl1}^{BCD}$	$\dfrac{n_{ijl}^{ABD} n_{jkl}^{BCD}}{n_{jl}^{BD}}$

[a] It is permissible in each case to let $m_{ijkl0} = 1$. In Model 1, $m_{ijkl3} = \hat{m}_{ijkl}$. In Model 3, 4, and 6, $m_{ijkl2} = \hat{m}_{ijkl}$.

estimates. Thus for four of the six models under study, this algorithm converges to the maximum–likelihood estimate after a single cycle. Maximum–likelihood estimates are listed in Table 4.8, together with one cycle of the Deming–Stephan algorithm.

Chi-Square Statistics

Chi-square statistics for these models are listed in Table 4.9, together with the corresponding degrees of freedom. Models 4, 5, and 6 all fit the table well enough to have F statistics less than 1. The differences between likelihood-ratio chi-squares for these models are all small. Thus no evidence exists that Model 4 does not hold. The change in L^2 between Model 3 and Model 4 is quite large for the number of degrees of freedom involved. The probability that a chi-square variable with four degrees of freedom exceeds 22.6 is about 0.00016. Thus Model 3 does not appear to hold. Consequently, Models 1 and 2, both of which are less general than Model 3, also appear not to fit the data.

Given that Model 4 appears to hold, at least approximately, the difference between L^2 statistics for Model 3 and Model 4 provides strong evidence that the interaction λ_{il}^{AD} of time of survey and abortion attitudes is not zero.

Table 4.9

Chi-Square Statistics for Table 4.6

Model	X^2	L^2	Degrees of freedom v	$F < L^2/v$	Decrease in L^2 from preceding model	Degrees of freedom of decrease
1	117.79	118.2	64	1.85	—	—
2	94.6	95.5	60	1.59	22.7	4
3	68.4	68.3	52	1.32	27.2	8
4	45.9	45.7	48	0.95	22.6	4
5	41.7	41.8	44	0.95	3.0	4
6	33.9	33.8	36	0.94	8.0	8

Given Model 4, one may test for the presence of the religion-by-education-by attitudes interaction λ_{jkl}^{BCD} by comparing Model 4 to the model

$$(2')\qquad \log m_{ijkl} = \lambda + \lambda_i^A + \lambda_j^B + \lambda_k^C + \lambda_l^D + \lambda_{ij}^{AD} + \lambda_{jk}^{BC} + \lambda_{jl}^{BD} + \lambda_{kl}^{CD}.$$

This model has a close relationship to Model 2 in terms of maximum-likelihood estimates. This relationship can be exploited to permit use of already available L^2 statistics in comparison of Model 3 and Model 2. The key feature is that both the likelihood-ratio chi-square for comparison of Models 4 and $2'$ and the likelihood-ratio chi-square for comparison of Models 3 and 2 reduce to the likelihood-ratio chi-square for the model of no three-factor interaction for the marginal table of counts n_{jkl}^{BCD}, $1 \leqslant j \leqslant 3$, $1 \leqslant k \leqslant 3$, $1 \leqslant l \leqslant 3$.

This claim follows from Goodman's (1971b) generalization of (4.7). Let b, d, $T(a)$, $1 \leq a \leq b$, $V(a)$, $2 \leq a \leq b$, and $c(a)$, $2 \leq a \leq b$, be defined as in (4.7). Let $c(a) < a$ for $a > e$. Let Z consist of all letters in some $T(a)$, $1 \leqslant a \leqslant e$. Then

$$\hat{\mathbf{m}} = \frac{1}{d}\,\hat{\mathbf{m}}^Z \times \frac{\mathbf{n}^{T(e+1)}}{\mathbf{n}^{V(e+1)}} \times \cdots \times \frac{\mathbf{n}^{T(b)}}{\mathbf{n}^{V(b)}}, \tag{4.10}$$

where $\mathbf{m}^Z$ is the maximum-likelihood estimate of the mean $\mathbf{m}^Z$ of the table $\mathbf{n}^Z$ for the hierarchical model for $\mathbf{n}^Z$ with generating class $T(a)$, $1 \leqslant a \leqslant e$.

In Model $2'$, one may let $T(1) = BC$, $T(2) = BD$, $T(3) = CD$, and $T(4) = AD$, so that $d = 1$, $V(4) = D$, $c(4) = 2$, $Z = BCD$, and $e = 3$. Thus

$$\hat{\mathbf{m}} = \hat{\mathbf{m}}^{BCD}(\mathbf{n}^{AD}/\mathbf{n}^D),$$

where $\hat{\mathbf{m}}^{BCD}$ is the maximum-likelihood estimate of $\mathbf{m}^{BCD}$, the table of marginal means m_{jkl}^{BCD}, under the hierarchical model for $\mathbf{n}^{BCD}$, the table of marginal totals n_{jkl}^{BCD}, with generating class consisting of BC, BD, and CD.

Equivalently,

$$\hat{m}_{ijkl} = \hat{m}_{jkl}^{BCD} n_{il}^{AD}/n_l^{D}.$$

Similarly, in Model 2, $T(1) = BC$, $T(2) = BD$, $T(3) = CD$, $T(4) = AD$, $d = 1$, $V(4)$ is empty, $c(4) = 1$, $Z = BCD$, $e = 3$, and

$$\hat{\mathbf{m}} = \hat{\mathbf{m}}^{BCD}(\mathbf{n}^A/N),$$

where $\hat{\mathbf{m}}^{BCD}$ is defined as in Model 2' Thus $\hat{m}_{ijkl} = \hat{m}_{jkl}^{BCD} n_i^A/N$.

In each case, the estimate $\hat{m}_{jkl}^{BCD}$ is the maximum-likelihood estimate of the mean m_{jkl}^{BCD} of the marginal count n_{jkl}^{BCD} under the marginal model of no three-factor interaction

$$\log m_{jkl}^{BCD} = \kappa + \kappa_j^{B} + \kappa_k^{C} + \kappa_l^{D} + \kappa_{jk}^{BC} + \kappa_{jl}^{BD} + \kappa_{kl}^{CD},$$

where

$$\sum_j \kappa_j^{B} = \sum_k \kappa_k^{C} = \sum_l \kappa_l^{D} = \cdots = \sum_k \kappa_{kl}^{CD} = \sum_l \kappa_{kl}^{CD} = 0.$$

Thus

$$\log \hat{m}_{jkl}^{BCD} = \hat{\kappa} + \hat{\kappa}_j + \hat{\kappa}_k^{C} + \hat{\kappa}_l^{D} + \hat{\kappa}_{jk}^{BC} + \hat{\kappa}_{jl}^{BD} + \hat{\kappa}_{kl}^{CD},$$
$$\hat{m}_{jk}^{BC} = n_{jk}^{BC},$$
$$\hat{m}_{jl}^{BD} = n_{jl}^{BD},$$

and

$$\hat{m}_{kl}^{CD} = n_{kl}^{CD}.$$

The difference between the likelihood-ratio chi-square statistics for Model 2 and Model 3 is thus

$$2\sum_i \sum_j \sum_k \sum_l n_{ijkl} \log\left[\frac{n_i^A n_{jkl}^{BCD}/n}{n_i^A \hat{m}_{jkl}^{BCD}/n}\right] = 2\sum_j \sum_k \sum_l n_{jk}^{BCD} \log\left[\frac{n_{jkl}^{BCD}}{\hat{m}_{jkl}^{BCD}}\right].$$

The difference between likelihood-ratio chi-square statistics for Model 2' and Model 4 is also

$$2\sum_i \sum_j \sum_k \sum_l n_{ijkl} \log\left[\frac{n_{il}^{AD} n_{jkl}^{BCD}/n_l^{D}}{n_i^{AD} \hat{m}_{jkl}^{BCD}/n_l^{D}}\right] = 2\sum_j \sum_k \sum_l n_{jkl}^{BCD} \log\left[\frac{n_{jkl}^{BCD}}{\hat{m}_{jkl}^{BCD}}\right].$$

Since the difference in L^2 between Model 2 and Model 3 is 27.2 and the corresponding degrees of freedom are 8, the significance level of the statistic is less than 0.001. Thus given that Model 4 holds, at least approximately, there is quite strong evidence that the λ_{jkl}^{BCD} are not 0. Given that λ_{jkl}^{BCD} is required in a hierarchical model, λ_{jk}^{BC} must also be included. Thus Model 4 appears to be the simplest hierarchical model consistent with the data.

The procedures used here to select Model 4 are by no means the only one available. Detailed discussion of the selection of hierarchical models can be found in Goodman (1971a, 1973c) and Bishop, Fienberg, and Holland (1975, pages 155–168). Some of these approaches are analogous to stepwise regression methods discussed in Draper and Smith (1966, pages 163–216).

Residual Analysis

Various residual analyses may be performed to check Model 4. As will be shown later in this subsection, the adjusted residuals may be found by the formula

$$r_{ijkl} = \frac{n_{ijkl} - \hat{m}_{ijkl}}{\left[(n_{jkl}^{BCD} n_{il}^{AD}(n_l{}^D - n_{jkl}^{BCD})(n_l{}^D - n_{il}^{AD})/(n_l{}^D)^3 \right]^{1/2}} .$$

The residuals shown in Table 4.10 do not appear excessively large, and patterns are hard to detect. Thus Model 4 does appear to provide a satisfactory fit to the data.

Table 4.10

Adjusted and Standardized Residuals for Table 4.6
for the Hierarchical Model with Generating
Class Consisting of BCD and AD[a]

Year	Religion	Education	Abortion Attitudes		
			Positive	Mixed	Negative
1972	N. Prot.	⩽ 8 yr	9	16	41
			13.93	16.27	41.63
			−1.60	−0.09	−0.13
			−1.32	−0.07	−0.10
		9–12 yr	85	52	105.
			83.29	49.52	100.27
			0.26	0.50	0.66
			0.19	0.35	0.47
		⩾ 13 yr	77	30	38
			69.36	23.34	36.20
			1.22	1.81	0.39
			0.92	1.38	0.30
	S. Prot.	⩽ 8 yr	8	8	46
			7.67	12.03	42.35
			0.14	−1.48	0.73
			0.12	−1.16	0.56
		9–12 yr	35	29	54
			38.09	34.66	60.45
			−0.63	−1.30	−1.11
			−0.50	−0.96	−0.83

Table 4.10 (*continued*)

Year	Religion	Education	Abortion Attitudes		
			Positive	Mixed	Negative
		⩾13 yr	37	15	22
			39.23	13.44	26.43
			−0.45	0.55	−1.11
			−0.36	0.43	−0.86
	Cath.	⩽8 yr	11	14	38
			7.11	14.15	31.86
			1.74	−0.05	1.41
			1.46	−0.04	1.09
		9–12 yr	47	35	115
			48.89	36.43	112.94
			−0.35	−0.32	0.28
			−0.27	−0.24	0.19
		⩾13 yr	25	21	42
			26.44	20.26	48.86
			−0.34	0.24	−1.30
			−0.28	0.19	−0.98
1973	N. Prot.	⩽8 yr	17	17	42
			17.85	13.68	37.64
			−0.26	1.11	0.90
			−0.20	0.90	0.71
		9–12 yr	102	38	84
			106.73	41.64	90.67
			−0.66	−0.76	−0.95
			−0.46	−0.56	−0.70
		⩾13 yr	88	15	31
			88.88	19.63	32.73
			−0.13	−1.32	−0.38
			−0.09	−1.05	−0.30
	S. Prot.	⩽8 yr	14	11	34
			9.83	10.11	38.30
			1.68	0.34	−0.78
			1.33	0.28	−0.69
		9–12 yr	61	30	59
			48.81	29.15	54.66
			2.32	0.20	0.76
			1.74	0.16	0.59
		⩾13 yr	49	11	19
			50.27	11.30	23.89
			−0.24	−0.11	−1.25
			−0.18	−0.09	−1.00
	Cath.	⩽8 yr	6	16	26
			9.11	11.90	28.80
			−1.30	1.47	−0.66
			−1.03	1.19	−0.52
		9–12 yr	60	29	108
			62.65	30.64	102.12
			−0.46	−0.39	0.81
			−0.34	−0.30	0.58

Table 4.10 (continued)

Year	Religion	Education	Positive	Mixed	Negative
			Abortion Attitudes		
		≥13 yr	31	18	50
			33.88	16.95	44.19
			−0.65	0.32	1.12
			−0.49	0.25	0.87
1974	N. Prot.	≤8 yr	23	13	32
			17.22	16.05	35.73
			1.77	−0.98	−0.78
			1.39	−0.76	−0.62
		9–12 yr	106	50	88
			103.00	48.84	86.06
			0.43	0.23	0.28
			0.30	0.17	0.21
		≥13 yr	79	21	31
			85.76	23.03	31.07
			−1.02	−0.55	−0.02
			−0.73	−0.42	−0.01
	S. Prot.	≤8 yr	5	15	37
			9.49	11.86	36.35
			−1.83	1.16	0.14
			−1.46	0.91	0.11
		9–12 yr	38	39	54
			47.10	34.19	51.89
			−1.75	1.11	0.38
			−1.33	0.82	0.29
		≥13 yr	52	12	32
			48.51	13.26	22.68
			0.66	−0.44	2.42
			0.50	−0.35	1.96
	Cath.	≤8 yr	8	10	24
			8.79	13.96	27.34
			−0.33	−1.36	−0.80
			−0.27	−1.06	−0.64
		9–12 yr	65	39	89
			60.46	35.93	96.94
			0.78	0.69	−1.10
			0.58	0.51	−0.81
		≥13 yr	37	18	43
			32.69	19.89	41.94
			0.98	−0.55	0.21
			0.75	−0.42	0.16

[a] First line is observed counted, second line is estimated expected count, third line is adjusted residual, and fourth line is standardized residual.

In general, adjusted residuals can be computed most easily when $\hat{\mathbf{m}}$ has a closed-form expression. If

$$\hat{\mathbf{m}} = \mathbf{n}^{T(1)} \times \left[\frac{\mathbf{n}^{T(2)}}{\mathbf{n}^{V(2)}}\right] \times \cdots \times \left[\frac{\mathbf{n}^{T(b)}}{\mathbf{n}^{V(b)}}\right],$$

then

$$r_{ijkl} = (n_{ijkl} - \hat{m}_{ijkl})/\hat{c}_{ijkl}^{1/2},$$

and one may symbolically write

$$\hat{\mathbf{c}} = \hat{\mathbf{m}} \left\{ 1 - \hat{\mathbf{m}} \sum_{a=1}^{b} \left[1/\mathbf{n}^{T(a)}\right] + \hat{\mathbf{m}} \sum_{a=2}^{b} \left[1/\mathbf{n}^{V(a)}\right] \right\}.$$

To verify this formula, one may use general formulas of Haberman (1974a, pages 138 and 211). In the hierarchical model under study,

$$n_{jkl}^{BCD} n_{il}^{AD} (n_l^D - n_{jkl}^{BCD})(n_l^D - n_{il}^{AD})/(n_l^D)^3 = \hat{m}_{ijkl}\left[1 - \hat{m}_{ijkl}\left(\frac{1}{n_{jkl}^{BCD}} + \frac{1}{n_{il}^{AD}}\right) + \frac{\hat{m}_{ijkl}}{n_l^D}\right].$$

Thus the formula for Model 4 is consistent with the general formula.

Programs such as C-TAB by Haberman (1973c) or MULTIQUAL by Bock and Yates (1973) report standardized residuals

$$s_{ijkl} = (n_{ijkl} - \hat{m}_{ijkl})/\hat{m}_{ijkl}^{1/2}$$

described in Haberman (1973b). These residuals are easily computed, whether or not $\hat{m}_{ijkl}$ can be expressed in closed form and whether or not the Newton–Raphson algorithm has been used. One has

$$X^2 = \sum_i \sum_j \sum_k \sum_l s_{ijkl}^2.$$

If the model under study holds, s_{ijk} has an approximate $N(0, v_{ijkl})$ distribution with $v_{ijkl} = c_{ijkl}/m_{ijkl} < 1$. If the v_{ijkl} are close to 1, then the s_{ijkl} are similar to the r_{ijkl}. This pattern can be seen in Table 4.10.

It should also be noted that a computer routine that uses the Newton–Raphson algorithm and calculates adjusted residuals can certainly do so in the case of hierarchical models, whether or not closed-form expressions are available for maximum-likelihood estimates. The computational inefficiency that results from use of the Newton–Raphson algorithm when closed-form expressions are available for $\hat{\mathbf{m}}$ is a minor issue unless the table has many cells. (Many might mean several hundred; however, this number is not clearly defined at the present time.)

Interpretation of Model 4

Model 4 implies that given response D_h, religion B_h and education C_h of respondent are conditionally independent of year A_h of survey. This result suggests that religion and education do not affect the relationship between year of survey and response. Similarly, year of survey does not affect the relationship between religion, education, and response.

To analyze the relationship between year of survey and response, consider the log cross-product ratios

$$\tau_{(ii')(ll')}^{AD} = \log\left[\frac{m_{il}^{AD}m_{i'l'}^{AD}}{m_{il'}^{AD}m_{i'l}^{AD}}\right].$$

Since

$$m_{il}^{AD} = n_i^A p_{l\cdot i}^{D\cdot A},$$

where $p_{l\cdot i}^{D\cdot A}$ is the probability that $D_h = l$ given that $A_h = i$, it follows that

$$\tau_{(ii')(ll')}^{AD} = \log\left(\frac{n_i^A p_{l\cdot i}^{D\cdot A} n_{i'}^A p_{l'\cdot i'}^{D\cdot A}}{n_i^A p_{l'\cdot i}^{D\cdot A} n_{i'}^A p_{l\cdot i'}^{D\cdot A}}\right) = \log\left(\frac{p_{l\cdot i}^{D\cdot A} p_{l'\cdot i'}^{D\cdot A}}{p_{l'\cdot i}^{D\cdot A} p_{l\cdot i'}^{D\cdot A}}\right).$$

Under Model 4, the conditional probability

$$p_{jk\cdot il}^{BC\cdot AD} = p_{jkl\cdot i}^{BCD\cdot A}/p_{l\cdot i}^{D\cdot A}$$

that $B_h = j$ and $C_h = k$ given $A_h = i$ and $D_h = l$ is independent of i. Thus

$$m_{ijkl} = n_i^A p_{jkl\cdot i}^{BCD\cdot A} = n_i^A p_{l\cdot i}^{D\cdot A} p_{jk\cdot il}^{BC\cdot AD} = n_i^A p_{l\cdot i}^{D\cdot A} p_{jk\cdot 1l}^{BC\cdot AD}.$$

The conditional log cross-product ratio for A_h and D_h given B_h and C_h is

$$\tau_{(ii')(ll')\cdot jk}^{AD\cdot BC} = \log\left(\frac{m_{ijkl}m_{i'jkl'}}{m_{ijkl'}m_{i'jkl}}\right)$$

$$= \log\left(\frac{n_i^A p_{l\cdot i}^{D\cdot A} p_{jk\cdot 1l}^{BC\cdot AD}}{n_i^A p_{l'\cdot i}^{D\cdot A} p_{jk\cdot 1l'}^{BC\cdot AD}} - \frac{n_{i'}^A p_{l\cdot i'}^{D\cdot A} p_{jk\cdot 1l'}^{BC\cdot AD}}{n_{i'}^A p_{l'\cdot i'}^{D\cdot A} p_{jk\cdot 1l}^{BC\cdot AD}}\right)$$

$$= \log\left(\frac{p_{l\cdot i}^{D\cdot A} p_{l'\cdot i'}^{D\cdot A}}{p_{l'\cdot i}^{D\cdot A} p_{l\cdot i'}^{D\cdot A}}\right)$$

$$= \tau_{(ii')(ll')}^{AD}.$$

Consequently, the conditional log cross-product ratio $\tau_{(ii')(ll')\cdot jk}^{AD\cdot BC}$ is independent of j and k, so that the interaction between year of survey and response is independent of the religion and education of the respondent. This result may be described in terms of the log odds

$$\tau_{ll'\cdot ijk}^{D\cdot ABC} = \log\left(\frac{p_{l\cdot ijk}^{D\cdot ABC}}{p_{l'\cdot ijk}^{D}}\right)$$

of response $D_h = l$ rather than $D_h = l'$ given that $A_h = i$, $B_h = j$, and $C_h = k$. Since

$$p_{jkl\cdot i}^{BCD\cdot A} = p_{l\cdot ijk}^{D\cdot ABC} p_{jk\cdot i}^{BC\cdot A},$$

where $p_{jk \cdot i}^{BC \cdot A}$ is the conditional probability that $B_h = j$ and $C_h = k$ if $A_h = i$, it follows that

$$\tau_{ll' \cdot ijk}^{D \cdot ABC} = \log\left(\frac{n_i^A p_{l \cdot ijk}^{D \cdot ABC} p_{jk \cdot i}^{BC \cdot A}}{n_i^A p_{l' \cdot ijk}^{D \cdot ABC} p_{jk \cdot i}^{BC \cdot A}}\right) = \log\left(\frac{m_{ijkl}}{m_{ijkl'}}\right).$$

Therefore, $\tau_{(ii')(ll') \cdot jk}^{AD \cdot BC}$ is the difference $\tau_{ll' \cdot ijk}^{D \cdot ABC} - \tau_{ll' \cdot i'jk}^{D \cdot ABC}$ in the log odds of response $D_h = l$ rather than $D_h = l'$ that results from a change $A_h = i'$ to $A_h = i$ in which B_h remains equal to j and C_h remains equal to k.

As an example, consider Catholic respondents ($j = 3$) with at least 13 years education ($k = 3$). For 1972 ($i = 1$), the estimated log odds that the response is positive rather than negative is

$$\tau_{13 \cdot 133}^{D \cdot ABC} = \log\left(\frac{26.44}{48.87}\right) = -0.614.$$

The corresponding odds in 1974 ($i = 3$) is

$$\hat{\tau}_{13 \cdot 333}^{D \cdot ABC} = \log\left(\frac{32.69}{41.94}\right) = -0.249.$$

The difference

$$\begin{aligned}
\hat{\tau}_{(13)(13) \cdot 33}^{AD \cdot BC} &= \hat{\tau}_{13 \cdot 133}^{D \cdot ABC} - \hat{\tau}_{13 \cdot 333}^{D \cdot ABC} \\
&= \hat{\tau}_{(13)(13)}^{AD} \\
&= \log\frac{\hat{m}_{11}^{AD} \hat{m}_{33}^{AD}}{\hat{m}_{13}^{AD} \hat{m}_{31}^{AD}} \\
&= \log\frac{n_{11}^{AD} n_{33}^{AD}}{n_{13}^{AD} n_{31}^{AD}} \\
&= \log\left(\frac{334 \times 430}{501 \times 413}\right) \\
&= -0.365.
\end{aligned}$$

In the case of Northern Protestant respondents ($j = 1$) with 9–12 years education ($k = 2$), the corresponding log odds for 1972 and 1974 are

$$\hat{\tau}_{13 \cdot 112}^{D \cdot ABC} = \log\left(\frac{83.29}{100.27}\right) = -0.186$$

and

$$\hat{\tau}_{13 \cdot 312}^{D \cdot ABC} = \log\left(\frac{102.99}{86.06}\right) = 0.180.$$

The difference

$$\hat{\tau}^{AD \cdot BC}_{(13)(13) \cdot 12} = \hat{\tau}^{D \cdot ABC}_{13 \cdot 112} - \hat{\tau}^{D \cdot ABC}_{13 \cdot 312}$$

remains $\hat{\tau}^{AD}_{(12)(13)} = -0.365$.
In general

$$\hat{\tau}^{AD}_{(ii')(ll')} = \log\left(\frac{n^{AD}_{il} n^{AD}_{i'l'}}{n^{AD}_{il'} n^{AD}_{i'l}}\right)$$

and

$$s^2(\hat{\tau}^{AD}_{(ii')(ll')}) = \frac{1}{n^{AD}_{il}} + \frac{1}{n^{AD}_{il'}} + \frac{1}{n^{AD}_{i'l}} + \frac{1}{n^{AD}_{i'l'}}.$$

Results are tabulated in Table 4.11. The estimated cross-product ratio $\hat{q}^{AD}_{(ii')(ll')}$ is

$$\frac{n^{AD}_{il} n^{AD}_{i'l'}}{n^{AD}_{il'} n^{AD}_{i'l}}.$$

Note that all but two of the estimated log cross-product ratios in Table 4.11 are negative. The overall pattern suggested is a tendency for respondents to have more favorable attitudes toward the types of abortion in question in later years. As shown by inspection of the EASD column, the magnitude of the change in attitudes is not well determined relative to its size; however, the changes do not appear negligible. For example, the estimate $\hat{q}^{AD}_{(13)(13)} =$

Table 4.11

Estimated Log Cross-Product Ratios and Cross-Product Ratios Based on Table 4.6
for Interaction of Year of Survey and Abortion Attitudes of Respondent

Year	i	Year	i'	Attitude	l	Attitude	l'	Estimated log cross product ratio $\hat{\tau}^{AD}_{(ii')(ll')}$	EASD $s(\hat{\tau}^{AD}_{(ii')(ll')})$	Estimated cross-product ratio $\hat{q}^{AD}_{(ii')(ll')}$
1972	1	1973	2	Positive	1	Mixed	2	−0.421	0.124	0.656
						Negative	3	−0.349	0.098	0.706
				Mixed	2	Negative	3	0.073	0.119	1.075
1972	1	1974	3	Positive	1	Mixed	2	−0.226	0.121	0.798
						Negative	3	−0.365	0.099	0.694
				Mixed	2	Negative	3	−0.139	0.116	0.870
1973	2	1974	3	Positive	1	Mixed	2	0.195	0.122	1.216
						Negative	3	−0.016	0.096	0.984
				Mixed	2	Negative	3	−0.212	0.121	0.809

0.694 corresponds to a increase in the odds ratio for positive versus negative responses of $100(0.694^{-1} - 1) = 44.1$ percent from 1972 to 1974.

Nonetheless, to simply describe the table as showing an increasing tendency for subjects to be favorable to a woman's ability to have a legal nontherapeutic abortion does not appear adequate. The estimated increase between 1973 and 1974 in the log odds of a positive rather than negative attitude is only -0.016, and for the same years, 0.195 is the estimated decrease in the log odds of a postive rather than a mixed response. Thus the table does show some changes in attitudes over a three-year period; however, it would be dangerous to extrapolate these results to other years.

The interaction between education, religion, and abortion attitudes is somewhat complex. By arguments quite similar to those used to analyze the interaction of attitude and year of survey, one finds that the conditional log cross-product ratio

$$\tau^{BD \cdot AC}_{(jj')(ll') \cdot ik} = \log\left(\frac{m_{ijkl}m_{ij'kl'}}{m_{ijkl'}m_{ij'kl}}\right)$$

$$= \log\left(\frac{p^{D \cdot ABC}_{l \cdot ijk} p^{D \cdot ABC}_{l' \cdot ij'k}}{p^{D \cdot ABC}_{l' \cdot ijk} p^{D \cdot ABC}_{l \cdot ij'k}}\right)$$

for religion and attitudes given year and education is independent of the year i and equal to

$$\tau^{BD \cdot C}_{(jj')(ll') \cdot k} = \log\left(\frac{m^{BCD}_{jkl}m^{BCD}_{j'kl'}}{m^{BCD}_{jkl'}m^{BCD}_{j'kl}}\right).$$

Thus

$$\hat{\tau}^{BD \cdot AC}_{(jj')(ll') \cdot ik} = \hat{\tau}^{BD \cdot C}_{(jj')(ll') \cdot k} = \log\left(\frac{n^{BCD}_{jkl}n^{BCD}_{j'kl'}}{n^{BCD}_{jkl'}n^{BCD}_{j'kl'}}\right),$$

$$\hat{q}^{BD \cdot AC}_{(jj')(ll') \cdot ik} = \hat{q}^{BD \cdot C}_{(jj')(ll') \cdot k} = \frac{n^{BCD}_{jkl}n^{BCD}_{j'kl'}}{n^{BCD}_{jkl'}n^{BCD}_{j'kl}},$$

and

$$s^2(\hat{\tau}^{BD \cdot AC}_{(jj')(ll') \cdot ik}) = s^2(\hat{\tau}^{BD \cdot C}_{(jj')(ll') \cdot k})$$

$$= \frac{1}{n^{BCD}_{jkl}} + \frac{1}{n^{BCD}_{jkl'}} + \frac{1}{n^{BCD}_{j'kl}} + \frac{1}{n^{BCD}_{j'kl'}}.$$

Computations are summarized in Table 4.12. In Table 4.13, the estimates

$$\hat{\tau}^{CD \cdot AB}_{(kk')(ll') \cdot ij} = \hat{\tau}^{CD \cdot B}_{(kk')(ll') \cdot j} = \log\left(\frac{n^{BCD}_{jkl}n^{BCD}_{jk'l'}}{n^{BCD}_{jkl'}n^{BCD}_{jk'l}}\right),$$

$$\hat{q}^{CD \cdot AB}_{(kk')(ll') \cdot ij} = \hat{\tau}^{CD \cdot B}_{(kk')(ll') \cdot j} = \frac{n^{BCD}_{jk'l}n^{BCD}_{jk'l'}}{n^{BCD}_{jkl'}n^{BCD}_{jk'l'}},$$

Table 4.12

Estimated Log Cross-Product Ratios Based on Table 4.6 for Interaction
of Religion and Abortion Attitudes Given Education

Education in years	k	Religion	j	Religion	j′	Attitude	l	Attitude	l′	$\hat{\tau}^{BD\text{-}C}_{(jj')(ll')\cdot k}$	$s(\hat{\tau}^{BD\text{-}C}_{(jj')(ll')\cdot k})$	$\hat{q}^{BD\text{-}C}_{(jj')(ll')\cdot k}$
≤8	1	N. Prot.	1	S. Prot.	2	Positive	1	Mixed	2	0.294	0.330	1.341
								Negative	3	0.613	0.273	1.846
						Mixed	2	Negative	3	0.320	0.262	1.376
				Cath.	3	Positive	1	Mixed	2	0.533	0.327	1.704
								Negative	3	0.405	0.284	1.500
						Mixed	2	Negative	3	−0.128	0.258	0.880
		S. Prot.	2	Cath.	3	Positive	1	Mixed	2	0.239	0.363	1.271
								Negative	3	−0.208	0.311	0.812
						Mixed	2	Negative	3	−0.447	0.273	0.639
9–12	2	N. Prot.	1	S. Prot.	2	Positive	1	Mixed	2	0.426	0.168	1.531
								Negative	3	0.276	0.143	1.318
						Mixed	2	Negative	3	−0.149	0.164	0.861
				Cath.	3	Positive	1	Mixed	2	0.226	0.161	1.253
								Negative	3	0.652	0.127	1.919
						Mixed	2	Negative	3	0.426	0.154	1.531
		S. Prot.	2	Cath.	3	Positive	1	Mixed	2	−0.200	0.182	0.819
								Negative	3	0.375	0.150	1.456
						Mixed	2	Negative	3	0.575	0.171	1.778
≥13	3	N. Prot.	1	S. Prot.	2	Positive	1	Mixed	2	0.018	0.230	1.018
								Negative	3	0.255	0.187	1.291
						Mixed	2	Negative	3	0.237	0.255	1.268
				Cath.	3	Positive	1	Mixed	2	0.818	0.218	2.266
								Negative	3	1.265	0.180	3.542
						Mixed	2	Negative	3	0.447	0.224	1.563
		S. Prot.	2	Cath.	3	Positive	1	Mixed	2	0.800	0.249	2.226
								Negative	3	1.009	0.198	2.744
						Mixed	2	Negative	3	0.209	0.255	1.233

Table 4.15

Estimated Log Cross-Product Ratios and Cross-Product Ratios Based on Table 4.6 for Interaction of Education and Abortion Attitudes Given Religion

Religion	j	Education in years	k	Education in years	k'	Attitude	l	Attitude	l'	$\hat{\tau}^{CD \cdot B}_{(kk'l)(l' \cdot j)}$	$s(\hat{\tau}^{CD \cdot B}_{(kk'l)(l' \cdot j)})$	$\hat{q}^{CD \cdot B}_{(kk'l)(l' \cdot j)}$
N. Prot.	1	≤8	1	9–12	2	Positive	1	Mixed	2	−0.675	0.230	0.509
								Negative	3	−0.909	0.190	0.403
						Mixed	2	Negative	3	−0.234	0.203	0.791
				≥13	3	Positive	1	Mixed	2	−1.244	0.248	0.288
								Negative	3	−1.745	0.208	0.175
						Mixed	2	Negative	3	−0.501	0.236	0.606
		9–12	2	≥13	3	Positive	1	Mixed	2	−0.569	0.173	0.566
								Negative	3	−0.836	0.145	0.434
						Mixed	2	Negative	3	−0.267	0.189	0.766
S. Prot.	2	≤8	1	9–12	2	Positive	1	Mixed	2	−0.543	0.290	0.581
								Negative	3	−1.246	0.243	0.288
				≥13	3	Mixed	2	Negative	3	−0.703	0.233	0.495
						Positive	1	Mixed	2	−1.520	0.316	0.219
								Negative	3	−2.103	0.258	0.122
		9–12	2	≥13	3	Mixed	2	Negative	3	−0.583	0.279	0.558
						Positive	1	Mixed	2	−0.977	0.226	0.377
								Negative	3	−0.857	0.185	0.424
						Mixed	2	Negative	3	0.120	0.237	1.127
Cath.	3	≤8	1	9–12	2	Positive	1	Mixed	2	−0.983	0.284	0.374
								Negative	3	−0.663	0.246	0.515
				≥13	3	Mixed	2	Negative	3	0.320	0.222	1.377
						Positive	1	Mixed	2	−0.960	0.305	0.383
								Negative	3	−0.886	0.264	0.412
		9–12	2	≥13	3	Mixed	2	Negative	3	0.074	0.248	1.077
						Positive	1	Mixed	2	0.023	0.209	1.023
								Negative	3	−0.223	0.165	0.800
						Mixed	2	Negative	3	−0.246	0.195	0.782

and

$$s^2(\hat{\tau}_{(kk')(ll')\cdot ij}^{CD\cdot AB}) = \frac{1}{n_{jkl}^{BCD}} + \frac{1}{n_{jkl'}^{BCD}} + \frac{1}{n_{jk'l}^{BCD}} + \frac{1}{n_{jk'l'}^{BCD}}$$

are provided.

Of the 27 coefficient estimates $\hat{\tau}_{(jj')(ll')\cdot k}^{BD\cdot C}$ in Table 4.12, 22 are positive. The principal pattern is that given education, Northern Protestants are more favorable to abortions than are Southern Protestants and Southern Protestants are more favorable to abortions than are Catholics.

Contrasts involving Catholics appear to be affected substantially by education. The sharpest contrast between Catholics and Protestants is found in the group with at least 13 years of education for comparisons between positive and negative answers. The odds of a positive relative than negative response are estimated to be 3.54 times higher for Northern Protestants than for Catholics and 2.74 times higher for Southern Protestants than for Catholics. These estimates are much larger than corresponding estimates in the other educational groups, none of which exceed 1.92. Contrasts between Northern and Southern Protestants are less clearly affected by education. For example, $\hat{q}_{(12)(13)\cdot j}^{BD\cdot C}$ ranges from 1.29 to 1.85. To further examine the influence of education on the partial association of religion and attitudes toward legal nontherapeutic abortions, consider the three-factor interactions

$$\hat{\tau}_{(jj')(kk')(ll')}^{BCD} = \hat{\tau}_{(jj')(ll')\cdot k}^{BC\cdot C} - \hat{\tau}_{(jj')(ll')\cdot k'}^{BD\cdot C}$$

$$= \log\left(\frac{n_{jkl}^{BCD} n_{j'kl'}^{BCD} n_{jk'l'}^{BCD} n_{j'k'l}^{BCD}}{n_{jkl'}^{BCD} n_{j'kl}^{BCD} n_{jk'l}^{BCD} n_{j'k'l'}^{BCD}}\right)$$

shown in Table 4.14. In this table,

$$s(\hat{\tau}_{(jj')(kk')(ll')}^{BCD}) = \left[s^2(\hat{\tau}_{(jj')(ll')\cdot k}^{BD\cdot C}) + s^2(\hat{\tau}_{(jj')(ll')\cdot k'}^{BD\cdot C})\right]^{1/2}$$

$$= \left(\frac{1}{n_{jkl}^{BCD}} + \frac{1}{n_{j'kl}^{BCD}} + \frac{1}{n_{jk'l}^{BCD}} + \frac{1}{n_{j'k'l}^{BCD}} + \frac{1}{n_{jkl'}^{BCD}} + \frac{1}{n_{j'kl'}^{BCD}} + \frac{1}{n_{jk'l'}^{BCD}} + \frac{1}{n_{j'k'l'}^{BCD}}\right)^{1/2},$$

and the standardized values are $\hat{\tau}_{(jj')(kk')(ll')}^{BCD}/s(\hat{\tau}_{(jj')(kk')(ll')}^{BCD})$. The largest standardized values involve contrasts with Catholics ($j = 3$). For contrasts between Northern and Southern Protestants ($j = 1, j' = 2$), the largest absolute value of a standardized value is only 1.52. In Chapter 6 of Volume 2, the question whether all $\tau_{(12)(kk')(ll')}^{BCD}$ can be assumed zero will be investigated.

In Table 4.13, 23 of 27 estimates $\hat{\tau}_{(kk')(ll')\cdot j}^{CD\cdot B}$ are negative. The basic pattern is a tendency given religion for attitudes to nontherapeutic abortions to become increasingly favorable as education increases. The odds of a positive rather than a negative response are estimated to be 0.175 times as large for Northern Protestants with no more than eight years schooling as for North-

Table 4.14

Estimated Three-Factor Interactions between Religion, Education,
and Abortion Attitudes for Table 4.6

Religion[a]		Education[b]		Attitudes[c]				Standardized
j	j'	k	k'	l	l'	$\hat{\tau}^{BCD}_{(jj')(kk')(ll')}$	$s(\hat{\tau}^{BCD}_{(jj')(kk')(ll')})$	value
1	2	1	2	1	2	−0.132	0.370	−0.36
					3	0.337	0.308	1.09
				2	3	0.469	0.309	1.52
			3	1	2	0.276	0.402	0.69
					3	0.358	0.331	1.08
				2	3	0.082	0.365	0.22
		2	3	1	2	0.408	0.285	1.43
					3	0.021	0.236	0.09
				2	3	−0.387	0.303	−1.27
1	3	1	2	1	2	0.307	0.365	0.84
					3	−0.246	0.311	−0.79
				2	3	−0.554	0.301	−1.84
			3	1	2	−0.285	0.393	−0.72
					3	−0.859	0.336	−2.56
				2	3	−0.575	0.342	−1.68
		2	3	1	2	−0.592	0.271	−2.18
					3	−0.613	0.220	−2.79
				2	3	−0.021	0.272	−0.08
2	3	1	2	1	2	0.439	0.406	1.08
					3	−0.583	0.346	−1.69
				2	3	−1.023	0.322	−3.18
			3	1	2	−0.561	0.440	−1.28
					3	−1.217	0.369	−3.30
				2	3	−0.657	0.373	−1.76
		2	3	1	2	−1.000	0.308	−3.24
					3	−0.634	0.248	−2.56
				2	3	0.366	0.307	1.19

[a] Codes for religion are (1) Northern Protestant, (2) Southern Protestant, (3) Catholic.
[b] Codes for education are (1) more than 8 years, (2) 9–12 years, (3) at least 13 years.
[c] Codes for attitudes are (1) positive, (2) mixed, (3) negative.

ern Protestants with at least 13 years education. The corresponding ratio
for Southern Protestants is 0.122. Catholics, however, present a somewhat
different picture, for $\hat{q}^{CD \cdot B}_{(13)(13) \cdot 3}$ is 0.412. Of four positive $\hat{\tau}^{CD \cdot B}_{(kk')(ll') \cdot j}$, three occur
in the Catholic group. Thus the association of education and abortion
attitudes appears weaker for Catholics than for the other groups. This
pattern is supported by the preponderance of negative three-factor interactions $\hat{\tau}^{BCD}_{(j3)(kk')(ll')}$ in Table 4.14.

The estimated asymptotic standard deviations in Tables 4.12, 4.13, and 4.14 are too large for a very accurate assessment of any parameter values. Some improvements are possible here if more assumptions can be made about the interaction terms by use of models that are not hierarchical. Attempts in this direction are discussed in Chapter 6 of Volume 2.

Limitations of the Analysis

The analysis in this section provides information concerning the extent to which education, time, and religion are helpful in the prediction of the attitudes of respondents toward the permissibility of legal abortions for nontherapeutic causes. The analysis does not imply that religion causally affects abortion attitudes or that education causally affects abortion attitudes. Even given the assumption that the response to the abortion questions cannot affect the other three variables, there still remain variables that may affect abortion attitudes and may be associated with religion and education. For example, consider the variables sex, marital status, and age. Thus caution is required in interpreting Table 4.6. The cautionary note expressed here is by no means new. As an example, see Yule (1900). The problems encountered arise whenever nonexperimental data are analyzed.

The reader may wonder why more variables have not been included in the analysis. Two answers are possible. The first is that the approximations of this chapter only apply when each cell has a sufficiently large number of observations. How large is sufficient is not well known, but inclusion of more variables would result in many cells with no more than one or two observations. In such cases, many approximations become questionable. Another difficulty is more a problem of human understanding than of statistical theory. Comprehension of the interactions between six or seven variables can be difficult. Some of these problems can be solved with the logit and multinomial response models discussed in the next two chapters, provided the interaction structure of the table is favorable and considerable care is taken.

Examination of Table 4.6 should still be regarded as informative, even if this one table cannot hope to describe all factors relevant to abortion. It is possible that some information from the table can remain unchanged even if more variables are considered.

This point is discussed in Yule (1900), Bishop (1971), and Bishop, Fienberg, and Holland (1975, pages 47–48). As an example, let A_h, D_h, and K_{he}, $1 \leqslant e \leqslant f$, be observed. Then the partial association $\tau_{(ii')(ll') \cdot k_1 \cdots k_f}^{AD \cdot K_1 \cdots K_f}$ of A_h and D_h given K_{he}, $1 \leqslant e \leqslant f$, can be estimated by

$$\hat{\tau}_{(ii')(ll')}^{AD} = \log\left(\frac{n_{il}^{AD} n_{i'l'}^{AD}}{n_{il'}^{AD} n_{i'l}^{AD}}\right)$$

as long as given D_h, the K_{he}, $1 \leq e \leq f$, are conditionally independent of A_h or as long as given A_h, the K_{he}, $1 \leq e \leq f$, are conditionally independent of D_h. This fact has been used already in the case of $K_{h1} = B_h$ and $K_{h2} = C_h$.

EXERCISES

4.1 Consider an $r \times s \times t \times u$ table of counts n_{ijkl} in which independent and identically distributed observations (A_h, B_h, C_h, D_h), $1 \leq h \leq N$, are cross-classified. Let n_{ijkl} be the number of h such that $A_h = i$, $B_h = j$, $C_h = k$, and $D_h = l$. Let $p_{ijkl} > 0$ be the probability that $A_h = i$, $B_h = j$, $C_h = k$, and $D_h = l$. Assume that $1 \leq A_h \leq r$, $1 \leq B_h \leq s$, $1 \leq C_h \leq t$, and $1 \leq D_h \leq l$. Find maximum-likelihood estimates for the expected value $m_{ijkl} = N p_{ijkl}$ of n_{ijkl} under the following hierarchical models:

(a) $\log m_{ijkl} = \lambda + \lambda_i^A + \lambda_j^B$,

(b) $\log m_{ijkl} = \lambda + \lambda_i^A + \lambda_j^B + \lambda_k^C + \lambda_l^D$,

(c) $\log m_{ijkl} = \lambda + \lambda_i^A + \lambda_j^B + \lambda_k^C + \lambda_l^D + \lambda_{ij}^{AB} + \lambda_{ik}^{AC} + \lambda_{il}^{AD}$,

(d) $\log m_{ijkl} = \lambda + \lambda_i^A + \lambda_j^B + \lambda_k^C + \lambda_l^D + \lambda_{ij}^{AB} + \lambda_{kl}^{CD}$.

Solution

(a) Note that the generating class consists of $T(1) = A$ and $T(2) = B$. Thus $V(2)$ is empty, $U = AB$, and $d = tu$. The solution

$$m_{ijkl} = \frac{n_i^A n_j^B}{tuN}$$

follows from (4.6).

(b) Here the generating class contains $T(1) = A$, $T(2) = B$, $T(3) = C$, and $T(4) = D$. Thus $V(2)$, $V(3)$, and $V(4)$ are empty, $u = ABCD$, $d = 1$, and $c(2) = c(3) = c(4) = 1$. By (4.7),

$$\hat{m}_{ijkl} = n_i^A(n_j^B/N)(n_k^C/N)(n_l^P/N) = \frac{n_i^A n_j^B n_k^C n_l^D}{N^3}.$$

(c) In this case, the generating class contains $T(1) = AB$, $T(2) = AC$, and $T(3) = AD$. Thus $V(2) = V(3) = A$, $u = ABCD$, $d = 1$, and $c(2) = c(3) = 1$. By (4.7),

$$\hat{m}_{ijkl} = n_{ij}^{AB}(n_{ik}^{AC}/n_i^A)(n_{il}^{AD}/n_i^A) = \frac{n_{ij}^{AB} n_{ik}^{AC} n_{il}^{AD}}{(n_i^A)^2}.$$

(d) Here the generating class contains $T(1) = AB$ and $T(2) = CD$, so that $V(2)$ is empty and $u = ABCD$. Thus $d = 1$ and by (4.6),

$$\hat{m}_{ijkl} = n_{ij}^{AB} n_{kl}^{CD}/N.$$

4.2   Let n_{ijkl} be defined as in Exercise 4.1. Which of the following hierarchical models have closed-form expressions for m_{ijkl}?

(a) $\log m_{ijkl} = \lambda + \lambda_i^A + \lambda_j^B + \lambda_k^C + \lambda_{ij}^{AB} + \lambda_{ik}^{AC} + \lambda_{jk}^{BC}$,

(b) $\log m_{ijkl} = \lambda + \lambda_i^A + \lambda_j^B + \lambda_k^C + \lambda_l^D + \lambda_k^{CD}$,

(c) $\log m_{ijkl} = \lambda + \lambda_i^A + \lambda_j^B + \lambda_k^C + \lambda_l^D + \lambda_{ij}^{AB} + \lambda_{il}^{AC} + \lambda_{jl}^{BC} + \lambda_{ijk}^{ABC}$,

(d) $\log m_{ijkl} = \lambda + \lambda_i^{iA} + \lambda_j^B + \lambda_k^C + \lambda_l^D + \lambda_{ij}^{AB} + \lambda_{ik}^{AC} + \lambda_{il}^{AD} + \lambda_{jk}^{BC}$,
$$+ \lambda_{jl}^{BD} + \lambda_{kl}^{CD} + \lambda_{ijk}^{ABC} + \lambda_{ijl}^{ABD} + \lambda_{jkl}^{BCD}.$$

Solution

(a)   The generating class contains AB, AC, and BC. No matter how these three combinations of letters are arranged, $c(3) = 3$. Thus $\hat{m}_{ijkl}$ cannot be expressed in closed form.

(b)   The generating class contains $T(1) = A$, $T(2) = B$, and $T(3) = CD$, so that $V(1)$ and $V(2)$ are empty and $c(2) = c(3) = 1$. Thus $\hat{m}_{ijkl}$ can be expressed in closed form.

(c)   The generating class consists of ABC and D. Since there are only two members, a closed-form expression is available for $\hat{m}_{ijkl}$.

(d)   The generating class contains ABC, ABD, and BCD. As in (a), $c(3) = 3$ under any definition of $T(1)$, $T(2)$, and $T(3)$.

4.3   Use the Deming–Stephan algorithm to obtain the maximum-likelihood estimates shown in Table 4.4. Verify that the degrees of freedom are 12.

Solution

Table 4.4 may be obtained using available computer programs such as those described in Fay and Goodman (1975) or Haberman (1972, 1973c). Standard program packages also supply degrees of freedom. Alternately, the degrees of freedom are given by the formula

$b(YE) + b(YR) + b(YSE) + b(YSR) + b(YER) + b(SER) + b(YSER) = 12.$

4.4   Show that in the iterative proportional fitting algorithm for the hierarchical model specified by (4.1), $m_{ijkl3} = \hat{m}_{ijkl}$. Show also that in the iterative proportional fitting algorithm for the model defined by (4.2), $m_{ijkl2} = \hat{m}_{ijkl}$. To simplify the algebra, assume that $m_{ijkl0} = 1$. Note that the order of members of the generating class does not affect the result.

Solution

In (4.1), one may have

$$m_{ijkl} = m_{ijkl0} n_{ij}^{YS} / m_{ij0}^{YS} = \tfrac{1}{6} n_{ij}^{YS},$$

$$m_{ijkl2} = m_{ijkl1} n_{jk}^{SE} / m_{jk1}^{SE}$$
$$= (\tfrac{1}{6} n_{ij}^{YS}) n_{jk}^{SE} / (\tfrac{1}{3} n_j^S)$$
$$= \tfrac{1}{2} n_{ij}^{YS} n_{jk}^{SE} / n_j^S,$$

$$m_{ijkl3} = m_{ijkl2}n_{kl}^{ER}/m_{kl2}^{ER}$$
$$= (\tfrac{1}{2}n_{ij}^{YS}n_{jk}^{SE}/n_j^{S})n_{kl}^{ER}/(\tfrac{1}{2}n_k^{E})$$
$$= n_{ij}^{YS}n_{jk}^{SE}n_{kl}^{ER}/(n_j^{S}n_k^{E}).$$

The same result occurs if

$$m_{ijkl1} = m_{ijkl0}n_{jk}^{SE}/m_{jk1}^{SE},$$
$$m_{ijkl2} = m_{ijkl1}n_{ij}^{YS}/m_{ij1}^{YS},$$

and

$$m_{ijkl3} = m_{ijkl2}n_{kl}^{ER}/m_{kl2}^{ER}$$

or if any other ordinary YS, SE, or ER is used.
  In (4.2), either

$$m_{ijkl1} = m_{ijkl0}n_{ij}^{YS}/m_{ijkl0} = \tfrac{1}{6}n_{ij}^{YS}$$

and

$$m_{ijkl2} = m_{ijkl1}n_{jkl}^{SER}/m_{jkl1}$$
$$= (\tfrac{1}{6}n_{ij}^{YS})n_{jkl}^{SER}/(\tfrac{1}{6}n_j^{S})$$
$$= n_{ij}^{YS}n_{jkl}^{SER}/n_j^{S}$$

or

$$m_{ijkl1} = m_{ijkl0}n_{jkl}^{SER}/m_{jkl0}^{SER} = \tfrac{1}{2}n_{jkl}^{SER}$$

and

$$m_{ijkl2} = m_{ijkl1}n_{ij}^{YS}/m_{ij1}^{YS}$$
$$= (\tfrac{1}{2}n_{jkl}^{SER})n_{ij}^{YS}/(\tfrac{1}{2}n_j^{S})$$
$$= n_{ij}^{YS}n_{jkl}^{SER}/n_j^{S}.$$

4.5 Interpret the models in Exercise 4.1 in terms of independence, conditional independence, and equiprobability.

Solution

(a) Let $U(1) = A$, $U(2) = B$, and $U(3) = CD$ and let W be empty. The corresponding model is

$$\log m_{ijkl} = \lambda + \lambda_i^A + \lambda_j^B + \lambda_k^C + \lambda_l^D + \lambda_{kl}^{CD}.$$

In (a), the added requirement that $\lambda_k^C = 0$, $\lambda_l^D = 0$, and $\lambda_{kl}^{CD} = 0$ holds if and only if p_{ijkl} is independent of k and l, so that

$$p_{kl \cdot ij}^{CD \cdot AB} = p_{ijkl}/p_{kl}^{CD} = 1/(tu).$$

Thus A_h and B_h are independent, and given A_h and B_h, all possible combinations of C_h and D_h are equally probable.

(b)   Let $U(1) = A, U(2) = B, U(3) = C$, and $U(4) = D$ and let W be empty. Then the model of mutual independence of A_h, B_h, C_h, and D_h corresponds to (b).

(c)   Let $U(1) = B$, $U(2) = C$, $U(3) = D$, and $W = A$. Then the model of conditional independence of B_h, C_h, and D_h given A_h corresponds to (c).

(d)   Let $U(1) = AB$ and $U(2) = CD$, and let W be empty. Then (d) corresponds to a model of independence of the pairs (A_h, B_h) and (C_h, D_h).

4.6   An alternate computational technique for Table 4.7 can be used to speed calculations. The method is derived from Good (1958). If the table is an $r_1 \times r_2 \times r_3 \times r_4$ table, then an $(r_1 + 1) \times (r_2 + 1) \times (r_3 + 1) \times (r_4 + 1)$ table of entries y_{ijkl0}, $0 \leqslant i \leqslant r_1, 0 \leqslant j \leqslant r_2, 0 \leqslant k \leqslant r_3, 0 \leqslant l \leqslant r_4$, is defined so that

$$y_{ijkl0} = \log(n_{ijkl} + \tfrac{1}{2}), \qquad 1 \leqslant i \leqslant r_1, \quad 1 \leqslant j \leqslant r_2, \quad 1 \leqslant k \leqslant r_3, \quad 1 \leqslant l \leqslant r_4,$$
$$= 0, \, i, j, k, \qquad\quad \text{or} \quad l = 0.$$

For $0 \leqslant v \leqslant 3$,

$$y_{jkli(v+1)} = \sum_{i'=1}^{r_v+1} y_{i'jklv}, \qquad\qquad i = 0,$$

$$= r_{v+1} y_{ijklv} - y_{jkl0(v+1)}, \qquad 1 \leqslant i \leqslant r_{v+1}.$$

Then

$$\hat{\lambda} = \frac{1}{r_1 r_2 r_3 r_4} y_{00003},$$

$$\hat{\lambda}_i^A = \frac{1}{r_1 r_2 r_3 r_4} y_{i0003},$$

$$\hat{\lambda}_j^B = \frac{1}{r_1 r_2 r_3 r_4} y_{0j003},$$

$$\hat{\lambda}_{ij}^{AB} = \frac{1}{r_1 r_2 r_3 r_4} y_{ij003},$$

etc.

Similarly, let

$$w_{ijkl0} = \frac{1}{n_{ijkl} + \tfrac{1}{2}}, \qquad 1 \leqslant i \leqslant r_1, \quad 1 \leqslant j \leqslant r_2, \quad 1 \leqslant k \leqslant r_3, \quad 1 \leqslant l \leqslant r_4,$$

$$= 0, \, i, j, k, \qquad\quad \text{or} \quad l = 0.$$

For $0 \leqslant v \leqslant 3$, let

$$w_{jkli(v+1)} = \sum_{i'=1}^{v+1} w_{i'jklv}, \qquad\qquad i = 0,$$

$$= r_{v+1}(r_{v+1} - 2)w_{ijklv} + w_{jkl0(v+1)}, \qquad 1 \leqslant i \leqslant r_{v+1}.$$

Then

$$s^2(\hat{\lambda}_i^A) = \left(\frac{1}{r_1 r_2 r_3 r_4}\right)^2 w_{i0003},$$

$$s^2(\hat{\lambda}_j^B) = \left(\frac{1}{r_1 r_2 r_3 r_4}\right)^2 w_{0j003},$$

$$s^2(\hat{\lambda}_{ij}^{AB}) = \left(\frac{1}{r_1 r_2 r_3 r_4}\right)^2 w_{ij003},$$

etc.

Use this algorithm to obtain Table 4.7.

Solution

A step-by-step worksheet cannot be provided given space requirements. Nonetheless, it may be helpful in checking work to note that in Table 4.6, $r_1 = r_2 = r_3 = r_4 = 3$, so that the tables of y_{ijklv} and w_{ijklv} are all $4 \times 4 \times 4 \times 4$ tables. Note that

$$y_{jkli(v+1)} = y_{ijklv} + y_{2jklv} + y_{3jklv}, \qquad i = 0,$$

$$= 3y_{ijklv} - y_{jkl(v+1)}, \qquad 1 \leqslant i \leqslant 3,$$

$$w_{jkli(v+1)} = w_{1jklv} + w_{2jklv} + w_{3jklv}, \qquad i = 0,$$

$$= 3w_{ijklv} + w_{jkl0(v+1)}, \qquad 1 \leqslant i \leqslant 3.$$

4.7 Verify Table 4.8 and Table 4.9.

Solution

Table 4.9 can be checked with the aid of available computer packages such as ECTA or C-TAB. The degrees of freedon can be found by hand as well as through these programs. For example, in Model 1, there are

$$b(AB) + b(AC) + b(AD) + b(BC) + b(ABC) + b(ABD) + b(ACD)$$
$$+ b(BCD) + b(ABCD)$$

degrees of freedom. Since

$$b(AB) = b(AC) = b(AD) = b(BC) = (3 - 1)(3 - 1) = 4,$$
$$b(ABC) = b(ABD) = b(ACD) = b(BCD) = (3 - 1)(3 - 1)(3 - 1) = 8,$$

Table 4.15

Attitudes toward Therapeutic Abortions among White Christian Subjects
in the 1972–1974 General Social Surveys[a]

Year	Religion[b]	Education in years	Attitudes[c] Positive	Attitudes[c] Mixed	Attitudes[c] Negative	Total
1972	N. Prot.	≤8	48	12	9	69
		9–12	197	43	13	253
		≥13	139	9	4	152
	S. Prot.	≤8	30	17	9	56
		9–12	97	10	6	113
		≥13	68	8	1	77
	Cath.	≤8	32	12	14	58
		9–12	131	50	18	199
		≥13	64	13	8	85
1973	N. Prot.	≤8	59	16	4	79
		9–12	197	24	6	227
		≥13	124	11	4	139
	S. Prot.	≤8	34	16	4	54
		9–12	118	29	6	153
		≥13	82	4	1	87
	Cath.	≤8	32	14	2	48
		9–12	141	45	16	202
		≥13	72	20	7	99
1974	N. Prot.	≤8	49	16	7	72
		9–12	219	26	10	255
		≥13	131	10	4	145
	S. Prot.	≤8	30	19	1	50
		9–12	106	21	5	132
		≥13	87	11	2	100
	Cath.	≤8	29	9	3	41
		9–12	149	30	15	194
		≥13	69	18	11	98

[a] From National Opinion Research Center, 1972, 1973, and 1974 General Social Surveys, University of Chicago.

[b] For an explanation of this category, see Table 4.6.

[c] For the general format of the question, see Table 4.6. The three conditions used in this table are the following:

A. If there is a strong chance of serious defect in the baby?

C. If the woman's own health is seriously endangered by the pregnancy?

E. If she became pregnant as a result of rape?

Inclusion rules for subjects are the same as in Table 4.6

and

$$b(ABCD) = (3 - 1)(3 - 1)(3 - 1)(3 - 1) = 16,$$

the sum is 64.

The explicit expressions for $\hat{m}_{ijkl}$ are also readily checked. In Model 1, $T(1) = A$, $T(2) = BD$, $T(3) = CD$, $U = ABCD$, $V(1)$ is empty, $V(2) = D$, $c(1) = 1$, $c(2) = 2$, $d = 1$, and

$$\hat{m}_{ijkl} = n_i^{A}(n_{jl}^{BD}/N)(n_{kl}^{CD}/n_l^{D}).$$

In Model 3, $T(1) = A$, $T(2) = BCD$, $U = ABCD$, and $V(2)$ is empty, so that

$$\hat{m}_{ijkl} = n_i^{A} n_{jkl}^{BCD}/N.$$

In Model 4, $T(1) = AD$, $T(2) = BCD$, $U = ABCD$, and $V(2) = D$, so that

$$\hat{m}_{ijkl} = n_{il}^{AD} n_{jkl}^{BCD}/n_l^{D}.$$

In Model 6, $T(1) = ABD$, $T(2) = BCD$, $U = ABCD$, and $V(2) = BD$, so that

$$\hat{m}_{ijkl} = n_{ijl}^{ABD} n_{jkl}^{BCD}/n_{jl}^{BD}.$$

4.8 Table 4.15 examines the relationship of year of survey and religion and education to respondent to attitudes toward abortions for reasons such as danger to the woman's health, a strong chance of a birth defect in the baby, or pregnancy due to rape. Analyse the data in this table in a manner parallel to that used in Table 4.16. Compare your conclusions for Table 4.15 to those reached in Table 4.6.

Solution

Space does not permit a detailed analysis; however, it should be noted that results for Table 4.15 are similar to those obtained for Table 4.6. The main difference is that there is more evidence in Table 4.15 of interaction between year of survey, education, and attitudes.

5 *Logit Models*

Logit models are a class of models used to explore the relationship of a dichotomous dependent variable to one or more independent variables. In these models, the logit, or log odds, that the dependent variable has a specified value is a linear function of the independent variables (or variables). Logit models are analogous to ordinary regression models in which the expected value of a continuous dependent variable is a linear function of one or more independent variables. The Newton–Raphson algorithm for logit models and formulas for computation of asymptotic variances of parameter estimates both exploit this analogy.

Logit models are log–linear models, just like the models examined in the first four chapters. They deserve special treatment among the class of log–linear models since the Newton–Raphson algorithm for logits is simpler than is the Newton–Raphson algorithm presented in earlier chapters and since interpretations of parameter estimates in logit models are more clearly analogous to similar interpretations in the case of ordinary regression models.

Logit analysis can be applied whether the independent variables X_{hk} are discrete or continuous; however, the methods used are not necessarily the same. Sections 5.1 and 5.2 deal with logit models for discrete indepenbent variables. In Section 5.1, Table 3.1 is reanalyzed by means of a logit model equivalent to the model of no three-factor interaction of Section 3.1. This section introduces the Newton–Raphson algorithm for logit analysis for the case of discrete independent variables. As usual, the algorithm corresponds to a series of weighted regressions. In Section 5.2, Table 3.4 is reanalyzed with a logit model not equivalent to any hierarchical log–linear model in Chapter 3. This model examines the influence of education and sex on attitudes toward women's roles by use of an arbitrary scoring of the education variable. This scoring system permits a more sensitive analysis than that provided by hierarchical log–linear models. In Section 5.3, the problems of continuous predictor variables are considered through a variant of Table 3.4 in which the education variable is recorded by number of years completed rather than by membership in one of three groups. This modest change in the data causes some chi-square approximations to become unacceptable and slows convergence of the Newton–Raphson algorithm. Residual analysis is also made more difficult.

Section 5.4 summarizes the general theory of logit models, while Section 5.5 considers a more general class of models for dichotomous responses, the class of quantal-response models. The most common member of this class is the probit model.

Logit and probit analysis have an extensive literature, much of which is concerned with biological assay. Finney (1964, 1971) provides a thorough discussion in this context and an extensive bibliography. For examples of early work in this area, see Fisher and Yates (1963 [1938], page 16) and Garwood (1941). Logit and probit models also have an extensive literature in psychology concerned with paired comparisons. Books by David (1963) and Bock and Jones (1968) can serve as guides to work in this area. An extensive bibliography and a more mathematically-sophisticated discussion of logit analysis appears in Cox (1970). Haberman (1974a, pages 303–373) provides a rigorous but mathematically-demanding discussion of logit and probit models. Treatments of logit and probit analysis are much less extensive in the social science literature. Exceptions include Theil (1970) and Nerlove and Press (1973). McFadden's (1976) review is helpful.

5.1 Logits and Table 3.1

Logit models are often equivalent to familiar hierarchical log–linear models. The logit formulation provides an alternate computational technique and a different perspective for interpretation. In Table 3.1, a log–linear model of no three-factor interaction was considered. In this section, this model is shown to be equivalent to an additive logit model. To show equivalence, let subject h have race A_h and sex B_h, and let the type of assault be C_h. To explore the relationship of sex and race to type of assault, it is helpful regard sex A_h and race B_h as independent variables and type of assault C_h as a dependent variable. Let $p_{k \cdot ij} = p_{k \cdot ij}^{C \cdot AB}$ denote the conditional probability that assault $C_h = k$ given that $A_h = i$ and $B_h = j$. Under the Poisson model,

$$p_{k \cdot ij}^{C \cdot AB} = m_{ijk}/m_{ij}^{AB}.$$

Let

$$\omega_{ij} = \tau_{12 \cdot ij}^{C \cdot AB} = \log\left(\frac{p_{1 \cdot ij}^{C \cdot AB}}{p_{2 \cdot ij}^{C \cdot AB}}\right) = \log\left(\frac{m_{ij1}}{m_{ij2}}\right)$$

be the log odds that C_h is 1 rather than 2 given that $A_h = i$ and $B_h = j$. This ratio (or sometimes $\frac{1}{2}\omega_{ij}$) is the logit of $p_{1 \cdot ij} = p_{1 \cdot ij}^{C \cdot AB}$. Knowledge of the logit ω_{ij} is equivalent to knowledge of the probabilities $p_{1 \cdot ij}$ and $p_{2 \cdot ij}$, for $p_{1 \cdot ij} + p_{2 \cdot ij} = 1$,

$$\omega_{ij} = \log\left(\frac{p_{1 \cdot ij}}{1 - p_{1 \cdot ij}}\right),$$

and

$$p_{1 \cdot ij} = (1 + \exp - \omega_{ij})^{-1}.$$

The logit ω_{ij} can be decomposed in the same manner as can the means of a two-way table in analysis of variance. Thus

$$\omega_{ij} = \eta + \eta_i^A + \eta_j^B + \eta_{ij}^{AB},$$

where

$$\sum_i \eta_i^A = \sum_j \eta_j^B = \sum_i \eta_{ij}^{AB} = \sum_j \eta_{ij}^{AB} = 0.$$

The restraints on the η-parameters imply that $\eta_2^A = -\eta_1^A$, $\eta_2^B = -\eta_1^B$, and $\eta_{22}^{AB} = -\eta_{21}^{AB} = -\eta_{12}^{AB} = \eta_{11}^{AB}$. The η-parameters for the logit ω_{ij} correspond to the λ-parameters of the log–linear representation

$$\log m_{ijk} = \lambda + \lambda_i^A + \lambda_j^B + \lambda_k^C + \lambda_{ij}^{AB} + \lambda_{ik}^{AC} + \lambda_{jk}^{BC} + \lambda_{ijk}^{ABC}$$

used in Chapter 3. To demonstrate the correspondence, an argument based on Bishop (1969) may be used. Observe that

$$\begin{aligned}
\omega_{ij} &= \log m_{ij1} - \log m_{ij2} \\
&= \lambda + \lambda_i^A + \lambda_j^B + \lambda_1^C + \lambda_{ij}^{AB} + \lambda_{i1}^{AC} + \lambda_{j1}^{BC} + \lambda_{ij1}^{ABC} \\
&\quad - \lambda - \lambda_i^A - \lambda_j^B - \lambda_2^C - \lambda_{ij}^{AB} - \lambda_{i2}^{AC} - \lambda_{j2}^{BC} - \lambda_{ij2}^{ABC} \\
&= (\lambda_1^C - \lambda_2^C) + (\lambda_{i1}^{AC} - \lambda_{i2}^{AC}) + (\lambda_{j1}^{BC} - \lambda_{j2}^{BC}) + (\lambda_{ij1}^{ABC} - \lambda_{ij2}^{ABC}) \\
&= 2(\lambda_1^C + \lambda_{i1}^{AC} + \lambda_{j1}^{BC} + \lambda_{ij2}^{ABC}).
\end{aligned}$$

Therefore, the η-parameters are related to the λ-parameters by the equations

$$\eta_i^A = 2\lambda_{i1}^{AC},$$
$$\eta_j^B = 2\lambda_{j1}^{BC},$$
$$\eta_{ij}^{AB} = 2\lambda_{ij1}^{ABC}.$$

The model that all λ_{ijk}^{ABC} are 0 holds if and only if the η_{ij}^{AB} are 0. Thus a log–linear model of no three-factor interaction holds if and only if the additive logit model

$$\omega_{ij} = \eta + \eta_i{}^A + \eta_j{}^B$$

holds. This result reflects the general observation in Haberman (1974a, pages 311–313) that a logit model is always equivalent to a log–linear model. For further examples on this point, see Exercises 5.1 and 5.2.

Maximum-likelihood estimates for the additive logit model can be obtained through versions of the Newton–Raphson algorithm or through iterative proportional fitting. Iterative proportional fitting has already been applied in Section 3.2 to the equivalent log–linear model of no three-factor interactions λ_{ijk}^{ABC}, so this algorithm will not be considered here. One version of the Newton–Raphson algorithm has been used in Exercise 3.3 with the log–linear model that all λ_{ijk}^{ABC} are zero. In this section, a different Newton–Raphson algorithm is explored in which the presence of binary responses is exploited. As usual, this algorithm reduces to a series of weighted regression problems, and estimated asymptotic variances of the η-parameters are by-products of the computations. The principal simplification is that the analogous weighted regression models now have an observation for each index i and j. In Chapter 3, the weighted regression models encountered required observations for each index $i, j,$ and k. Thus the number of observations in the regression models is cut in half. Relevant references to the algorithm presented here include Fisher and Yates (1963[1938], page 16), Dyke and Patterson (1952), and Yates (1955). The procedure used here is essentially that of Yates (1955), although a few differences in approach are adopted here to permit easy generalization. These references are also helpful for estimation of asymptotic variances and for computation of X^2.

The analogies to weighted regression can be clarified if the additive logit model is reparametrized. The same procedure is used here that was adopted earlier in Section 2.6 in connection with additive log–linear models for two-way tables. Let

$$\beta_1 = \eta_1{}^A,$$
$$\beta_2 = \eta_1{}^B,$$
$$x_{ij1} = 1, \qquad i = 1,$$
$$\phantom{x_{ij1}} = -1, \qquad i = 2,$$

and

$$x_{ij2} = 1, \qquad j = 1,$$
$$\phantom{x_{ij2}} = -1, \qquad j = 2,$$

so that the additive logit model can be written as

$$\omega_{ij} = \eta + \beta_1 x_{ij1} + \beta_2 x_{ij2}.$$

Let m_{ijk0} be an initial approximation to $\hat{m}_{ijk}$, say $n_{ijk} + \frac{1}{2}$. Then one may obtain an initial approximation

$$y_{ij0} = \log(m_{ij10}/m_{ij20})$$

for ω_{ij} and an initial approximation

$$p_{1 \cdot ij0} = m_{ij10}/m_{ij0}^{AB} = m_{ij10}/(m_{ij10} + m_{ij20})$$

for

$$\hat{p}_{1 \cdot ij} = \hat{p}_{1 \cdot ij}^{C \cdot AB}.$$

Initial approximations η_0 for $\hat{\eta}$, β_{10} for $\hat{\beta}_1 = \hat{\eta}_1{}^A$, and β_{20} for $\hat{\beta}_2 = \eta_1{}^B$ are obtained as in a weighted regression analysis

$$y_{ij0} = \eta + \beta_1 x_{ij1} + \beta_2 x_{ij2} + \varepsilon_{ij}$$

in which the ε_{ij} are independent observations with respective known means of 0 and variances of u_{ij0}^{-1}. Here

$$u_{ij0} = m_{ij0}^{AB} p_{1 \cdot ij0}(1 - p_{1 \cdot ij0}) = m_{ij10} m_{ij20}/(m_{ij10} + m_{ij20})$$

is the variance of a binomial variable with sample size m_{ij0}^{AB} and probability $p_{1 \cdot ij0}$. Unlike the earlier versions of the Newton–Raphson algorithm, the constant term must be computed. The regression parameters β_{10} and β_{20} are determined by the simultaneous equations

$$S_{110}\beta_{10} + S_{120}\beta_{20} = w_{10}$$

and

$$S_{210}\beta_{10} + S_{220}\beta_{20} = w_{20},$$

while the constant η_0 satisfies the equation

$$\eta_0 = \bar{y}_0 - \beta_{10}\theta_{10} - \beta_{20}\theta_{20}.$$

In these formulas,

$$S_{kl0} = \sum_i \sum_j (x_{ijk} - \theta_{k0})(x_{ijl} - \theta_{l0})u_{ij0},$$

$$w_{k0} = \sum_i \sum_j (x_{ijk} - \theta_{k0})y_{ij0}u_{ij0},$$

$$\theta_{k0} = \sum_i \sum_j x_{ijk}u_{ij0} \bigg/ \sum_i \sum_j u_{ij0},$$

and

$$\bar{y}_0 = \sum_i \sum_j y_{ij0}u_{ij0} \bigg/ \sum_i \sum_j u_{ij0}.$$

Given η_0, β_{10}, and β_{20}, $\hat{\omega}_{ij}$ has an initial approximation

$$\omega_{ij0} = \eta_0 + \beta_{10}x_{ij1} + \beta_{20}x_{ij2}.$$

The corresponding approximation of $\hat{p}_{ij}$ is

$$p_{1 \cdot ij1} = (1 + \exp - \omega_{ij0})^{-1},$$

and the corresponding approximation of $\hat{m}_{ijk}$ is

$$\begin{aligned}
m_{ijk1} &= n_{ij}^{AB}p_{1 \cdot ij1}, & k &= 1, \\
&= n_{ij}^{AB}(1 - p_{1 \cdot ij1}), & k &= 2.
\end{aligned}$$

For all practical purposes, η_0, β_{10}, β_{20}, ω_{ij0}, $p_{1 \cdot ij1}$, and m_{ijk1} are all quite adequate approximations for the corresponding maximum-likelihood estimates. Just as in other versions of the Newton–Raphson algorithm, this situation is common when all counts n_{ijk} are large. Nonetheless, an added iteration is presented here for illustrative purposes.

In the new iteration, working logits

$$y_{ij1} = \omega_{ij0} + (n_{ij1} - m_{ij11})/u_{ij1}$$

and new weights

$$u_{ij1} = n_{ij}^{AB}p_{1 \cdot ij1}(1 - p_{1 \cdot ij1})$$

are computed. Computations proceed as in the weighted regression problem

$$y_{ij1} = \eta + \beta_1 x_{ij1} + \beta_2 x_{ij2} + \varepsilon_{ij},$$

where each ε_{ij} is an independent random variable with mean 0 and variance u_{ij1}^{-1}. Thus β_{11} and β_{21} satisfy the simultaneous equations

$$S_{111}\beta_{11} + S_{121}\beta_{21} = w_{11}$$

and

$$S_{211}\beta_{11} + S_{221}\beta_{21} = w_{21},$$

and η_1 is determined by the formula

$$\eta_1 = \bar{y}_2 - \beta_{11}\theta_{11} - \beta_{21}\theta_{21}.$$

In these formulas,

$$S_{kl1} = \sum_i \sum_j (x_{ijk} - \theta_{k1})(x_{ijl} - \theta_{l1})u_{ij1},$$

$$w_{k1} = \sum_i \sum_j (x_{ijk} - \theta_{k1})y_{ij1}u_{ij1},$$

$$\theta_{k1} = \sum_i \sum_j x_{ijk}u_{ij1} \bigg/ \sum_i \sum_j u_{ij1},$$

and

$$\bar{y}_1 = \sum_i \sum_j y_{ij1} u_{ij1} \bigg/ \sum_i \sum_j u_{ij1},$$

In practice, computational labor can be reduced by noting that

$$\eta_1 = \eta_0 + \gamma_1,$$
$$\beta_{11} = \beta_{10} + \delta_{11},$$

and

$$\beta_{21} = \beta_{20} + \delta_{21}.$$

Here δ_{11} and δ_{21} are determined by the simultaneous equations

$$S_{11}\delta_{11} + S_{121}\delta_{21} = a_{11},$$
$$S_{211}\delta_{11} + S_{221}\delta_{21} = a_{21}.$$

In these equations, for $1 \leqslant k \leqslant 2$,

$$a_{k1} = \sum_i \sum_j (x_{ijk} - \theta_{k1})(n_{ij1} - m_{ij11}).$$

The parameter γ_1 is determined by the equation

$$\gamma_1 = \left[\sum_i \sum_j (n_{ij1} - m_{ij11}) \bigg/ \sum_i \sum_j u_{ij1} \right] - \delta_{11}\theta_{11} - \delta_{21}\theta_{21}.$$

Given η_1, β_{11}, and β_{21},

$$\omega_{ij1} = \eta_1 + \beta_{11}x_{ij1} + \beta_{21}x_{ij2},$$
$$p_{1 \cdot ij2} = (1 + \exp - \omega_{ij1})^{-1},$$
$$m_{ijk2} = n_{ij}^{AB} p_{1 \cdot ij2}, \qquad k = 1,$$
$$= n_{ij}^{AB}(1 - p_{1 \cdot ij2}), \qquad k = 2.$$

Further iterations are obtained in a similar manner. Calculations are summarized in Table 5.1.

The close relationship of $p_{1 \cdot ij1}$ to $\hat{p}_{1 \cdot ij}$ and of the β_{j0} to $\hat{\beta}_j$, $1 \leqslant j \leqslant 2$, is not surprising. This feature has been emphasized at least since Berkson (1944, 1953, 1955). References to a rather polemical literature on whether the approximations to maximum-likelihood estimates are better or worse than the maximum-likelihood estimates themselves can be found in Finney (1971, pages 51–52 and 93–97). Due to problems encountered in Section 5.3, the author recommends use of maximum-likelihood estimates whenever the initial approximations such as $p_{1 \cdot ij1}$ are likely to differ noticeably from the maximum-likelihood estimates. In practice, differences are likely whenever some n_{ijk} are modest in size, say less than 10.

Table 5.1

Computation of Maximum-Likelihood Estimates for the Additive Logit Model for Table 3.1

i	j	n_{ij1}	n_{ij2}	m_{ij10}	m_{ij20}	y_{ij0}	m_{ij0}^{AB}	u_{ij0}	x_{ij1}	x_{ij2}
1	1	3910	808	3910.5	808.5	1.5762	4719.0	669.98	1	1
	2	1050	234	1050.5	234.5	1.4996	1285.0	191.71	1	−1
2	1	5218	1385	5218.5	1385.5	1.3261	6604.0	1094.81	−1	1
	2	929	298	˙929.5	298.5	1.1359	1228.0	225.94	−1	−1
Total		11,107	2725	11,109	2727		13,386	2182.5	0	0

i	j	$y_{ij0}u_{ij0}$	$x_{ij1}u_{ij0}$	$x_{ij2}u_{ij0}$	$(x_{ij1} - \theta_{10})^2 u_{ij0}$	$(x_{ij1} - \theta_{10})(x_{ij2} - \theta_{20})u_{ij0}$
1	1	1056.06	669.98	669.98	981.49	310.36
	2	287.48	191.71	−191.71	280.84	−375.26
2	1	1451.90	−1094.8	1094.8	682.67	−330.88
	2	256.64	−225.94	−225.94	140.88	288.54
Total		3052.1	−459.08	1347.2	2085.9	−107.23

i	j	$(x_{ij2} - \theta_{20})^2 u_{ij0}$	$(x_{ij1} - \theta_{10})y_{ij0}u_{ij0}$	$(x_{ij2} - \theta_{20})y_{ij0}u_{ij0}$
1	1	98.141	1278.2	404.18
	2	501.42	347.95	−464.92
2	1	160.37	−1146.5	555.69
	2	590.96	−202.66	−415.06
Total		1350.9	276.99	79.890

$\theta_{10} = -459.08/2182.5 = -0.21035,$

$\theta_{20} = 1347.2/2182.5 = 0.61727,$

$\bar{y}_0 = 3052.1/2182.5 = 1.3985,$

$2085.9\beta_{10} - 107.23\beta_{20} = 276.99; \qquad -107.23\beta_{10} + 1350.9\beta_{20} = 79.890.$

$\beta_{10} = \dfrac{(276.99)(1350.9) - (79.890)(-107.23)}{(2085.9)(1350.9) - (-107.23)^2} = 0.13639,$

$\beta_{20} = \dfrac{(2085.9)(79.890) - (107.23)(276.99)}{(2085.9)(1350.9) - (-107.23)^2} = 0.069965,$

$\eta_0 = 1.3985 - 0.13639(-0.21035) - 0.069965(0.61727) = 1.3840,$

Table 5.1 (continued)

i	j	$(x_{ij2} - \theta_{21})(n_{ij1} - m_{ij11})$	$x_{ij1}u_{ij1}$	$x_{ij2}u_{ij1}$	$(x_{ij1} - \theta_{11})^2u_{ij1}$	$(x_{ij1} - \theta_{11})(x_{ij2} - \theta_{21})u_{ij1}$
1	1	-3.4718	663.65	663.65	972.79	308.15
	2	-15.982	197.56	-197.56	289.59	-385.65
2	1	3.7378	-1100.1	1100.1	685.36	-333.01
	2	14.648	-220.90	-220.90	137.62	281.84
Total		-1.0680		-459.82	1345.3	2085.4	-129.67

i	j	ω_{ij0}	$p_{1 \cdot ij1}$	n_{ijk}^{AB}	m_{ij11}	m_{ij21}	u_{ij1}	$n_{ij1} - m_{ij1}$	$(x_{ij1} - \theta_{11})(n_{ij1} - m_{ij1})$
1	1	1.5903	0.83066	4718	3919.1	798.95	663.65	-9.0527	-10.960
	2	1.4504	0.81006	1284	1040.1	243.89	197.56	9.8870	11.970
2	1	1.3175	0.78877	6603	5208.3	1394.7	1100.1	9.7461	-7.6925
	2	1.1776	0.76452	1227	938.06	288.94	220.90	-9.0618	7.1523
Total				13,832	11,105.	2726.5	2182.2	1.5186	0.46987

$\theta_{11} = -459.82/2182.2 = -0.21071; \qquad \theta_{21} = 1345.3/2182.2 = 0.61649,$

$\sum_i \sum_j (n_{ij1} - m_{ij11})/\sum_i \sum_j u_{ij1} = 1.5186/2182.2 = 0.00069587,$

$2085.4\delta_{11} - 129.67\delta_{21} = 0.46987; \qquad -129.67\delta_{11} + 1352.9\delta_{21} = -1.0680,$

$$\delta_{11} = \frac{(0.46987)(1352.9) - (-1.0680)(-129.67)}{(2085.4)(1352.9) - (-129.67)^2} = 0.00017729,$$

$$\delta_{21} = \frac{(2085.4)(-1.0680) - (-129.67)(0.46987)}{(2085.4)(1352.9) - (-129.67)^2} = -0.00077244,$$

$\gamma_1 = 0.00069587 - 0.00017729(-0.21071) + 0.00077244(0.61649) = 0.0012094,$

$\beta_{11} = 0.13639 + 0.00017729 = 0.13657; \qquad \beta_{21} = 0.069965 - 0.00077244 = 0.069193$

$\eta_1 = 1.3840 + 0.0012094 = 1.3852.$

i	j	$(x_{ij2} - \theta_{21})^2u_{ij1}$	w_{ij2}	$p_{21 \cdot ij1}$	m_{ij12}	m_{ij22}	u_{ij2}	$x_{ij1}u_{ij2}$	$x_{ij2}u_{ij2}$
1	1	97.612	1.5909	0.83075	3919.5	798.54	663.38	663.38	663.38
	2	516.24	1.4525	0.81039	1040.5	243.46	197.30	197.30	-197.30
2	1	161.81	1.3178	0.78881	5208.5	1394.5	1100.0	-1100.0	1100.0
	2	577.21	1.1794	0.76484	938.46	288.54	220.69	-220.69	-220.69
Total		1352.9			11,107.	2727.0	2181.3	-459.98	1345.4

i	j	$(x_{ij1} - \theta_{12})^2u_{ij2}$	$(x_{ij1} - \theta_{12})(x_{ij2} - \theta_{22})u_{ij2}$	$(x_{ij2} - \theta_{22})^2u_{ij2}$
1	1	972.65	307.84	97.432
	2	289.28	-386.25	515.72
2	1	684.98	-332.66	161.55
	2	137.43	281.56	576.86
Total		2084.3	-129.50	1351.6

$\theta_{12} = -459.781/2181.3 = -0.21087; \qquad \theta_{22} = 1345.4/2181.3 = 0.61676,$

$\hat{S}_{11}\hat{S}_{22} - (\hat{S}_{12})^2 = (2084.3)(1351.5) - (-129.50)^2 = 2,800,400,$

$\hat{S}^{11} = 1351.6/2,800,400 = 0.00048264; \qquad \hat{S}^{12} = \hat{S}^{21} = 129.50/2,800,400 = 0.000046245,$

$\hat{S}^{22} = 2084.3/2,800,400 = 0.00074431; \qquad s(\hat{\eta}_1^A) = 0.021969; \qquad s(\hat{\eta}_1^B) = 0.027282.$

Given maximum-likelihood estimates $\hat{m}_{ijk}$, one can obtain Pearson chi-square statistics and likelihood-ratio chi-square statistics just as in Section 3.1. The degrees of freedom can now also be obtained by noting that there are 4 logits ω_{ij} and 3 parameters η, β_1, and β_2, so that there is $4 - 3 = 1$ degree of freedom. It is also interesting to note that

$$X^2 = \sum_i \sum_j (n_{ij1} - \hat{m}_{ij1})^2 / \hat{u}_{ij},$$

where

$$\hat{u}_{ij} = n_{ij}^{AB} \hat{p}_{1 \cdot ij}(1 - \hat{p}_{1 \cdot ij}).$$

This formula can be used to simplify calculation of the Pearson chi-square statistic.

The estimated asymptotic covariance matrix of $\hat{\boldsymbol{\beta}}$, the vector with co-ordinates $\hat{\beta}_1$ and $\hat{\beta}_2$, is

$$\hat{S}^{-1} = \begin{bmatrix} \hat{S}^{11} & \hat{S}^{12} \\ \hat{S}^{21} & \hat{S}^{22} \end{bmatrix},$$

the inverse of

$$\hat{S} = \begin{bmatrix} \hat{S}_{11} & \hat{S}_{12} \\ \hat{S}_{21} & \hat{S}_{22} \end{bmatrix},$$

where

$$\hat{S}_{kl} = \sum_i \sum_j (x_{ijk} - \hat{\theta}_k)(x_{ijk} - \hat{\theta}_l)\hat{u}_{ij}$$

and

$$\hat{\theta}_k = \sum_i \sum_j x_{ijk}\hat{u}_{ij} \Big/ \sum_i \sum_j \hat{u}_{ij}.$$

Thus $\hat{\eta}_1^A = \hat{\beta}_1$ has $EASD$ $(\hat{S}^{11})^{1/2}$ and $\hat{\eta}_1^B = \hat{\beta}_2$ has $EASD$ $(\hat{S}^{22})^{1/2}$. The expressions derived in Table 5.1 are based on approximation of $\hat{S}_{kl}$ by S_{kl2}. To compare these results to those of Section 3.1, note that

$$\hat{\eta}_1^A = \tfrac{1}{2}(\hat{\omega}_{11} - \hat{\omega}_{22}) = \tfrac{1}{2}\hat{\tau}_{(12)(12)}^{AC}$$

and

$$\hat{\eta}_1^B = \tfrac{1}{2}(\hat{\omega}_{11} - \hat{\omega}_{12}) = \tfrac{1}{2}\hat{\tau}_{(12)(12)}^{BC}$$

under the additive model.

Logit analysis for this example has made relatively little difference. All conclusions reached here have all been reached in Section 3.1. The main change is that attention has been concentrated on a limited number of

η-parameters, namely η, η_i^A, $1 \leqslant i \leqslant 2$, and η_j^B, $1 \leqslant j \leqslant 2$, rather than on the many λ-parameters λ, λ_i^A, $1 \leqslant i \leqslant 2$, λ_j^B, $1 \leqslant j \leqslant 2$, λ_k^C, $1 \leqslant k \leqslant 2$, λ_{ij}^{AB}, $1 \leqslant i \leqslant 2$, $1 \leqslant j \leqslant 2$, λ_{ik}^{AC}, $1 \leqslant i \leqslant 2$, $1 \leqslant k \leqslant 2$, and λ_{jk}^{BC}, $1 \leqslant j \leqslant 2$, $1 \leqslant k \leqslant 2$. Normally this concentration on fewer parameters simplifies the Newton–Raphson algorithm and simplifies estimation of asymptotic variances. In the next section, a more important gain is achieved by logit analysis, for a logit model is used that does not correspond to any hierarchical log–linear model of Chapter 3.

5.2 Logits and Women's Roles

In this section, another example is presented of a dichotomous dependent variable and polytomous independent variables. Changes occur, however, since an attempt is made to exploit the fact that one independent variable is ordinal and has three categories. Hierarchical models cannot be used for this purpose. Consequently, the logit model which results does not correspond to a hierarchical log–linear model. Nonetheless, the Newton–Raphson algorithm for logits is easily applied, and chi-square statistics and estimated asymptotic standard deviations are readily found. The data analyzed are summarized in Table 5.2. This table has been formed by combining results of the 1974 and 1975 General Social Surveys. Given the observation in Section 4.1 that conditional on sex of respondent, education and response have not been shown to have any relationship to year of survey, such a procedure is justifiable, for the probability of a given response appears dependent on sex and education rather than on year of survey.

Since the value of sex and education in prediction of response is under study, logits are defined in terms of the conditional probabilities $p_{k \cdot ij} = p_{k \cdot ij}^{R \cdot SE}$ that response $R_h = k$ given that sex $S_h = i$ and education $E_h = j$. Let n_{ijk} denote the number of subjects h with $S_h = i$, $E_h = j$, and $R_h = k$. Given that n_{ij}^{SE} subjects have sex $S_h = i$ and $E_h = j$, the conditional expected value of n_{ijk} is $n_{ij}^{SE} p_{k \cdot ij}$. Thus the logit ω_{ij} of $p_{1 \cdot ij}$ satisfies the equation

$$\omega_{ij} = \log\left(\frac{p_{1 \cdot ij}}{1 - p_{1 \cdot ij}}\right) = \log\left(\frac{p_{1 \cdot ij}}{p_{2 \cdot ij}}\right) = \log\left(\frac{m_{ij1}}{m_{ij2}}\right).$$

As in Section 5.1, one may write

$$\omega_{ij} = \eta + \eta_i^S + \eta_j^E + \eta_{ij}^{SE},$$

where

$$\sum \eta_i^S = \sum \eta_j^E = \sum_i \eta_{ij}^{SE} = \sum_j \lambda_{ij}^{SE} = 0.$$

Table 5.2

Subjects in the 1974 and 1975 General Social Surveys, Cross-Classified by
Attitude Toward Women Staying at Home, Sex of Respondent,
and Education of Respondent[a]

		Response[b]				
		Agree		Disagree		
Sex	Education in years	No.	Percent	No.	Percent	Total
Male	$\leqslant 8$	161	64.1	90	35.9	251
	9–12	212	35.9	378	64.1	590
	$\geqslant 13$	92	19.8	372	80.2	464
	Total	465	35.6	840	64.4	1305
Female	$\leqslant 8$	169	71.6	67	28.4	236
	9–12	325	36.4	567	63.6	892
	$\geqslant 13$	61	13.9	377	86.1	438
	Total	555	35.4	1011	64.6	1566
Total	$\leqslant 8$	330	67.8	157	32.2	487
	9–12	537	36.2	945	63.8	1482
	$\geqslant 13$	153	17.0	749	83.0	902
	Total	1020	35.5	1851	64.5	2871

[a] From Table 3.4 and Table 3.20.
[b] For details on subject selection and the question asked, see 3.4.

To exploit the fact that education is an ordinal variable, one may assign scores t_1, t_2, and t_3 to the three educational groups. Many scoring methods can be adopted. The simplest assumes equal spacing of categories, so that $t_1 = -1$, $t_2 = 0$, and $t_3 = 1$. The only real requirement for the methods of this section is that

$$t_1 + t_2 + t_3 = 0.$$

A simple logit model based on these scores assumes that

$$\eta_j^E = \gamma^E t_j$$

for some unknown γ^E and

$$\eta_{ij}^{SE} = \gamma^{SE} t_j, \qquad i = 1,$$
$$= -\gamma^{SE} t_j, \qquad i = 2,$$

for some unknown γ^{SE}. Equivalently, for each i, ω_{ij} is a linear function

$$\omega_{ij} = \alpha_i + \beta_i t_j$$

of the education score t_j, where

$$\alpha_i = \eta + \eta_i^S$$

and

$$\beta_i = \gamma^E + \gamma^{SE}, \qquad i = 1,$$
$$= \gamma^E - \gamma^{SE}, \qquad i = 2.$$

Models with this or similar structures are very common in the literature on biological assay. For example, see Finney (1964, 1971).

The proposed logit model does not correspond to a hierarchical log-linear model. One may write

$$\log m_{ijk} = \lambda + \lambda_i^B + \lambda_j^E + \lambda_k^R + \lambda_{ij}^{SE} + \lambda_{ik}^{SR} + \lambda_{jk}^{ER} + \lambda_{ijk}^{SER},$$

where the λ-parameters are uniquely defined by the constraints

$$\sum \lambda_i^S = \sum \lambda_j^E = \sum \lambda_k^R = \sum_i \lambda_{ij}^{SE} = \sum_j \lambda_{ij}^{SE} = \sum_i \lambda_{ik}^{SR} = \sum_k \lambda_{ik}^{SR}$$
$$= \sum_j \lambda_{jk}^{ER} = \sum_k \lambda_{jk}^{ER} = \sum_i \lambda_{ijk}^{SER} = \sum_j \lambda_{ijk}^{SER} = \sum_k \lambda_{ijk}^{SER} = 0.$$

As in Section 5.1, the η-parameters and λ-parameters are related. One has

$$\eta = 2\lambda_1^R,$$
$$\eta_i^S = 2\lambda_{i1}^{SR},$$
$$\eta_j^E = 2\lambda_{j1}^{ER},$$

and

$$\eta_{ij}^{SE} = 2\lambda_{ij1}^{SER}.$$

As in evident from Haberman (1974a, pages 311–313), the general rule relating η-parameters and λ-parameters is that for any superscript T and subscript t, $\eta_t^T = 2\lambda_{t1}^{TR}$. Under the proposed model, $\eta_j^E = \gamma^E t_j$. Therefore,

$$\lambda_{jk}^{ER} = \tfrac{1}{2}\gamma^E t_j, \qquad k = 1,$$
$$= -\tfrac{1}{2}\gamma^E t_j, \qquad k = 2.$$

Similarly,

$$\lambda_{ijk}^{SER} = \tfrac{1}{2}\gamma^{SE} t_j, \qquad i = k,$$
$$= -\tfrac{1}{2}\gamma^{SE} t_j, \qquad i \neq k.$$

Thus linear constraints are imposed on the λ-parameters by the proposed logit model; however, no set of λ-parameters is set to 0. Thus the logit model is not equivalent to a hierarchical model.

The Newton–Raphson Algorithm

As usual, the Newton–Raphson algorithm can be described as a series of weighted regression problems. Analysis can be based on the parametrization

$$\omega_{ij} = \eta + \eta_1{}^S x_{ij1} + \gamma^E x_{ij2} + \gamma^{SE} x_{ij3},$$

where

$$x_{ij1} = 1, \qquad i = 1,$$
$$= -1, \qquad i = 2,$$
$$x_{ij2} = t_j,$$

and

$$x_{ij3} = t_j, \qquad i = 1,$$
$$= -t_j, \qquad i = 2;$$

however, calculations are much similar if the parametrization

$$\omega_{ij} = \alpha_i + \beta_i t_j$$

is employed. This parametrization is used in this section. The first parametrization results in an algorithm in which solutions of three simultaneous linear equations is needed. The second parametrization involves the same kinds of computations needed to perform two simple weighted regression analyses, each of which only involves a single independent variable. Either version of the Newton–Raphson algorithm for logits is simpler than a Newton–Raphson algorithm based on a parametrization in terms of $\log m_{ijk}$.

To begin computations, let $m_{ijk0} > 0$ be an initial approximation for $\hat{m}_{ijk}$ such as $n_{ijk} + \frac{1}{2}$. Define an empirical logit

$$y_{ij0} = \log\left(\frac{m_{ij10}}{m_{ij20}}\right)$$

and an empirical weight

$$u_{ij0} = \frac{m_{ij10} m_{ij20}}{m_{ij10} + m_{ij20}}.$$

Estimate the α_i and β_i as in the weighted regression problem

$$y_{ij0} = \alpha_i + \beta_i t_j + \varepsilon_{ij},$$

where the ε_{ij} are independent random variables with known means 0 and known variances u_{ij0}^{-1}.

Thus one obtains approximations α_{i0} and β_{i0} for the maximum-likelihood estimates $\hat{\alpha}_i$ and $\hat{\beta}_i$, where

$$\beta_{i0} = \sum_j (t_j - \theta_{i0}) y_{ij0} u_{ij0} \bigg/ \sum_j (t_j - \theta_{i0})^2 u_{ij0},$$

$$\theta_{i0} = \sum_j t_{j0} u_{ij0} \bigg/ \sum_j u_{ij0},$$

$$\bar{y}_{i0} = \sum_j y_{ij0} u_{ij0} \bigg/ \sum_j u_{ij0}$$

$$\alpha_{i0} = \bar{y}_{i0} - \beta_{i0}\theta_{i0}.$$

These approximations yield an approximation

$$\omega_{ij0} = \alpha_{i0} + \beta_{i0} t_j.$$

for $\hat{\omega}_{ij}$ and approximations

$$p_{1 \cdot ij1} = [1 + \exp(-\omega_{ij0})]^{-1}$$

for $\hat{p}_{1 \cdot ij1}$ and

$$m_{ijk1} = n_{ij}^{SE} p_{1 \cdot ij1}, \qquad k = 1,$$
$$= n_{ij}^{SE}(1 - p_{1 \cdot ij1}), \qquad k = 2,$$

for $\hat{m}_{ijk}$. For a summary of calculations, see Table 5.3.

Table 5.3

Computation of Maximum-Likelihood Estimates for the Logit Model for Table 5.2

i	j	n_{ij1}	n_{ij2}	n_{ij}^{SE}	t_j	m_{ij10}	m_{ij20}	y_{ij0}	u_{ij0}	$t_j u_{ij0}$	$y_{ij} u_{ij0}$
1	1	161	90	251	-1	161.5	90.5	0.5792	58.09	-57.999	33.590
	2	212	378	590	0	212.5	378.5	-0.5773	136.09	0	-78.561
	3	92	372	464	1	92.5	372.5	-1.3930	74.10	74.099	-103.22
Total		465	840	1305	0	466.5	841.5		268.19	16.100	-148.20
2	1	169	67	236	-1	169.5	67.5	0.9207	48.28	-48.275	44.448
	2	325	567	892	0	325.5	567.5	-0.5559	206.85	0	-114.99
	3	61	377	438	1	61.5	337.5	-1.8145	52.88	52.884	-95.960
Total		555	1011	1566	0	556.5	1012.5		308.01	4.6091	-166.50

$\theta_{10} = 16.100/268.19 = 0.060033; \qquad \bar{y}_{10} = -148.20/268.19 = -0.55257,$

$\theta_{20} = 4.6091/308.01 = 0.014964; \qquad \bar{y}_{11} = -166.50/308.01 = -0.54055.$

Table 5.3 (continued)

i	j	$(t_j - \theta_{io})y_{ij0}u_{ij0}$	$(t_j - \theta_{io})^2u_{ij0}$	ω_{ij0}	$p_{1 \cdot ij1}$	m_{ij11}	m_{ij21}
1	1	-35.607	65.172	0.48147	0.61809	155.14	95.858
	2	4.716	0.4164	-0.49401	0.37895	223.58	366.42
	3	-97.026	65.470	-1.4695	0.18702	86.777	377.22
Total		-127.92	131.13			465.50	839.50
2	1	-45.113	49.751	0.84415	0.69934	165.04	70.956
	2	1.721	0.046	-0.52014	0.37282	332.56	559.44
	3	-94.524	51.313	-1.8844	0.13188	57.764	380.24
Total		-139.92	101.09			555.36	1010.6

$\beta_{10} = -127.92/131.13 = -0.97548;$ $\alpha_{10} = -0.55257 - (-0.97548)(0.060033) = -0.49401,$

$\beta_{20} = -99.157/101.69 = -1.3643;$ $\alpha_{20} = -0.53609 - (-1.3643)(0.014964) = -0.52014.$

i	j	$n_{ij1} - m_{ij11}$	u_{ij1}	t_ju_{ij1}	$(t_j - \theta_{i1})(n_{ij1} - m_{ij11})$	$(t_j - \theta_{i1})^2u_{ij1}$
1	1	5.8583	59.25	-59.249	-6.1047	64.338
	2	-11.580	138.85	0	0.4870	0.246
	3	5.2227	70.55	70.548	5.0030	64.739
Total		-0.49907	268.65	11.299	-0.61466	129.32
2	1	3.9559	49.62	-49.622	-3.9626	49.791
	2	-7.5552	208.57	0	0.0128	0.001
	3	3.2362	50.15	50.146	3.2307	49.976
Total		-0.36308	308.34	0.52359	-0.71906	99.767

$\theta_{11} = 11.299/268.65 = 0.042057;$ $\theta_{21} = 0.52359/308.34 = 0.0016981,$

$\delta_{11} = -0.61466/129.32 = -0.0047529;$ $\delta_{21} = -0.71906/99.767 = -0.0072074,$

$\beta_{11} = -0.97548 - 0.0047529 = -0.98023;$ $\beta_{21} = -1.3643 - 0.0072074 = -1.3715,$

$\gamma_{11} = -0.49907/268.65 - (-0.0047529)0.042057 = -0.0016578,$

$\gamma_{21} = -0.36308/308.34 - (-0.0072074)0.0016981 = -0.0011653,$

$\alpha_{11} = -0.49401 - 0.0016578 = -0.49667;$ $\alpha_{21} = -0.52014 - 0.0011653 = -0.52130.$

i	j	ω_{ij1}	$p_{1 \cdot ij2}$	m_{ij12}	m_{ij22}	u_{ij2}	t_ju_{ij2}	$(t_j - \theta_{i2})^2u_{ij2}$
1	1	0.48456	0.61882	155.32	95.675	59.21	-59.206	64.188
	2	-0.49567	0.37856	223.35	366.65	138.80	0	0.236
	3	-1.4759	0.18605	86.326	377.67	70.27	70.265	64.591
Total				465.00	840.00	268.27	11.059	129.02
2	1	0.85020	0.70061	165.34	70.656	49.50	-49.502	49.610
	2	-0.52130	0.37255	332.31	559.69	208.51	0	0.000
	3	-1.8928	0.13093	57.345	380.65	49.84	49.837	49.729
Total				555.00	1011.0	307.85	0.33479	99.399

$\theta_{12} = 11.059/268.27 = 0.041224;$ $\theta_{22} = 0.33479/307.85 = 0.0010875,$

$s(\hat{\beta}_1) = (1/129.02)^{1/2} = 0.088;$ $s(\hat{\beta}_2) = (1/99.339)^{1/2} = 0.100,$

$s(\hat{\alpha}_1) = \left[\dfrac{1}{268.27} + (0.041224)^2/129.02\right]^{1/2} = 0.061;$

$s(\hat{\alpha}_2) = \left[\dfrac{1}{307.85} + (0.0010875)^2/99.339\right]^{1/2} = 0.057.$

Although not really needed here, given the large values of n_{ijk}, a further iteration is supplied for illustrative purposes. In this iteration, α_{i1} and β_{i1} correspond to the weighted-regression problem

$$y_{ij1} = \alpha_i + \beta_i t_j + \varepsilon_{ij},$$

where the ε_{ij} are independent random variables with respective means 0 and variances u_{ij1}^{-1},

$$y_{ij1} = \omega_{ij0} + (n_{ij1} - m_{ij11})/m_{ij11},$$

and

$$u_{ij1} = n_{ij}^{SE} p_{ij1}(1 - p_{ij1}).$$

One has

$$\beta_{i1} = \beta_{i0} + \delta_{i1},$$

where

$$\delta_{i1} = \sum_{j} (t_j - \theta_{i1})(n_{ij1} - m_{ij11}) \bigg/ \sum_{j} (t_j - \theta_{i1})^2 u_{ij1},$$

and

$$\theta_{i1} = \sum_{j} t_j u_{ij1} \bigg/ \sum_{j} u_{ij1}.$$

Similarly,

$$\alpha_{i1} = \alpha_{i0} + \gamma_{i1},$$

where

$$\gamma_{i1} = \sum_{j} (n_{ij1} - m_{ij11}) \bigg/ \sum_{j} u_{ij1} - \delta_{i1}\theta_{i1}.$$

Thus

$$
\begin{aligned}
\omega_{ij1} &= \alpha_{i1} + \beta_{i1} t_j \\
p_{1 \cdot ij2} &= [1 + \exp(-\omega_{ij1})]^{-1}, \\
m_{ijk2} &= n_{ij}^{SE} p_{1 \cdot ij2}, & k = 1, \\
&= n_{ij}^{SE}(1 - p_{1 \cdot ij2}), & k = 2.
\end{aligned}
$$

The chi-square statistics are $X^2 = 2.72$ and $L^2 = 2.73$. There are $6 - 4 = 2$ degrees of freedom, so the model fits the data quite satisfactorily.

The estimated asymptotic standard deviations for $\hat{\alpha}_i$ and $\hat{\beta}_i$ are

$$s(\hat{\beta}_i) = \left[1 \Big/ \sum_j (t_j - \hat{\theta}_i)^2 \hat{u}_{ij} \right]^{1/2}$$

and

$$s(\hat{\alpha}_i) = \left\{ \left(1 \Big/ \sum_j \hat{u}_{ij} \right) + \left[\hat{\theta}_i^{\,2} \Big/ \sum_j (t_j - \hat{\theta}_i)^2 \right] \right\}^{1/2},$$

where

$$\hat{u}_{ij} = \hat{m}_{ij1}\hat{m}_{ij2}/n_{ij}^{SE}$$

and

$$\hat{\theta}_i = \sum_j t_j \hat{u}_{ij} \Big/ \sum_j \hat{u}_{ij}.$$

These formulas correspond to the variance estimates obtained from the corresponding regression problem

$$Y_{ij} = \alpha_i + \beta_i x_{ij} + \varepsilon_{ij},$$

where the Y_{ij} are hypothetical dependent variables and the ε_{ij} are independent and have respective means 0 and variance $\hat{u}_{ij}^{-1}$.

The slopes β_1 and β_2 corresponding to males and females, respectively, appear to differ. Since the estimates $\hat{\beta}_1$ and $\hat{\beta}_2$ are obtained from distinct 3×2 subtables n_{1jk}, $1 \leqslant j \leqslant 3$, $1 \leqslant k \leqslant 2$, and n_{2jk}, $1 \leqslant j \leqslant 3$, $1 \leqslant k \leqslant 2$, respectively, $\hat{\beta}_1$ and $\hat{\beta}_2$ are independent. Thus

$$\hat{\beta} - \hat{\beta}_2 = 0.391$$

has an *EASD* of

$$[s^2(\hat{\beta}_1) + s^2(\hat{\beta}_2)]^{1/2} = \left[\frac{1}{\displaystyle\sum_j (t_{1j} - \hat{\theta}_1)^2 \hat{u}_{1j}} + \frac{1}{\displaystyle\sum_j (t_{2j} - \hat{\theta}_2)^2 \hat{u}_{2j}} \right]^{1/2}$$

$$= \left[\frac{1}{129.02} + \frac{1}{99.339} \right]^{1/2} = 0.133.$$

The corresponding standardized value of $0.391/0.133 = 2.93$ is rather large. The probability that a standard normal deviate exceeds 2.93 in absolute value is only 0.003. Since the logit model is consistent with the data, this difference in slope provides strong evidence that the interactions η_{ij}^{SE} between sex and education are not all zero. In the log–linear scale, this interaction corresponds to an interaction λ_{ijk}^{SER}, for

$$\lambda_{ij1}^{SER} = \tfrac{1}{2}\eta_{ij}^{SE}.$$

Thus the logit model provides strong evidence that the three-factor interactions λ_{ijk}^{SER} are not all zero. The more definitive result obtained with a logit model than with hierarchical log–linear models illustrates the gains in precision of inferences that are achievable by exploiting ordering of categories.

The slope estimate $\hat{\beta}_i$ measures the assumed common difference between the logits ω_{ij} and $\omega_{i(j+1)}$ for $j = 1$ or 2. For example, it is estimated that the odds that a male respondent agrees with the statement are

$$\exp(-0.980) = 0.375$$

times as large if he has at least 13 years education as they are if he has 9–12 years education. The estimate is determined with only modest accuracy. An approximate 95 percent confidence interval for $\exp(-\beta_1)$ has lower bound

$$\exp(-0.980 - 1.96 \times 0.088) = 0.316$$

and upper bound

$$\exp(-0.980 + 1.96 \times 0.088) = 0.446.$$

Nonetheless, it is clear that a strong relationship exists among males between education and response.

Since $\hat{\beta}_2$ has been shown to be significantly smaller than $\hat{\beta}_1$, the relationship between education and response is stronger for females than for males. Note that

$$\exp(-1.37) = 0.254$$

is somewhat smaller than the corresponding figure of 0.375 for males. The corresponding approximate 95 percent confidence bounds for the ratio $\exp(-\beta_2)$ between the odds of a female with at least 13 years of education agreeing and the odds of a female with nine to 12 years of education aggreeing are

$$\exp(-1.37 - 1.96 \times 0.100) = 0.208.$$

and

$$\exp(-1.37 + 1.96 \times 0.100) = 0.309.$$

The coefficient $\hat{\alpha}_1$ is an estimate of the log odds ω_{12} of agreement for males with 9–12 years of education, while $\hat{\alpha}_2$ is an estimate of ω_{22}. These two coefficients differ by a negligible amount. Thus there is no evidence that male respondents with 9–12 years education differ from the corresponding female group in the probability of agreement.

This section has illustrated the use of logit models to exploit ordered classifications in a manner not possible with hierarchical log–linear models. The logit model considered in this section is also important since it can easily be generalized for use with continuous independent variables. An example of such a generalization is considered in the next section.

5.3 Logits, Women's Roles, and Small Frequency Counts

So far in this book, tables which have been studied generally have had large frequency counts in each cell. For example, in Table 3.1 each count n_{ijk} is at least 234, while in Table 3.4 each n_{ijk} is at least 28. When such large frequency counts are available, large-sample approximations considered in analysis are very satisfactory. These approximations become much more uncertain when the total number N of observations is large but most or all of the individual cell frequencies are small. In the case of logit models, it is often the case that large-sample approximations for the distribution of parameter estimates are quite satisfactory; however, some approximations for the distributions of chi-square statistics and adjusted residuals become unacceptable.

To illustrate problems encountered when many cell counts are small, consider Table 5.4. This table is similar to Table 3.4, except that education is now recorded in years completed. Thus a more thorough investigation is possible of the relationship of education and sex to attitudes toward women's roles. However, this finer division of educational level leads to a table in which seven of 84 cell counts are 0, four are 1, six are 2, four are 3, and four are 4. In all, 25 counts are less than 5. Thus questions arise concerning the validity of large-sample approximations when so many cell counts are small.

Despite these small cell counts, an analysis similar to that in Section 5.2 is possible. For $1 \leqslant i \leqslant 2$, $0 \leqslant j \leqslant 20$, $1 \leqslant k \leqslant 2$, let n_{ijk} be the number of subjects h, $1 \leqslant h \leqslant N$, with sex $S_h = i$, years of education $C_h = j$, and response $R_h = k$. Let $p_{k \cdot ij} = p_{k \cdot ij}^{R \cdot SC} > 0$ be the probability that response $R_h = k$, given that the subject has sex $S_h = i$ and education $C_h = j$, and let

$$\omega_{ij} = \log\left(\frac{p_{1 \cdot ij}}{1 - p_{1 \cdot ij}}\right)$$

be the logit of $p_{1 \cdot ij}$. As in Section 5.2, a logit model

$$\omega_{ij} = \alpha_i + \beta_i t_j$$

Table 5.4

Subjects in the 1974 and 1975 General Social Surveys, Cross-Classified
by Attitude Toward Women Staying at Home, Sex of Respondent, and Years
of Education of Respondent[a,b]

Years of education completed	Responses of males		Responses of females	
	Agree	Disagree	Agree	Disagree
0	4	2	4	2
1	2	0	1	0
2	4	0	0	0
3	6	3	6	1
4	5	5	10	0
5	13	7	14	7
6	25	9	17	5
7	27	15	26	16
8	75	49	91	36
9	29	29	30	35
10	32	45	55	67
11	36	59	50	62
12	115	245	190	403
13	31	70	17	92
14	28	79	18	81
15	9	23	7	34
16	15	110	13	115
17	3	29	3	28
18	1	28	0	21
19	2	13	1	2
20	3	20	2	4
Total	465	840	555	1011

[a] Data tapes from National Opinion Research Center, 1974 and 1975 General
Social Surveys, University of Chicago.
[b] The responses and the excluded subjects are those indicated in Table 3.4,
Table 3.20, and Table 5.2. Subjects are asked whether they agree that women
should run their homes and leave men run the country.

may be considered. Here α_i and β_i, $1 \leqslant i \leqslant 2$, are unknown parameters,
while t_j is a score for education. Since C_h has range from 0 to 20, it is convenient to set $t_j = j - 10$. Then t_j has ranges -10 to 10 and $\sum t_j = 0$. Other
choices of scores t_j are possible. The logit model of the Section 5.2 holds if

$$t_j = -1, \quad 0 \leqslant j \leqslant 8,$$
$$= 0, \quad 9 \leqslant j \leqslant 12,$$
$$= 1, \quad 13 \leqslant j \leqslant 20.$$

As in Section 5.2, an alternate parametrization can be used if $\sum t_j = 0$. For some unknown $\eta, \eta_i^S, 1 \leqslant i \leqslant 2, \eta_j^C, 0 \leqslant j \leqslant 20, \eta_{ij}^{SC}, 1 \leqslant i \leqslant 2, 0 \leqslant j \leqslant 20$, γ^C, and γ^{SC}, the logits $\omega_{ij}, 1 \leqslant i \leqslant 2, 0 \leqslant j \leqslant 20$, satisfy the equations

$$\omega_{ij} = \eta + \eta_i^S + \eta_j^C + \eta_{ij}^{SC},$$

$$\sum_i \eta_i^S = 0,$$

$$\eta_j^C = \gamma^C t_j,$$

and

$$\eta_{ij}^{SC} = -\eta_{2j}^{SC} = \gamma^{SC} t_j.$$

Observe that these equations imply that

$$\sum_j \eta_j^C = \sum_i \eta_{ij}^{SC} = \sum_j \eta_{ij}^{SC} = 0.$$

As in Section 5.2,

$$\eta = \tfrac{1}{2}(\alpha_1 + \alpha_2),$$
$$\eta_i^S = \alpha_i - \eta,$$
$$\gamma^C = \tfrac{1}{2}(\beta_1 + \beta_2),$$

and

$$\gamma^{SC} = \tfrac{1}{2}(\beta_1 - \beta_2).$$

On the other hand, the logit model of this section does not lead to a log–linear model of exactly the same type as considered in Section 5.2. Given that n_{ij}^{SC} subjects have sex i and years j of education, the mean of n_{ijk} is $m_{ijk} = n_{ij}^{SC} p_{k \cdot ij}$. Since $n_{22}^{SC} = 0, m_{221} = m_{222} = 0$, so that $\log m_{ijk}$ cannot be found for all i, j, and k. Instead of an ordinary log–linear model for a $2 \times 21 \times 2$ table, one obtains a log–linear model for an incomplete $2 \times 21 \times 2$ table. Such models will be discussed in Chapter 7 of Volume 2. For now, it suffices to note that for i or j not equal to 2,

$$\log m_{ijk} = \lambda + \lambda_i^S + \lambda_j^C + \lambda_k^R + \lambda_{ij}^{SC} + \lambda_{ik}^{SR} + \lambda_{jk}^{CR} + \lambda_{ijk}^{SCR},$$

where

$$\sum \lambda_i^S = \sum \lambda_j^C = \sum \lambda_k^R = \sum_i \lambda_{ij}^{SC} = \sum_j \lambda_{ij}^{SC} = \sum_i \lambda_{ik}^{SR} = \sum_k \lambda_{ik}^{SR} = 0,$$

$$\lambda_{j1}^{CR} = -\lambda_{j2}^{CR} = \frac{1}{2}\gamma^C t_j,$$

and

$$\lambda_{1j1}^{SCR} = -\lambda_{2j1}^{SCR} = -\lambda_{1j2}^{SCR} = \lambda_{2j2}^{SCR} = \tfrac{1}{2}\gamma^{SC} t_j.$$

Note that these equations imply that

$$\sum_j \lambda_{jk}^{CR} = \sum_k \lambda_{jk}^{CR} = \sum_i \lambda_{ijk}^{SCR} = \sum_j \lambda_{ijk}^{SCR} = \sum_k \lambda_{ijk}^{SCR} = 0.$$

The Newton–Raphson Algorithm

Computation of maximum-likelihood estimates can proceed in almost the same manner as in Section 5.2. The only significant change involves the initial estimates m_{ijk0} of the $\hat{m}_{ijk}$. Since many counts n_{ij}^{SC} are small, it appears advisable to let

$$m_{ijk0} = n_{ij}^{SC} p_{k \cdot 2j0} = n_{ij}^{SC}(n_{ijk} + \tfrac{1}{2})/(n_{ij}^{SC} + 1),$$

where $p_{k \cdot ij0} = (n_{ijk} + \tfrac{1}{2})/(n_{ij}^{SC} + 1)$ is an initial estimate of $\hat{p}_{k \cdot ij}$. This choice leads to an empirical logit

$$y_{ij0} = \log\left(\frac{p_{1 \cdot ij0}}{1 - p_{2 \cdot ij0}}\right) = \log\left(\frac{n_{ij1} + \tfrac{1}{2}}{n_{ij2} + \tfrac{1}{2}}\right),$$

just as in Section 5.2. The empirical weight u_{ij0} of

$$u_{ij0} = \frac{m_{ij10} m_{ij20}}{m_{ij10} + m_{ij20}}$$

is then equal to

$$n_{ij}^{SC} \frac{(n_{ij1} + \tfrac{1}{2})(n_{ij2} + \tfrac{1}{2})}{(n_{ij}^{SC} + 1)^2} = n_{ij}^{SC} p_{1 \cdot ij0}(1 - p_{1 \cdot ij0}).$$

The corresponding value in Section 5.2 was

$$(n_{ij1} + \tfrac{1}{2})(n_{ij2} + \tfrac{1}{2})/(n_{ij}^{SC} + 1).$$

The weighting system used here has the advantage that when n_{ij}^{SC} is 0, $u_{ij0} = 0$, so that unobserved combinations of predicting variables receive no weight.

Given the y_{ij0} and u_{ij0}, approximations α_{i0} for the $\hat{\alpha}_i$ and β_{i0} for the $\hat{\beta}_2$ are found as in the weighted-regression problem

$$y_{ij0} = \alpha_i + \beta_i t_j + \varepsilon_{ij} \qquad \text{if} \quad u_{ij0} > 0,$$

where for i and j such that $u_{ij0} > 0$, each ε_{ij} is an independent random variable with mean 0 and variance u_{ij0}^{-1}. Thus

$$\beta_{i0} = \sum_j (t_j - \theta_{i0}) y_{ij0} u_{ij0} \bigg/ \sum_j (t_j - \theta_{i0})^2 u_{ij0}$$

and

$$\alpha_{i0} = \bar{y}_{i0} - \beta_{i0}\bar{t}_{i0},$$

where

$$\theta_{i0} = \sum_j t_j u_{ij0} \Big/ \sum_j u_{ij0}$$

and

$$\bar{y}_{i0} = \sum_j y_{ij0} u_{ij0} \Big/ \sum_j u_{ij0}.$$

The corresponding approximation for $\hat{\omega}_{ij}$ is

$$\omega_{ij0} = \alpha_{i0} + \beta_{i0} t_j.$$

The new approximation $p_{1 \cdot ij1}$ for $\hat{p}_{1 \cdot ij}$ is $p_{1 \cdot ij1} = (1 + \exp -\omega_{ij0})^{-1}$. Subsequent iterations procede in similar fashion. For example, in the next iteration, the working logit is

$$y_{ij1} = \omega_{ij0} + (n_{ij1} - n_{ij}^{SC} p_{1 \cdot ij1}) / u_{ij1},$$

where

$$u_{ij1} = n_{ij}^{SC} p_{1 \cdot ij1} (1 - p_{1 \cdot ij1}).$$

The corresponding weighted-regression model used to compute new approximations α_{i1} and β_{i1} is

$$y_{ij1} = \alpha_i + \beta_i t_j + \varepsilon_{ij} \qquad \text{if} \quad u_{ij1} > 0,$$

where for i and j such that $u_{ij1} > 0$, each ε_{ij} is an independent random variable with mean 0 and variance u_{ij1}^{-1}. The α_{i1} and β_{i1} may be found from α_{i0} and β_{i0} by the equations

$$\alpha_{i1} = \alpha_{i0} + \gamma_{i1}$$

and

$$\beta_{i1} = \beta_{i0} + \delta_{i1},$$

where

$$\delta_{i1} = \sum_j (t_j - \theta_{i1})(n_{ij1} - n_{ij}^{SC} p_{1 \cdot ij1}) \Big/ \sum_j (t_j - \theta_{i1})^2 u_{ij1},$$

$$\gamma_{i1} = \sum_j (n_{ij1} - n_{ij}^{SC} p_{1 \cdot ij1}) \Big/ \sum_j u_{ij1} - \delta_{i1} \bar{t}_{i1},$$

and

$$\theta_{i1} = \sum_j t_j u_{ij1} \Big/ \sum_j u_{ij1}.$$

New approximations ω_{ij1} and $p_{1 \cdot ij2}$ are found from the equations

$$\omega_{ij1} = \alpha_{i1} + \beta_{i1} t_j$$

and

$$p_{1 \cdot ij2} = (1 + \exp - \omega_{ij1})^{-1}.$$

Computations are summarized in Table 5.5.

Convergence here is rapid, but not as fast as in Table 5.3. Note that differences between $p_{1 \cdot ij1}$ and $p_{1 \cdot ij2}$ are large enough so that a third iteration is used to obtain $p_{1 \cdot ij3}$. Changes from $p_{1 \cdot ij2}$ to $p_{1 \cdot ij3}$ are small enough so that further iteration is unnecessary.

Ungrouped Data and the Newton–Raphson Algorithm

Computations in Table 5.5 are based on the limited number of values assumed by the education variable C_h. The Newton–Raphson algorithm can still be applied if C_h assumes hundreds of values; however, an alternate approach then should be used which does not require that data be summarized in a contingency table. To apply this approach to the data in Table 5.4, it is helpful to make some small changes in notation. Let subjects of sex i be numbered from 1 to n_i^S, the number of subjects of sex i. Let R_{gi} denote the response of subject g of sex i, let C'_{gi} denote the years of education completed by that subject, let $T_{gi} = C'_{gi} - 10$, and let $P_{k \cdot gi}$ denote the probability that $R'_{gi} = k$. The log–linear model under study assumes that the logit

$$\Omega_{gi} = \log\left(\frac{P_{1 \cdot gi}}{1 - P_{1 \cdot gi}}\right) = \log\left(\frac{P_{1 \cdot gi}}{P_{2 \cdot gi}}\right)$$

satisfies the equation

$$\Omega_{gi} = \alpha_i + \beta_i T_{gi}.$$

Note that if $C'_{gi} = j$, then $\Omega_{gi} = \omega_{ij} = \alpha_i + \beta_i t_j$.

As usual, the Newton–Raphson algorithm reduces to a series of weighted regressions. Initial approximations $P_{1 \cdot gi0}$ for the estimate $\hat{P}_{1 \cdot gi}$ are required such that $0 < P_{1 \cdot gi0} < 1$. In Table 5.4, the fact that many T_{gi} are equal can be used to estimate $\hat{P}_{1 \cdot gi}$; however, cruder methods are generally used. Two possibilities are the approximations

$$P_{1 \cdot gi0} = e^3/(e^3 + 1), \qquad R'_{gi} = 1,$$
$$= 1/(e^3 + 1), \qquad R'_{gi} = 2,$$

Table 5.5

Computation of Maximum-likelihood Estimates for the Logit Model for Table 5.4—Grouping Used

j	y_{1j0}	y_{2j0}	u_{ij0}	u_{2j0}	ω_{1j0}	ω_{2j0}	$p_{1.1j1}$	$p_{1.2j1}$
0	0.58779	0.58779	1.3776	1.3776	1.8672	2.5487	0.86613	0.92749
1	1.8094	1.0986	0.27777	0.18750	1.6546	2.2744	0.83951	0.90673
2	2.1972	0.00000	0.36000	0.00000	1.4420	2.0000	0.80877	0.88080
3	0.61904	1.4663	2.0475	1.0664	1.2294	1.7257	0.77372	0.84886
4	0.00000	3.0445	2.5000	0.43388	1.0168	1.4514	0.73435	0.81021
5	0.58779	0.65925	4.5918	4.7185	0.80423	1.1770	0.69088	0.76441
6	0.98739	1.1574	6.7237	4.0028	0.59163	0.90268	0.64374	0.71150
7	0.57335	0.47378	9.6823	9.9321	0.37904	0.62834	0.59364	0.65211
8	0.42216	0.91903	29.659	25.888	0.16645	0.35400	0.54152	0.58759
9	0.00000	−0.15181	14.500	16.157	−0.04615	0.07966	0.48847	0.51990
10	−0.33647	−0.19574	18.715	30.210	−0.25874	−0.19468	0.43567	0.45148
11	−0.48866	−0.21319	22.387	27.684	−0.47136	−0.46902	0.38430	0.38485
12	−0.75403	−0.75052	78.329	129.19	−0.68393	−0.74336	0.33538	0.32227
13	−0.80563	−1.6650	21.559	14.582	−0.89652	−1.0177	0.28977	0.26548
14	−1.0258	−1.4828	20.785	14.927	−1.1091	−1.2920	0.24804	0.21551
15	−0.90571	−1.5260	6.5601	6.0140	−1.3217	−1.5664	0.21053	0.17273
16	−1.9642	−2.1466	13.485	11.994	−1.5343	−1.8407	0.17736	0.13697
17	−2.1316	−2.0971	3.0340	3.0198	−1.7469	−2.1151	0.14844	0.10764
18	−2.9444	−3.7612	1.3775	0.46643	−1.9595	−2.3894	0.12352	0.08398
19	−1.6864	−0.51083	1.9775	0.70312	−2.1721	−2.6637	0.10229	0.06515
20	−1.7677	−0.58779	2.8650	1.3776	−2.3847	−2.9381	0.08435	0.05030

$\alpha_{10} = -0.25874;$ $\quad \alpha_{20} = -0.19468;$ $\quad \beta_{10} = -0.21259;$ $\quad \beta_{20} = 0.27434.$

j	u_{1j1}	u_{1j2}	ω_{1j1}	ω_{1j2}	$p_{1.2j1}$	$p_{1.2j2}$
0	0.69568	0.40352	2.0910	2.9791	0.89003	0.95162
1	0.26946	0.08457	1.8577	2.6659	0.86502	0.93498
2	0.61866	0.00000	1.6243	2.3526	0.83539	0.91314
3	1.57757	0.89807	1.3910	2.0394	0.80075	0.88487
4	1.9508	1.5377	1.1576	1.7262	0.76090	0.84892
5	4.2713	3.7818	0.92429	1.4129	0.71592	0.80423
6	7.7975	4.5159	0.69095	1.0997	0.66618	0.75021
7	10.132	9.5282	0.45761	0.78648	0.61245	0.68707
8	30.786	30.176	0.22426	0.47324	0.55583	0.61615
9	14.492	16.224	−0.09082	0.16001	0.49773	0.53992
10	18.931	30.213	−0.24243	−0.15323	0.43969	0.46177
11	22.478	26.515	−0.47577	−0.46646	0.38325	0.38545
12	80.245	129.52	−0.70911	−0.7797	0.32979	0.31439
13	20.786	21.255	−0.94246	−1.0929	0.28040	0.25107
14	19.957	16.737	−1.1758	−1.4062	0.23581	0.19684
15	5.3187	5.8587	−1.4091	−1.7194	0.19637	0.15195
16	18.238	15.130	−1.6425	−2.0326	0.16213	0.11582
17	4.0450	2.9777	−1.8758	−2.3459	0.13287	0.08739
18	3.1397	1.6155	−2.1092	−2.6591	0.10821	0.06543
19	1.3773	0.18271	−2.3425	−2.9723	0.08766	0.04869
20	1.7764	0.28663	−2.5759	−3.2856	0.07071	0.03607

$\alpha_{11} = -0.24243;$ $\quad \alpha_{21} = -0.15323;$ $\quad \beta_{11} = -0.23335;$ $\quad \beta_{21} = -0.31324.$

Table 5.5 (continued)

i	u_{1j2}	u_{2j2}	ω_{1j2}	ω_{2j2}	$p_{1.1j3}$	$p_{1.2j3}$
0	0.58728	0.27623	2.0982	3.0029	0.89073	0.95270
1	0.23351	0.06079	1.8642	2.6875	0.86578	0.93628
2	0.55005	0.00000	1.6301	2.3721	0.03619	0.91467
3	1.4360	0.71311	1.3961	2.0567	0.80157	0.88662
4	1.8193	1.2825	1.1621	1.7412	0.76171	0.85085
5	4.0676	3.3063	0.92806	1.4258	0.71668	0.80625
6	7.5611	4.1227	0.69403	1.1104	0.66686	0.75221
7	9.9690	9.0301	0.46000	0.79504	0.61302	0.68891
8	30.613	30.037	0.22598	0.47963	0.55626	0.61766
9	14.500	16.146	−0.00805	0.16423	0.49799	0.54096
10	18.970	30.322	−0.24208	−0.15118	0.43978	0.46228
11	22.455	26.530	−0.47610	−0.46659	0.38317	0.38542
12	79.571	127.82	−0.71013	−0.78199	0.32957	0.31389
13	20.380	20.496	−0.94416	−1.0974	0.28006	0.25023
14	19.282	15.651	−1.1782	−1.4128	0.23538	0.19579
15	5.0499	5.2833	−1.4122	−1.7282	0.19589	0.15082
16	16.980	13.108	−1.6462	−2.0436	0.16162	0.11470
17	3.6869	2.4725	−1.8803	−2.3590	0.13236	0.08635
18	2.7985	1.2841	−2.1143	−2.6744	0.10772	0.06450
19	1.1997	0.13896	−2.3483	−2.9898	0.08720	0.04789
20	1.5113	0.20861	−2.5823	−3.3052	0.07028	0.03539

$\alpha_{12} = -0.24208$; $\alpha_{22} = -0.15118$; $\beta_{12} = -0.23403$; $\beta_{22} = -0.31541$.

$s(\hat{\alpha}_1) = 0.0669$; $s(\hat{\alpha}_2) = 0.06436$; $s(\hat{\beta}_1) = 0.02019$; $s(\hat{\beta}_2) = 0.02365$.

of Cox (1970, page 90) and

$$P_{1\cdot gi0} = e^2/(e^2 + 1), \qquad R'_{gi} = 1,$$
$$= 1/(e^2 + 1), \qquad R'_{gi} = 2,$$

of Goodman (1975). In Table 5.6, the Cox approximation is used. Given the $P_{1\cdot gi0}$, a logit

$$Y_{gi0} = \log\left(\frac{P_{1\cdot gi0}}{1 - P_{1\cdot gi0}}\right)$$

and a weight

$$U_{gi0} = P_{1\cdot gi0}(1 - P_{1\cdot gi0})$$

are computed. Under the Cox approximation, Y_{gi0} is 3 or -3 and U_{gi0} is always $e^3/(e^3 + 1)^2$. Approximations α_{i0} and β_{i0} for $\hat{\alpha}_i$ and $\hat{\beta}_i$ are then found as in the weighted-regression problem

$$Y_{gi0} = \alpha_i + \beta_i T_{gi} + \varepsilon_{gi},$$

Table 5.6

Computation of Maximum-Likelihood Estimates for the Logit Model
for Table 5.4—Grouping Not Used

j	y_{1j0}	y_{2j0}	$u_{1j0}{}^a$	$u_{2j0}{}^a$	ω_{1j0}	ω_{2j0}	$p_{1.1j1}$	$p_{1.2j1}$
0	1.0000	1.0000	0.27106	0.27106	2.4938	3.4171	0.92370	0.96824
1	3.0000	3.0000	0.09035	0.04518	2.2099	3.0502	0.90113	0.95479
2^b	3.0000	—	0.18071	0.00000	1.9260	2.6832	0.87280	0.93603
3	1.0000	2.1429	0.40659	0.31624	1.6421	2.3162	0.83782	0.91021
4	0.00000	3.0000	0.45177	0.45177	1.3582	1.9492	0.79546	0.87536
5	0.9000	1.0000	0.90353	0.94871	1.0743	1.5823	0.74541	0.82953
6	1.4118	1.6364	1.5360	0.99389	0.79040	1.2153	0.68792	0.77124
7	0.85714	0.71429	1.8974	1.8974	0.50650	0.84834	0.62399	0.70022
8	0.62903	1.2992	5.6019	5.7374	0.22261	0.48137	0.55542	0.61807
9	0.00000	−0.23077	2.6202	2.9365	−0.06129	0.11440	0.48468	0.52857
10	−0.50649	−0.29508	3.4786	5.5116	−0.34518	−0.25257	0.41455	0.43719
11	−0.72632	−0.32143	4.2918	5.0598	−0.62908	−0.61954	0.34772	0.34989
12	−1.0833	−1.0776	16.264	26.790	−0.91297	−0.98651	0.28639	0.27160
13	−1.1584	−2.0642	4.5628	4.9243	−1.1969	−1.3535	0.23203	0.20530
14	−1.4299	−1.9091	4.8339	4.4725	−1.4808	−1.7205	0.18531	0.15181
15	−1.3125	−1.9756	1.4457	1.8522	−1.7647	−2.0874	0.14621	0.11033
16	−2.2800	−2.3906	5.6471	5.7826	−2.0486	−2.4544	0.11420	0.07912
17	−2.4375	−2.4194	1.4457	1.4005	−2.3324	−2.8214	0.08847	0.05618
18	−2.7931	−3.0000	1.3101	0.94871	−2.6163	−3.1883	0.06809	0.03961
19	−2.2000	−1.0000	0.67765	0.13553	−2.9002	−3.5553	0.05214	0.02778
20	−2.2174	−1.0000	1.0391	0.27106	−3.1841	−3.9223	0.03977	0.01941

$\alpha_{10} = -0.34518;$ $\alpha_{20} = -0.25257;$ $\beta_{10} = -0.28390;$ $\beta_{20} = -0.36697.$

[a] The first iteration is unchanged if $u_{ij0} = n_{ij}^{SC}$ or $\frac{1}{4}n_{ij}^{SC}$ instead of $n_{ij}^{SC}e^3/(e^3 + 1)^2$.

[b] Since $n_{22}^{SC} = 0$, y_{220} is not defined; however the weight u_{220} is 0. Thus computations are not affected.

j	u_{1j1}	u_{2j1}	ω_{1j1}	ω_{2j1}	$p_{1.1j2}$	$p_{1.2j2}$
0	0.42285	0.18453	2.0363	2.9413	0.88455	0.94985
1	0.17819	0.04317	1.8090	2.6324	0.85924	0.93292
2	0.44407	0.00000	1.5818	2.3235	0.82946	0.91081
3	1.2229	0.57209	1.3546	2.0147	0.79487	0.88233
4	1.6270	1.0910	1.1273	1.7058	0.75534	0.84629
5	3.7955	2.9697	0.90006	1.3969	0.71096	0.80170
6	7.2994	3.8815	0.67282	1.0881	0.66213	0.74802
7	9.8544	8.8163	0.44558	0.77920	0.60959	0.68551
8	30.619	29.980	0.21833	0.47033	0.55437	0.61546
9	14.486	16.197	−0.00891	0.16147	0.49777	0.54028
10	18.688	30.019	−0.23616	−0.14740	0.44123	0.46322
11	21.547	25.476	−0.46340	−0.45626	0.38618	0.38787
12	73.574	117.32	−0.69064	−0.76513	0.33389	0.31753
13	17.998	17.784	−0.91789	−1.0740	0.28539	0.25464
14	16.154	12.748	−1.1451	−1.3829	0.24138	0.20055
15	3.9946	4.0243	−1.3724	−1.6917	0.20224	0.15555
16	12.645	9.3258	−1.5996	−2.0006	0.16804	0.11914
17	2.5806	1.6437	−1.8269	−2.3095	0.13861	0.09034
18	1.8403	0.79880	−2.0541	−2.6183	0.11364	0.06797
19	0.74135	0.08102	−2.2814	−2.9272	0.09268	0.05083
20	0.87827	0.11421	−2.5086	−3.2360	0.07526	0.03783

$\alpha_{11} = -0.23616;$ $\alpha_{21} = -0.14740;$ $\beta_{11} = -0.22724;$ $\beta_{21} = -0.30887.$

Table 5.6 *(continued)*

j	u_{1j2}	u_{2j2}	ω_{1j2}	ω_{2j2}	$p_{1.1j3}$	$p_{1.2j3}$
0	0.61271	0.28582	2.0973	3.0022	0.89064	0.95267
1	0.24188	0.06258	1.8634	2.6869	0.86569	0.93625
2	0.56583	0.00000	1.6294	2.3715	0.83609	0.91463
3	1.4674	0.72678	1.3955	2.0562	0.80147	0.88657
4	1.8480	1.3008	1.1616	1.7409	0.76162	0.85080
5	4.1099	3.3386	0.92764	1.4255	0.71660	0.80620
6	7.6062	4.1467	0.69370	1.1102	0.66679	0.75216
7	9.9956	9.0547	0.45976	0.79484	0.61296	0.68887
8	30.633	30.057	0.22583	0.47950	0.55622	0.61763
9	14.500	16.145	−0.00811	0.16416	0.49797	0.54095
10	18.984	30.335	−0.24205	−0.15118	0.43978	0.46228
11	22.519	26.592	−0.47598	−0.46652	0.38320	0.38544
12	80.067	128.51	−0.70992	−0.78186	0.32962	0.31392
13	20.598	20.688	−0.94386	−1.0972	0.28012	0.25027
14	19.593	15.873	−1.1778	−1.4125	0.23545	0.19582
15	5.1628	5.3855	−1.4117	−1.7279	0.19596	0.15086
16	17.475	13.433	−1.6457	−2.0432	0.16170	0.11474
17	3.8208	2.5476	−1.8796	−2.3586	0.13243	0.00639
18	2.9210	1.3303	−2.1135	−2.6739	0.10779	0.06452
19	1.2613	0.14473	−2.3475	−2.9892	0.08727	0.04791
20	1.6007	0.21840	−2.5814	−3.3046	0.07034	0.0354

$$\alpha_{12} = -0.24205; \quad \alpha_{22} = -0.15118; \quad \beta_{12} = -0.23395; \quad \beta_{22} = -0.31534.$$
$$s(\hat{\alpha}_1) = -0.06669; \quad s(\hat{\alpha}_2) = 0.06436; \quad s(\hat{\beta}_1) = 0.02019; \quad s(\hat{\beta}_2) = 0.02365.$$

where each ε_{gi} is an independent random variable with mean 0 and variance U_{gi0}^{-1}. Thus

$$\beta_{i0} = \sum_g (T_{gi} - \theta_{i0}) Y_{gi0} U_{gi0} \Big/ \sum_g (T_{gi} - \theta_{i0})^2 U_{ig0}$$

and

$$\alpha_{i0} = \bar{y}_{i0} - \beta_{i0}\theta_{i0},$$

where

$$\bar{y}_{i0} = \sum_g Y_{gi0} U_{gi0} \Big/ \sum_g U_{gi0}$$

and

$$\theta_{i0} = \sum_g T_{gi} U_{gi0} \Big/ \sum_g U_{gi0}.$$

If

$$y_{ij0} = 6(n_{ij1}/n_{ij}^{SC} - \tfrac{1}{2})$$

and

$$u_{ij0} = n_{ij}^{SC} e^3 / (e^3 + 1)^2,$$

then one obtains similar expressions to those used in Table 5.5, for

$$\beta_{i0} = \sum_j (t_j - \theta_{i0}) y_{ij0} u_{ij0} \Big/ \sum_j (t_j - \theta_{i0})^2 u_{ij0},$$

$$\bar{y}_{i0} = \sum_j y_{ij0} u_{ij0} \Big/ \sum_j u_{ij0},$$

and

$$\theta_{i0} = \sum_j t_j u_{ij0} \Big/ \sum_j u_{ij0}.$$

The approximations α_{i0} and β_{i0} lead to approximations

$$\Omega_{gi0} = \alpha_{i0} + \beta_{i0} T_{gi}$$

for the logits Ω_{gi0} and approximation

$$P_{1 \cdot gi1} = (1 + \exp -\Omega_{gi0})^{-1}$$

for the probabilities $P_{1 \cdot gi}$. In Table 5.6, $\omega_{ij0} = \Omega_{gi0}$ and $p_{1 \cdot ij1} = P_{1 \cdot gi1}$ if $C'_{gi} = j$.

Subsequent iterations can be effectively illustrated by the computations leading to $P_{1 \cdot gi2}$. These computations begin with the working logit

$$Y_{gi1} = \Omega_{gi0} + (X_{gi} - P_{1 \cdot gi1})/U_{gi1}$$

and the working weight

$$U_{gi1} = P_{1 \cdot gi1}(1 - P_{1 \cdot gi1}),$$

where $X_{gi} = 1$ if $R'_{gi} = 1$ and $X_{gi} = 0$ if $R'_{gi} = 2$. The approximations α_{i1} and β_{i1} are found as in the weighted-regression models

$$Y_{gi1} = \alpha_i + \beta_i T_{gi} + \varepsilon_{gi},$$

where each ε_{gi} is an independent random variable with mean 0 and variance U_{gi0}^{-1}, Thus

$$\beta_{i1} = \beta_{i0} + \delta_{i1}$$

and

$$\alpha_{i1} = \alpha_{i0} + \gamma_{i1},$$

where

$$\delta_{i1} = \sum_g (T_{gi} - \theta_{i1})(X_{gi} - P_{1 \cdot gi1}) \Big/ \sum_g (T_{gi} - \theta_{i1})^2 U_{gi1},$$

$$\gamma_{i1} = \sum_g (X_{gi} - P_{1 \cdot gi1}) \Big/ \sum_g U_{gi1} - \delta_{i1}\theta_{i1},$$

and

$$\theta_{i1} = \sum_g T_{gi} U_{gi1} \Big/ \sum_g U_{gi1}.$$

If y_{ij1} is the average of all Y_{gi1} such that $C'_{gi} = j$ and $u_{ij1} = n_{ij}^{SC} U_{gi0}$ if $C'_{gi} = j$, then, as in Table 5.5,

$$\delta_{i1} = \sum_j (t_j - \theta_{i1})(n_{ij1} - n_{ij}^{SC} p_{1 \cdot ij1}) \Big/ \sum_j (t_j - \theta_{i1})^2 u_{ij1},$$

$$\gamma_{i1} = \sum_j (n_{ij1} - n_{ij}^{SC} p_{1 \cdot ij1}) \Big/ \sum_j u_{ij1} - \delta_{i1} \theta_{i1},$$

and

$$\theta_{i1} = \sum_j t_j u_{ij1} \Big/ \sum_j u_{ij1}.$$

Given α_{i1} and β_{i1}, one then obtains approximations

$$\Omega_{gi1} = \alpha_{i1} + \beta_{i1} T_{gi}$$

for the estimated logits $\hat{\Omega}_{gi}$ and approximations

$$P_{1 \cdot gi2} = (1 + \exp -\Omega_{gi1})^{-1}$$

for the estimated probabilities $\hat{P}_{1 \cdot gi}$. In Table 5.6, $\omega_{ij1} = \Omega_{gi1}$ and $p_{1 \cdot ij2} = P_{1 \cdot gi2}$. As is evident from Table 5.6, the convergence of the Newton–Raphson algorithm has slowed to only a limited degree by ignoring grouping.

Large-Sample Properties of Estimates

Even though many of the cell counts n_{ijk} in Table 5.4 are small, the standard normal approximations for parameters such as $\hat{\alpha}$ and $\hat{\beta}_i$ continue to apply, provided the model is satisfied. The analogous weighted-regression problem assumes that hypothetical dependent variables Y_{gi}, $1 \leqslant g \leqslant n_i^S$, $1 \leqslant i \leqslant 2$, satisfy the model

$$Y_{gi} = \alpha_i + \beta_i T_{gi} + \varepsilon_{gi},$$

where the ε_{gi} are independent $N(0, U_{gh}^{-1})$ random variables and

$$U_{gi} = P_{1 \cdot gi} P_{2 \cdot gi} = P_{1 \cdot gi}(1 - P_{1 \cdot gi}).$$

The weighted-least-squares estimate b_i of β_i satisfies the equation

$$b_i = \sum_g (T_{gi} - \theta_i) Y_{gi} U_{gi} \Big/ \sum_g (T_{gi} - \theta_i)^2 U_{gi},$$

where

$$\theta_i = \sum_g T_{gi} U_{gi} \Big/ \sum_g U_{gi},$$

so that b_i has a normal distribution with mean β_i and variance

$$1 \Big/ \sum_g (T_{gi} - \theta_i)^2 U_{gi}.$$

Thus if the logit model holds, then $\hat{\beta}_i$, the maximum-likelihood estimate of β_i, has an approximate $N(\beta_i, \sigma^2(\hat{\beta}_i))$ distribution, where

$$\sigma^2(\hat{\beta}_i) = 1 \Big/ \sum_g (T_{gi} - \theta_i)^2 U_{gi}.$$

The accuracy of the approximation improves as n_i^S, the number of respondents of sex i, become large. It is not necessary that individual cell counts be large. Nonetheless, grouping does have some computational advantage. If

$$u_{ij} = n_{ij}^{SC} p_{1 \cdot ij}(1 - p_{1 \cdot ij}),$$

then

$$\sigma^2(\hat{\beta}_i) = 1 \Big/ \sum_j (t_j - \theta_i)^2 u_{ij}.$$

The maximum-likelihood estimate of $\sigma^2(\hat{\beta}_i)$ is

$$s^2(\hat{\beta}_i) = 1 \Big/ \sum_g (T_{gi} - \hat{\theta}_i)^2 \hat{U}_{gh} = 1 \Big/ \sum_j (t_j - \hat{\theta}_i)^2 \hat{u}_{ij}.$$

Here

$$\hat{U}_{gi} = \hat{P}_{1 \cdot gi}(1 - \hat{P}_{1 \cdot gi}),$$
$$\hat{u}_{ij} = n_{ij}^{SC} \hat{p}_{1 \cdot ij}(1 - \hat{p}_{1 \cdot ij}),$$

and

$$\hat{\theta}_i = \sum_g T_{gi} \hat{U}_{gi} \Big/ \sum_g \hat{U}_{gi} = \sum_j t_j \hat{u}_{ij} \Big/ \sum_j \hat{u}_{ij}.$$

Similarly, the weighted-least-squares estimate a_i of α_i satisfies the equation

$$a_i = \sum_g Y_{gi} U_{gi} \Big/ \sum_g U_{gi} - b_i \theta_i,$$

so that a_i has a normal distribution with mean α_i and variance

$$\left(1 \Big/ \sum_g U_{gi}\right) + \left[\theta_i^2 \Big/ \sum_g (T_{gi} - \theta_i)^2 U_{gi} \right].$$

Thus the maximum-likelihood estimate $\hat\alpha_i$ of α_i has an approximate $N(\alpha_i, \sigma(\hat\alpha_i))$ distribution, where

$$\sigma^2(\hat\alpha_i) = \left(1 \Big/ \sum_g U_{gi}\right) + \left[\theta_i^2 \Big/ \sum_g (T_{gi} - \theta_i)^2 U_{gi} \right]$$

$$= \left(1 \Big/ \sum_j u_{ij}\right) + \left[\theta_i^2 \Big/ \sum_j (t_j - \theta_i)^2 u_{ij} \right]$$

$$= \left(1 \Big/ \sum_j u_{ij}\right) + \theta_i^2 \sigma^2(\hat\beta_i).$$

As in the case of $\hat\beta_i$, the approximation improves as n_i^S increases. The estimated asymptotic standard deviation of $\hat\alpha_i$ is

$$s^2(\hat\alpha_i) = \left(1 \Big/ \sum_g U_{gh}\right) + \hat\theta_i^2 s^2(\hat\beta_i)$$

$$= \left(1 \Big/ \sum_j \hat u_{ij}\right) + \hat\theta_i^2 s^2(\hat\beta_i).$$

Estimates $s(\hat\alpha_i)$ and $s(\hat\beta_i)$ may be found in Tables 5.5 and 5.6.

To illustrate use of these estimates, consider $\hat\beta_1$ and $s(\hat\beta_1)$. An approximate 95 percent confidence interval for β_1 has lower bound

$$\hat\beta_1 + 1.96s(\hat\beta_1) = -0.274$$

and upper bound

$$\hat\beta_1 + 1.96s(\hat\beta_1) = -0.194.$$

Among male respondents, it is estimated that an increase of one year in education of respondent corresponds to a decrease of from 0.194 to 0.274 in the logit of the probability of agreeing that women should run their homes and leave running of the country for men. Equivalently, the percentage decrease in odds of agreement is estimated to range from

$$100(1 - e^{-0.194}) = 17.7$$

to

$$100(1 - e^{-0.274}) = 23.9$$

when educational level increases by one year. Thus the estimated association of response and education is quite strong.

Chi-Square Tests

Although maximum-likelihood estimates of parameters retain standard large-sample properties even when data are not grouped, the large-samples approximations for chi-square statistics are often unacceptable.

In the logit model for Table 5.4, the standard chi-square statistics

$$X^2 = \sum_i \sum_j \sum_k (n_{ijk} - \hat{m}_{ijk})^2 / \hat{m}_{ijk}$$

$$= \sum_i \sum_j (n_{ij1} - n_{ij}^{SC} \hat{p}_{1 \cdot ij})^2 / \hat{u}_{ij}$$

and

$$L^2 = 2 \sum_i \sum_j \sum_k n_{ijk} \log\left(\frac{n_{ijk}}{\hat{m}_{ijk}}\right)$$

$$= 2 \sum_i \sum_j \sum_k n_{ijk} \log\left(\frac{n_{ijk}}{n_{ij}^{SC} \hat{p}_{k \cdot ij}}\right)$$

may be considered. As usual $0/0$ and $0\log(0/0)$ are set to 0. In this example, $X^2 = 53.3$ and $L^2 = 72.7$. The large difference between X^2 and L^2 provides a strong indication that the chi-square approximations cannot be adequate, for the chi-square approximation leads to substantially different significance levels for the two statistics. There are $41 - 4 = 37$ degrees of freedom, for there are 41 logits ω_{ij} for which n_{ij}^{SC} is positive and there are 4 parameters α_i and β_i, $1 \leqslant i \leqslant 2$, to be estimated. The logit ω_{22} is ignored in this computation since the corresponding component

$$(n_{222} - n_{22}^{SC} \hat{p}_{1 \cdot 22})^2 / \hat{U}_{22}$$

of X^2 is always 0 and the component

$$2 \sum_k n_{22k} \log\left(\frac{n_{22k}}{n_{22}^{SC} \hat{p}_{k \cdot 22}}\right)$$

of L^2 also is always 0. The chi-square approximation for X^2 yields an approximate significance level of 0.04, while the corresponding approximation for L^2 yields an approximate significance level of 0.0005. One significance level suggests inadequacy in the model, while the other indicates that there is overwhelming evidence against the model. Clearly, both pictures cannot be correct.

In fact, the chi-square approximations for both X^2 and L^2 are questionable whenever many of the n_{ijk} are small, even if the marginal totals n_i^S are large. The approximation for the likelihood-ratio chi-square statistic L^2 is especially unreliable. Thus the test statistics X^2 and L^2 have little value in this problem, at least when taken individually.

The problems in use of X^2 and L^2 become even more extreme if grouping is ignored. In this case, the Pearson chi-square statistic reduces to

$$X_u^2 = \sum_i \sum_g (X_{gi} - \hat{P}_{1 \cdot gi})^2/\hat{U}_{gi}$$

and the likelihood-ratio chi-square statistic becomes

$$L_u^2 = -2 \sum_i \sum_g [X_{gi} \log \hat{P}_{1 \cdot gi} + (1 - X_{gi}) \log(1 - \hat{P}_{1 \cdot gi})].$$

Now there is a summand for each of the 2871 logits Ω_{gi}, and there are still the 4 estimated parameters α_i, α_2, β_1, and β_2. Thus there are $2871 - 4 = 2867$ degrees of freedom. To compute X_u^2 and L_u^2 from Table 5.5, note that

$$X_u^2 = \sum_i \sum_j [n_{ij1}(1 - \hat{p}_{1 \cdot ij})^2 + n_{ij2}\hat{p}_{1 \cdot ij}^2]/\hat{u}_{ij} = 2929.3$$

and

$$L_u^2 = -2 \sum_i \sum_j [n_{ij1} \log \hat{p}_{1 \cdot ij} + n_{ij2} \log(1 - \hat{p}_{1 \cdot ij})] = 3341.4.$$

The chi-square approximations for X_u^2 and L_u^2 imply that these statistics have an approximate normal distribution with mean 2867 and variance $2(2867) = 5734$. Thus X_u^2 is quite consistent with the logit model, and L_u^2 is quite inconsistent with the model. Again the difficulty is with the chi-square approximation. A more accurate normal approximation for X_u^2 has the same mean but a different variance, while a more accurate normal approximation for L_u^2 has both a different mean and different variance. To help understand the basis for this claim, see Exercise 5.4.

Despite the problems observed with X^2, L^2, X_u^2, and L_u^2, chi-square tests can still be employed to examine the logit model for Table 5.4. The key observation is that differences between two likelihood-ratio chi-square statistics have approximate chi-square distributions under quite general conditions.

For example, consider an alternate model for Table 5.4 with a quadratic term, so that

$$\Omega_{gi} = \alpha_i + \beta_{i1} T_{gi} + \beta_{i2} T_{gi}^2$$

or, equivalently,

$$\omega_{ij} = \alpha_i + \beta_{i1} t_j + \beta_{i2} t_j^2.$$

The likelihood-ratio chi-square statistic L'^2 for this model is 71.1, while the likelihood-ratio chi-square statistic $L_u'^2$ for ungrouped data is 3339.8

(see Exercise 5.6). The difference

$$L'^2 - L^2 = 1.60$$

in ordinary likelihood-ratio chi-square statistics is the same as the difference

$$L_u'^2 - L_u^2 = 1.60$$

in likelihood-ratio chi-square statistics for ungrouped data. The linear logit model is a special case of the quadratic logit model in which $\beta_{i2} = 0$ and $\beta_{i1} = \beta_i$. Since six parameters are estimated in the quadratic case and four parameters in the linear case, there are $6 - 4 = 2$ degrees of freedom associated with the difference in likelihood-ratio chi-squares. The chi-square approximation for this difference is satisfactory, even though the chi-square approximations for L^2, L'^2, L_u^2, and $L_u'^2$ are all unsatisfactory. Since 1.60 is not an unusually large value for a chi-square variable with two degrees of freedom, this test of the linear logit model provides little evidence against it.

A second possible alternative model divides subjects into three educational groups, just as in Section 5.2. It is now assumed that a separate linear logit model applies for each group and each sex, so that

$$
\begin{aligned}
\Omega_{gi} &= \alpha_{i1} + \beta_{i1} T_{gi}, & 0 \leqslant C'_{gi} \leqslant 8, \\
&= \alpha_{i2} + \beta_{i2} T_{gi}, & 9 \leqslant C'_{gi} \leqslant 12, \\
&= \alpha_{i3} + \beta_{i3} T_{gi}, & 13 \leqslant C'_{gi} \leqslant 20.
\end{aligned}
$$

Under the linear logit models, $\alpha_{i1} = \alpha_{i2} = \alpha_{i3}$ and $\beta_{i1} = \beta_{i2} = \beta_{i3}$.

Under the new model, the likelihood-ratio chi-square $L_u'^2$ is 3321.3. Thus the reduction from the likelihood-ratio chi-square for the linear model is 20.2. There are 12 parameters α_{il} and β_{il}, $1 \leqslant i \leqslant 2, 1 \leqslant l \leqslant 3$, in this model, compared to 4 parameters in (5.2). Thus the degrees of freedom for the difference $L_u^2 - L_u'^2$ are 8. Therefore, the significance level is the difference $L_u^2 - L_u'^2$ is about 0.01. Consequently, substantial evidence does exist that the linear logit model is unsatisfactory.

The evidence against the simple linear model comes primarily from female respondents. Female respondents contribute $L_{2u}^2 = 1805.58$ to the total L_u^2 and $L_{2u}'^2 = 1790.40$ to the total $L_u'^2$. If $\alpha_{21} = \alpha_{22} = \alpha_{23}$ and $\beta_{21} = \beta_{22} = \beta_{23}$, then $L_{2u}^2 - L_{2u}'^2$ has an approximate chi-square distribution on four degrees of freedom. Since $L_{2u}^2 - L_{2u}'^2$ is 15.18, there is rather strong evidence that for female respondents, the logit Ω_{g2} is not a linear function of years C'_{g2} of education. For males, the contribution to L_u^2 is $L_{1u}^2 = 1535.84$, the contribution to $L_u'^2$ is $L_{1u}'^2 = 1530.86$, and the difference is 4.98. Since the corresponding degrees of freedom are 4, this difference is not unusually large.

Thus the latest test does not provide much evidence that for males, Ω_{g1} is not a linear function of years C_{g1} of education.

Residual Analysis

Further tests of the linear logit model may be based on residual analysis. Normal approximations for adjusted residuals or generalized adjusted residuals are generally satisfactory if the raw residual is based on a substantial number of observations.

The simplest residuals to consider are ordinary adjusted residuals

$$(n_{ij1} - \hat{m}_{ij1})/\hat{c}_{ij1}^{1/2} = (n_{ij1} - n_{jj}^{SC}\hat{p}_{1\cdot ij})/\hat{c}_{ij1}^{1/2}.$$

Here $\hat{c}_{ij1}$ is the variance of $\hat{u}_{ij}r_{ij}$, where r_{ij} is the residual

$$r_{ij} = Y_{ij} - a_i - b_i t_j$$

in the weighted-regression model

$$Y_{ij} = \alpha_i + \beta_i t_j + \varepsilon_{ij}, \qquad \hat{u}_{ij} > 0.$$

Here the Y_{ij} are hypothetical dependent variables, and the ε_{ij} are independent $N(0, \hat{u}_{ij}^{-1})$ random variables. The coefficients a_i and b_i are the weighted-least-squares estimates of α_i and β_i under this model. Thus

$$\hat{c}_{ij1} = \hat{u}_{ij}[1 - (\hat{u}_{ij}/\hat{M}_i) - \hat{u}_{ij}(t_j - \bar{t}_i)^2 s^2(\hat{\beta}_i)],$$

where

$$\hat{M}_i = \sum_j \hat{u}_{ij}.$$

If the model holds for sex i and u_{ij} is large, then the adjusted residuals have approximate $N(0, 1)$ distributions.

More general residuals consider weighted combinations

$$d_i = \sum_j e_j(n_{ij1} - \hat{m}_{ij1}).$$

The corresponding adjusted residual is

$$z_i = d_i/s(d_i),$$

where $s^2(d_i)$ is the variance of

$$\sum_j e_j r_{ij}$$

in the corresponding weighted regression. Thus

$$s^2(d_i) = \sum_j e_j^2 \hat{u}_{ij} - \left[\left(\sum_j e_j \hat{u}_{ij}\right)^2 \bigg/ \hat{M}_i\right] - \left[\sum_j e_j(t_j - \hat{\theta}_i)\right]^2 s^2(\hat{\beta}_i).$$

Here z_i has an approximate standard normal distribution if the model holds for sex i and if each c_j is small relative to $s(d_i)$.

Analyses may also be described for ungrouped data through the weighted-regression model

$$Y_{gi} = \alpha_i + \beta_i T_{gi} + \varepsilon_{gi},$$

where the ε_{gi} are independent $N(0, \hat{U}_{gi}^{-1})$ random variables and the Y_{gi} are hypothetical dependent variables. If

$$d_i = \sum_g E_{gi}(X_{gi} - \hat{P}_{1 \cdot gi}),$$

then the corresponding adjusted residual is

$$z_i = d_i / s(d_i),$$

where

$$s^2(d_i) = \sum_g E_{gi}^2 \hat{U}_{gi} - \left[\left(\sum_g E_{gi} \hat{U}_{gi}\right)^2 \middle/ \hat{M}_i\right] - \left[\sum_g E_{gi}(T_{gh} - \hat{\theta}_i)\right]^2 s^2(\hat{\beta}_i).$$

Note that

$$\hat{M}_i = \sum_g \hat{U}_{gi}.$$

The normal approximation for z_i applies if the model holds for sex i and each E_{gi} is small relative to $s(d_i)$. Note that the formula for $\hat{c}_{ij1}$ corresponds to

$$e_{j1} = 1, \quad j = j',$$
$$= 0, \quad j \neq j',$$

or to

$$E_{g1} = 1, \quad T_{g1} = t_j,$$
$$= 0, \quad T_{g1} \neq t_j.$$

Results are summarized in Table 5.7. Since values of $\hat{u}_{ij}$ are small for subjects with no more than five years of education or at least 17 years of education, combined residuals are provided for these years. For example, one can obtain a difference

$$d_i = \sum_{j=0}^{5} n_{ij} - \sum_{j=0}^{5} N_{ij} \hat{p}_{1 \cdot ij}.$$

Here

$$s^2(d_i) = \sum_{j=0}^{5} \hat{u}_{ij} - \left[\left(\sum_{j=0}^{5} \hat{u}_{ij}\right)^2 \middle/ \hat{M}_i\right] - \left[\sum_{j=0}^{5} (t_j - \hat{\theta}_i)\hat{u}_{ij}\right]^2 s^2(\hat{\beta}_i).$$

Table 5.7

Residuals for the Logit Model for Table 5.4

Years of education completed	Male			Female		
	Raw residual	EASD of raw residual	Adjusted residual	Raw residual	EASD of raw residual	Adjusted residual
0	−1.34	0.75	−1.79	−1.72	0.51	−3.33
1	0.27	0.48	0.56	0.06	0.24	0.26
2	0.66	0.73	0.89	—	—	—
3	−1.21	1.17	−1.04	−0.21	0.83	−0.25
4	−2.62	1.32	−1.99	1.49	1.10	1.35
5	−1.33	1.93	−0.69	−2.93	1.73	−1.69
6	2.33	2.59	0.90	0.45	1.95	0.23
7	1.25	2.98	0.42	−2.93	2.81	−1.04
8	6.02	4.80	1.26	12.56	4.66	2.69
9	0.12	3.64	0.03	−5.16	3.81	−1.35
10	−1.86	4.17	−0.45	−1.40	5.15	−0.27
11	−0.40	4.53	−0.09	6.83	4.92	1.39
12	−3.64	7.35	−0.50	3.86	8.35	0.46
13	2.71	4.27	0.63	−10.27	4.29	−2.40
14	2.81	4.09	0.69	−1.38	3.71	−0.37
15	2.73	2.19	1.25	0.82	2.22	0.37
16	−5.20	3.64	−1.43	−1.68	3.21	−0.52
17	−1.24	1.86	−0.67	0.32	1.52	0.21
18	−2.12	1.62	−1.31	−1.35	1.10	−1.23
19	0.69	1.07	0.64	0.86	0.37	2.32
20	1.38	1.19	1.15	1.79	0.45	3.97
0–5	−5.59	2.69	−2.07	−3.30	1.98	−1.66
17–20	−1.28	2.92	−0.44	1.61	1.88	0.86

Table 5.7 does not provide much evidence against the linear logit model for male respondents; however, several adjusted residuals for females are rather large. Given the small number of observations on women with no education or 20 years of education, the very large standardized values for females with no years of education completed and for females with at least 20 years of education completed are difficult to interpret precisely; nonetheless, they are disturbing. The large residuals for eight years and 13 years of education are also troublesome. Given these residuals and results from the model with separate linear logit relationships for each educational group, it appears that certain changes in the amount of education have disproportionately large effects. For example, the estimated logits

$$\log\left(\frac{n_{2j} + \frac{1}{2}}{N_{2j} - n_{2j} + \frac{1}{2}}\right)$$

for female respondents with nine to 11 years education range from -0.21 to -0.15. For female respondents with 12 years of education, the estimated logit is -0.75. For female respondents with 13 to 15 years of education, the range of logits is from -1.67 to -1.48. These results are not surprising given that 12 years of education corresponds to completion of high school. Nonetheless, it is worth noting that comparable results are not observed among male respondents. There the logits range from 0.00 for nine years of education to -0.49 for 11 years of education to -0.75 for 12 years to -0.81 for 13 years to -0.91 for 15 years. Thus the change in the proportion of positive responses is much more regularly seen among male respondents than among female respondents.

In addition to providing a detailed analysis of the association of sex and education to attitudes toward women running homes rather than the country, this section has illustrated the possibility of using logit models when cell counts are small or when data are completely ungrouped. The next section provides a more general discussion of applications of logit models to grouped or ungrouped data.

5.4   Properties of Logit Models

In general, the logit models considered in this section involve N observations h, $1 \leqslant h \leqslant N$, on a dichotomous dependent variable Z_h and on q independent variables T_{hk}, $1 \leqslant k \leqslant q$. The probability is $P_{k \cdot h} > 0$ that $Z_h = k$, $1 \leqslant k \leqslant 2$. These probabilities $P_{1 \cdot h}$ and $P_{2 \cdot h}$ depend only on the independent variables T_{hk}, $1 \leqslant k \leqslant q$. For some unknown constants η and β_k, $1 \leqslant k \leqslant q$, the logits

$$\Omega_h = \log\left(\frac{P_{1 \cdot h}}{1 - P_{1 \cdot h}}\right)$$

are assumed to satisfy the equation

$$\Omega_h = \eta + \sum_k \beta_k T_{hk}, \qquad 1 \leqslant h \leqslant N. \tag{5.1}$$

For example, in Section 5.3, the logits Ω_{gi} for individuals of sex i satisfy a relationship like (5.1) with $q = 1$ independent variables.

In many common applications, many observations h have the same observed independent variables T_{hk}, $1 \leqslant k \leqslant q$. In such cases, it is often helpful to divide the observations h, $1 \leqslant h \leqslant N$, into groups j, $1 \leqslant j \leqslant s$, such that if h is in group j, then $\mathbf{T}_h = \mathbf{t}_j$, where $\mathbf{T}_h$ is the vector with co-ordinates T_{hk}, $1 \leqslant k \leqslant q$. For example, male respondents in Table 5.4 have possible responses T_{g1} of $t_j = j$, where j is an integer from 0 to 20. The data may be summarized in a 2×2 table by letting n_{ij} be the number of subjects h with $Z_h = i$ and $\mathbf{T}_h = t_j$. A contingency table with column-multinomial sampling is obtained. The count n_{1j} has a binomial distribution with same size $N_j = n_{1j} + n_{2j}$ and probability $p_{1 \cdot j}$. Here

$$p_{1 \cdot j} = (1 + \exp - \omega_j)^{-1}$$

and

$$\omega_j = \eta + \sum_k \beta_k t_{jk}. \tag{5.2}$$

This logit model for the n_{ij}, $1 \leqslant i \leqslant 2$, $1 \leqslant j \leqslant s$, is also a log–linear model. Let

$$p_{2 \cdot j} = 1 - p_{1 \cdot j}$$

and

$$m_{ij} = N_j p_{i \cdot j}.$$

Then the count n_{ij} has mean m_{ij}, and (5.2) is equivalent to the model

$$\log m_{ij} = \gamma_j + \sum_{k=0}^{q} \beta_k x_{ijk}$$

$$= \alpha + \sum_{k=0}^{q+s-1} \beta_k x_{ijk}, \tag{5.3}$$

where

$$\alpha = \log m_{11},$$
$$\gamma_j = \log m_{2j},$$
$$\beta_k = \gamma_{k-q+1} - \alpha, \qquad q + 1 \leqslant k \leqslant q + s - 1,$$
$$\beta_0 = \eta,$$

and

$$\begin{aligned}
x_{ijk} &= 1, & k = 0, \quad i = 1, \\
&= t_j, & 1 \leqslant k \leqslant q, \quad i = 1, \\
&= 0, & 0 \leqslant k \leqslant q, \quad i = 2, \\
&= 1, & j = k - q + 1, \quad q + 1 \leqslant k \leqslant q + s - 1, \\
&= 0, & j \neq k - q + 1, \quad q + 1 \leqslant k \leqslant q + s - 1.
\end{aligned}$$

For verification, see Exercise 5.2. Note that examples of this relationship of logits and log–linear models appear in Sections 5.1, 5.2, and 5.3, and in Exercise 5.1.

Maximum-Likelihood Equations

As Cox (1970, page 87) and others have noted, the model specified by (5.1) has maximum-likelihood equations

$$\hat{\Omega}_h = \hat{\eta} + \sum_k \hat{\beta}_k T_{hk}, \tag{5.4}$$

$$\hat{P}_{1\cdot h} = (1 + \exp -\hat{\Omega}_h)^{-1}, \tag{5.5}$$

$$\sum_h \hat{P}_{1\cdot h} = \sum_h X_h, \tag{5.6}$$

and

$$\sum_h T_{hk}\hat{P}_{1\cdot h} = \sum_h T_{hk}X_h, \qquad 1 \leqslant k \leqslant q, \tag{5.7}$$

where $X_h = 1$ if $Z_h = 1$ and $X_h = 0$ if $Z_h = 2$. Note that

$$\sum_h \hat{P}_{1\cdot h}$$

is the estimated expected number of $X_h = 1$, while

$$\sum_h X_h$$

is the observed number of $X_h = 1$. Note also that

$$\sum_h T_{hk}\hat{P}_{1\cdot h}$$

is the estimated expected sum of all T_{hk} such that $X_h = 1$, while

$$\sum_k T_{hk}X_h$$

is the observed sum of all such T_{hk}.

These equations are easily converted to formulas for grouped data. If

$$\hat{p}_{1\cdot j} = (1 + \exp -\hat{\omega}_j)^{-1} \tag{5.8}$$

and

$$\hat{\omega}_j = \hat{\eta} + \sum_k \hat{\beta}_k t_{jk}, \tag{5.9}$$

then

$$\sum_j N_j \hat{p}_{1\cdot j} = \sum_j n_{1j} \tag{5.10}$$

and

$$\sum_j t_{jk} N_j \hat{p}_{1\cdot j} = \sum_j t_{jk} n_{ij}. \tag{5.11}$$

To verify these formulas, note that

$$\hat{P}_{k \cdot h} = \hat{p}_{k \cdot j}$$

and

$$\hat{\Omega}_h = \hat{\omega}_j$$

if $\mathbf{T}_h = \mathbf{t}_j$.

The Newton–Raphson Algorithm

The Newton–Raphson algorithm may be used with grouped or ungrouped data. The grouped case will be considered first. This approach is preferable if many of the totals N_j are large. It has been used in Sections 5.1, 5.2, and 5.3.

In the grouped case, an initial estimate $m_{ij0} > 0$ for $\hat{m}_{ij} = N_j \hat{p}_{i \cdot j}$ is given. Common choices are $m_{ij0} = n_{ij}$, $m_{ij0} = n_{ij} + \frac{1}{2}$, and $m_{ij0} = N_j(n_{ij} + \frac{1}{2})/(N_j + 1)$. The empirical logit

$$y_{j0} = \log\left(\frac{m_{1j0}}{m_{2j0}}\right) \tag{5.12}$$

and the empirical weight

$$u_{j0} = m_{1j0}m_{2j0}/(m_{1j0} + m_{2j0}) \tag{5.13}$$

are then computed.

Approximations η_0 and β_{k0}, $1 \leqslant k \leqslant q$, are then found as in the weighted-regression model

$$y_{j0} = \eta + \sum_h \beta_k t_{jk} + \varepsilon_{j0},$$

where each ε_{j0} is an independent random variable with mean 0 and variance u_{j0}^{-1}. Thus the β_{k0}, $1 \leqslant k \leqslant q$, satisfy the simultaneous equations

$$\sum_l S_{kl0}\beta_{l0} = \omega_{k0}, \qquad 1 \leqslant k \leqslant q, \tag{5.14}$$

where

$$\omega_{k0} = \sum_j (t_{jk} - \theta_{k0})y_{j0}u_{ij0}, \tag{5.15}$$

$$S_{kl0} = \sum_j (t_{jk} - \theta_{k0})(t_{jl} - \theta_{l0})u_{ij0}, \tag{5.16}$$

and

$$\theta_{k0} = \sum_j t_{jk}u_{j0} \Big/ \sum_j u_{j0}. \tag{5.17}$$

In matrix terminology,

$$\boldsymbol{\beta}_0 = S_0^{-1}\boldsymbol{\omega}_0.$$

The approximation η_0 satisfies the equation

$$\eta_0 = \left(\sum_j y_{j0}u_{j0} \Big/ \sum_j u_{j0}\right) - \sum_k \beta_{k0}\theta_{k0}. \tag{5.18}$$

Given η_0 and β_{k0}, $1 \leqslant k \leqslant q$, an approximation

$$\omega_{j0} = \eta_0 + \sum_k \beta_{k0}t_j \tag{5.19}$$

for $\hat{\omega}_j$ is obtained. The next iteration begins with the new probability estimates

$$p_{1 \cdot j1} = (1 + \exp -\omega_{j0})^{-1} \tag{5.20}$$

and

$$p_{2 \cdot j1} = 1 - p_{1 \cdot j1} \tag{5.21}$$

and the new estimated means

$$m_{ij1} = N_j p_{i \cdot j1}. \tag{5.22}$$

At iteration $v \geqslant 1$, the working logit y_{jv} is defined as

$$y_{jv} = \omega_{j(v-1)} + (n_{1j} - m_{1jv})/u_{jv}, \tag{5.23}$$

where the working weight u_{jv} satisfies

$$u_{jv} = N_j p_{1 \cdot jv} p_{2 \cdot jv} = N_j p_{1 \cdot jv}(1 - p_{1 \cdot jv}). \tag{5.24}$$

The corresponding weighted-regression model is

$$y_{jv} = \eta + \sum_k \beta_k t_j + \varepsilon_j,$$

where each ε_j has an independent distribution with mean 0 and variance u_{jv}^{-1}. Thus approximations β_{kv}, $1 \leqslant k \leqslant q$, are determined by the simultaneous equations

$$\sum_l S_{klv}\beta_{lv} = w_{k0}, \qquad 1 \leqslant k \leqslant q,$$

where

$$w_{k0} = \sum_j (t_{jk} - \theta_{kv})y_{jv}u_{jv},$$

$$S_{kl0} = \sum_j (t_{jk} - \theta_{kv})(t_{jl} - \theta_{lv})u_{jv}. \tag{5.25}$$

and

$$\theta_{kv} = \sum_j t_{jk} u_{jv} \Big/ \sum_j u_{jv}. \tag{5.26}$$

Given the β_{kv}, $1 \leqslant k \leqslant q$, η_v satisfies

$$\eta_v = \left(\sum_j y_{jv} u_{jv} \Big/ \sum_j u_{jv} \right) - \sum_k \beta_{kv} \theta_{kv}.$$

Computations are simplified by letting

$$\beta_{kv} = \beta_{k(v-1)} + \delta_{kv}, \qquad 1 \leqslant k \leqslant q, \tag{5.27}$$

and

$$\eta_v = \eta_{v-1} + \gamma_v. \tag{5.28}$$

Here the δ_{kv}, $1 \leqslant k \leqslant q$, are determined from the simultaneous equations

$$\sum_l S_{klv} \delta_{lv} = a_{kv}, \qquad 1 \leqslant k \leqslant q, \tag{5.29}$$

where

$$a_{kv} = \sum_j (t_{jk} - \theta_{jv})(n_{1j} - m_{2jv}), \qquad 1 \leqslant k \leqslant q, \tag{5.30}$$

and

$$\gamma_v = \left[\sum_j (n_{1j} - m_{1jv}) \Big/ \sum_j u_{jv} \right] - \sum_k \delta_{kv} \theta_{kv}. \tag{5.31}$$

In matrix terms,

$$\boldsymbol{\delta}_v = S_v^{-1} \mathbf{a}_v.$$

The new estimate of $\hat{\omega}_j$ is

$$\omega_j = \eta_v + \sum_k \beta_{kv} t_{jk}, \tag{5.32}$$

so that

$$p_{1 \cdot j(v+1)} = (1 + \exp - \omega_{jv})^{-1}, \tag{5.33}$$

$$p_{2 \cdot j(v+1)} = 1 - p_{1 \cdot j(v+1)}, \tag{5.34}$$

and

$$m_{ij(v+1)} = N_j p_{i \cdot j(v+1)}. \tag{5.35}$$

If the N_j are generally large, convergence is generally very rapid.

In the ungrouped case, initial approximations $P_{k \cdot h0}$ for $\hat{P}_{k \cdot h}$ are needed. Since $P_{2 \cdot h0} = 1 - P_{1 \cdot h0}$, only $P_{1 \cdot h0}$ need be specified. Two simple possible choices are

$$P_{1 \cdot h0} = \exp(3X_h)/(e^3 + 1), \tag{5.36}$$

which corresponds to a proposal of Cox (1970, page 90), and

$$P_{1 \cdot h0} = \exp(2X_h)/(e^2 + 1), \tag{5.37}$$

which is based on some analysis by Goodman (1975). More complex procedures for initial values do not seem necessary. The former choice was used in Section 5.3. Readers interested in a more complex approach to initial values based on discriminant analysis should see Haberman (1974a, p. 372) or Cornfield, Kannel, and Truett (1967).

Given initial values $P_{1 \cdot h0}$, $0 < P_{1 \cdot h0} < 1$, the algorithm is quite similar to that used with grouped data. Let

$$U_{h0} = P_{1 \cdot h0}(1 - P_{1 \cdot h0}) \tag{5.38}$$

and

$$Y_{h0} = \log\left(\frac{P_{1 \cdot h0}}{1 - P_{1 \cdot h0}}\right). \tag{5.39}$$

In the model defined by (5.1), approximations η_0 and β_{k0}, $1 \leqslant k \leqslant q$, are found as in the regression model

$$Y_{h0} = \eta + \sum_k \beta_k T_{hk} + \varepsilon_h,$$

where the ε_h are independently distributed and ε_h has mean 0 and variance U_{h0}^{-1}. Thus the β_{k0}, $1 \leqslant k \leqslant q$, are determined by the simultaneous equations

$$\sum_l S_{kl0}\beta_{k0} = w_{k0}, \qquad 1 \leqslant k \leqslant q, \tag{5.40}$$

and

$$\eta_0 = \left(\sum_h Y_{h0}U_{k0} \bigg/ \sum_n U_{h0}\right) - \sum_h \beta_{k0}\theta_{k0}. \tag{5.41}$$

Here

$$\theta_{k0} = \sum_h T_{hk}U_{h0} \bigg/ \sum_h U_{h0}, \tag{5.42}$$

$$w_{k0} = \sum_h (T_{hk} - \theta_{k0})Y_{h0}U_{h0}, \tag{5.43}$$

and

$$S_{kl0} = \sum_h (T_{hk} - \theta_{k0})(T_{hl} - \theta_{l0})U_{h0}. \tag{5.44}$$

Given η_0 and β_{h0}, $1 \leqslant k \leqslant q$, one may then approximate $\hat{\Omega}_h$ and $\hat{P}_h$ by

$$\Omega_{h0} = \eta_0 + \sum_k \beta_{k0} T_{hk} \tag{5.45}$$

and

$$P_{1 \cdot h1} = (1 + \exp -\Omega_{h0})^{-1}. \tag{5.46}$$

At subsequent iterations such as iteration $v \geqslant 1$, a working logit such as

$$Y_{hv} = \Omega_{h(v-1)} + (X_h - P_{hv})/U_{hv} \tag{5.47}$$

is considered, where

$$U_{hv} = P_{hv}(1 - P_{hv}). \tag{5.48}$$

New approximations η_v and β_{kv}, $1 \leqslant k \leqslant q$, correspond to the regression model

$$Y_{hv} = \eta + \sum_k \beta_k T_{hk} + \varepsilon_h,$$

where the ε_h are independent and each ε_h has mean 0 and variance U_{hv}^{-1}. In practice, one has

$$\eta_v = \eta_{v-1} + \gamma_v \tag{5.49}$$

and

$$\beta_v = \beta_{k(v-1)} + \delta_{kv}. \tag{5.50}$$

To define γ_v and the δ_{kv}, $1 \leqslant k \leqslant q$, let

$$\theta_{kv} = \sum_h T_{hk} U_h \bigg/ \sum U_{hv}, \tag{5.51}$$

$$a_{kv} = \sum_h (T_{hk} - \theta_{kv})(X_h - P_{1 \cdot hv}), \tag{5.52}$$

and

$$S_{klv} = \sum_h (T_{hk} - \theta_{kv})(T_h - \theta_{lv})U_{hv}. \tag{5.53}$$

Then the δ_{kv} are determined by the simultaneous equations

$$\sum_l S_{klv} \delta_{lv} = a_{kv}, \qquad 1 \leqslant k \leqslant q, \tag{5.54}$$

and γ_v is determined by the equation

$$\gamma_v = \left[\sum_h (X_h - P_{1 \cdot hv}) \bigg/ \sum_h U_{hv}\right] - \sum_k \delta_{kv}\theta_{kv}. \tag{5.55}$$

The new approximations Ω_{hv} and $P_{1 \cdot h(v+1)}$ are then

$$\Omega_{hv} = \eta_v + \sum_k \beta_{kv} T_{hk} \tag{5.56}$$

and

$$P_{1 \cdot h(v+1)} = (1 + \exp -\Omega_{hv})^{-1}. \tag{5.57}$$

Convergence for the ungrouped case is generally slower than in the grouped case; however, convergence is still relatively rapid. A related but different computational technique for ungrouped data is described by Walher and Duncan (1967).

Large-Sample Properties of Maximum-Likelihood Estimates

If the model holds, then the large-sample distribution of parameter estimates such as $\hat{\eta}$ and $\hat{\beta}_k$, $1 \le k \le q$, is approximately the same as the distribution of the weighted least-squares estimates of η and $\hat{\beta}_k$, $1 \le k \le q$, in the regression problem

$$Y_h = \eta + \sum_k \beta_k T_{hk} + \varepsilon_h,$$

where the Y_h are hypothetical observed dependent variables, the ε_h are independent $N(0, U_h^{-1})$ random variables, and

$$U_h = P_{1 \cdot h}(1 - P_{1 \cdot h}).$$

If $\hat{\boldsymbol{\beta}}$ is the vector with coordinates $\hat{\beta}_k$, $1 \le k \le q$, and $\boldsymbol{\beta}$ is the vector with coordinates β_k, $1 \le k \le q$, then $\hat{\boldsymbol{\beta}}$ has an approximate $N(\boldsymbol{\beta}, S^{-1})$ distribution. Here S^{-1} is the inverse of S, a $q \times q$ matrix with elements

$$S_{kl} = \sum_h (T_{hk} - \theta_k)(T_{hl} - \theta_l)U_h$$

$$= \sum_j (t_{jk} - \theta_k)(t_{jl} - \theta_l)u_j, \qquad 1 \le k \le q, \quad 1 \le l \le q. \tag{5.58}$$

In the equation for S_{kl},

$$\theta_k = \sum_h T_{hk}U_h \bigg/ \sum_h U_h = \sum t_{jk}u_j \bigg/ \sum_j u_j \tag{5.59}$$

and

$$u_j = N_j p_{1 \cdot j}(1 - p_{1 \cdot j}).$$ (5.60)

The asymptotic covariance matrix S^{-1} has maximum-likelihood estimate $\hat{S}^{-1}$, where $\hat{S}$ has elements

$$\hat{S}_{kl} = \sum_h (T_{hk} - \hat{\theta}_k)(T_{hl} - \hat{\theta}_l)\hat{U}_h$$

$$= \sum_j (t_{jk} - \hat{\theta}_k)(t_{jl} - \hat{\theta}_l)\hat{u}_j,$$ (5.61)

$$\hat{\theta}_k = \sum_h T_{hk}\hat{U}_h \bigg/ \sum_h \hat{U}_h = \sum_j t_{jk}\hat{u}_j \bigg/ \sum_j \hat{u}_j,$$ (5.62)

$$\hat{u}_j = N_i \hat{p}_{1 \cdot j}(1 - \hat{p}_{1 \cdot j}),$$ (5.63)

and

$$\hat{U}_h = \hat{P}_{1 \cdot h}(1 - \hat{P}_{1 \cdot h}).$$ (5.64)

The estimate $\hat{\eta}$ has an approximate $N(0, \sigma^2(\hat{\eta}))$ distribution, where

$$\sigma^2(\hat{\eta}) = \left(1 \bigg/ \sum_n \hat{U}_h\right) + \sum_k \sum_l \theta_k \theta_l S^{kl}.$$ (5.65)

Note that S^{-1} has elements S^{kl}, $1 \leqslant k \leqslant q$, $1 \leqslant l \leqslant q$, and

$$\sum_h U_h = \sum_j u_j.$$ (5.66)

To estimate $\sigma^2(\hat{\eta})$, let

$$s^2(\hat{\eta}) = \left(1 \bigg/ \sum_h \hat{U}_h\right) + \sum_k \sum_l \hat{\theta}_k \hat{\theta}_l S^{kl}.$$ (5.67)

Here the $\hat{S}^{kl}$, $1 \leqslant k \leqslant q$, $1 \leqslant l \leqslant q$, are the elements of $\hat{S}^{-1}$ and

$$\sum_h \hat{U}_h = \sum_j \hat{u}_j.$$ (5.68)

The large-sample properties considered here have been assumed at least since Garwood (1941) for both grouped and ungrouped data; however, rigorous early proofs appear hard to find. Haberman (1974a, pages 358–369) does provide a proof under the assumption that as N becomes large, the distribution of the $\mathbf{T}_h$, $1 \leqslant h \leqslant N$, converges to a distribution with a non-singular covariance matrix. The exact statement of conditions is beyond the scope of this book.

Chi-Square Tests

The chi-square statistics for grouped data are the Pearson chi-square

$$X^2 = \sum_i \sum_j (n_{ij} - \hat{m}_{ij})^2 / \hat{m}_{ij} = \sum_j (n_{1j} - N_j \hat{p}_{1 \cdot j})^2 / \hat{u}_j \qquad (5.69)$$

and the likelihood-ratio chi-square

$$L^2 = 2 \sum_i \sum_j n_{ij} \log(n_{ij}/\hat{m}_{ij}). \qquad (5.70)$$

If all N_j are large and if (5.1) holds, then both X^2 and L^2 have approximate chi-square distributions with $s - q - 1$ degrees of freedom.

For ungrouped data, the chi-square statistics are

$$X_u^{\,2} = \sum_h (X_h - \hat{P}_{1 \cdot h})^2 / \hat{U}_h \qquad (5.71)$$

and

$$L_u^{\,2} = -2 \sum_h \left[X_h \log \hat{P}_{1 \cdot h} + (1 - X_h) \log(1 - \hat{P}_{1 \cdot h}) \right]. \qquad (5.72)$$

As suggested by Exercises 5.4 and 5.5, neither of these statistics has an approximate chi-square distribution; however, $L_u^{\,2}$ does have value for tests of (5.1) against restricted alternatives. Consider a model

$$\Omega_h = \alpha + \sum_{k=1}^{q'} \beta_k T_{hk}, \qquad (5.73)$$

where the T_{hk}, $1 \leqslant k \leqslant q$, are defined as in (5.1) and the T_{hk}, $q + 1 \leqslant k \leqslant q'$ are additional independent variables. Let $\hat{P}_{1 \cdot h}$ be the maximum-likelihood estimate of $P_{1 \cdot h}$ under (5.73), and let the corresponding likelihood-ratio chi-square for ungrouped data be

$$L_u'^{\,2} = -2 \sum_h \left[X_h \log \hat{P}'_{1 \cdot h} + (1 - X_h) \log(1 - \hat{P}'_{1 \cdot h}) \right]. \qquad (5.74)$$

As noted in Haberman (1974a, page 372), under quite general conditions, the difference in likelihood-ratio chi-square $L_u^{\,2} - L_u'^{\,2}$ has an approximate chi-square distribution with $q' - q$ degrees of freedom, provided (5.1) holds. The essential requirement is that the distribution of the vectors $\mathbf{T}_h'$, $1 \leqslant h \leqslant N$, approach the distribution of a random vector with nonsingular covariance matrix as N becomes large. Here $\mathbf{T}_h'$ has coordinates T_{hk}, $1 \leqslant k \leqslant q'$.

Changes in likelihood-ratio chi-squares may also be studied using grouped data. Let $T_{hk}' = t_{jk}$, $1 \leqslant k \leqslant q'$, if observation h is in group j. Thus (5.73) assumes that

$$\omega_j = \alpha + \sum_{k=1}^{q'} \beta_k t_{jk}. \qquad (5.75)$$

Let m_{ij} have maximum-likelihood estimate $\hat{m}'_{ij}$ (under 5.73), and let

$$L'^2 = 2 \sum_i \sum_j n_{ij} \log(n_{ij}/\hat{m}_{ij}) \qquad (5.76)$$

be the corresponding likelihood-ratio chi-square. Then

$$L^2 - L'^2 = L_u^2 - L_u'^2. \qquad (5.77)$$

Residual Analysis

Residual analysis involves ordinary adjusted residuals for individual frequency counts with grouped data and generalized residuals for both grouped and ungrouped data. Material presented here is based on Haberman (1973b; 1974a, pages 349–351; 1976; 1977a; 1977b). Different approaches to residuals are found in Cox and Snell (1968) and Cox (1970, pages 94–99). The application in this chapter in Section 5.3 illustrates the case of $q = 1$.

In the case of grouped data, the adjusted residual r_{ij} for model (5.3) has the form

$$r_{1j} = -r_{2j} = (n_{1j} - \hat{m}_{1j})/\hat{c}_{1j}^{1/2} = (n_{1j} - N_j \hat{p}_{1 \cdot j})/\hat{c}_{1j}^{1/2}, \qquad (5.78)$$

where

$$\hat{c}_{1j} = \hat{u}_j[1 - \hat{u}_j/\hat{M} - \sum_k \sum_l (t_{jk} - \hat{\theta}_k)(t_{jl} - \hat{\theta}_l)\hat{S}^{kl}] \qquad (5.79)$$

and

$$\hat{M} = \sum_j \hat{u}_j. \qquad (5.80)$$

Here the $\hat{S}^{kl}$ are the elements of $\hat{S}^{-1}$, and $\hat{\theta}_k$ and $\hat{S}$ are defined as in (5.61) and (5.62).

The estimated asymptotic variance $\hat{c}_{1j}$ of $n_{1j} - \hat{m}_{1j}$ is the variance of $\hat{u}_j R_j$ where R_j is the residual

$$Y_j - a - \sum_k b_k t_{jk}$$

under the regression model

$$Y_j = \alpha + \sum_k \beta_k t_{jk} + \varepsilon_j.$$

Here the Y_j are hypothetical dependent variables, and the ε_j are independent random variables with respective means 0 and variances $\hat{u}_j^{-1}$. The coefficients a and b_k, $1 \leq k \leq q$, are the respective weighted-least-squares estimates of α and β_k, $1 \leq k \leq q$. If the model holds, if the regularity conditions hold for approximate normality of parameter estimates, and if N_j is large, then r_{1j} has a distribution that is approximately the standard normal distribution.

More generally, let e_j, $1 \leqslant j \leqslant s$, be constants associated with each group j of observations. Consider the residual

$$d = \sum_j e_j(n_{1j} - \hat{m}_{1j}) = \sum_j e_j(n_{1j} - N_j\hat{p}_1 \cdot_j).$$ (5.81)

Then the estimated asymptotic variance of d is

$$s^2(d) = \sum_j e_j{}^2\hat{u}_j - \left[\left(\sum_j e_j\hat{u}_j\right)^2 \Big/ \hat{M}\right] - \sum_k \sum_l \hat{f}_k\hat{f}_l S^{kl},$$ (5.82)

where

$$\hat{f}_k = \sum_j e_j(t_{jk} - \hat{\theta}_k).$$ (5.83)

This formula for $s^2(d)$ is also the formula for the variance of

$$\sum_j e_j\hat{u}_j R_j$$

in the corresponding weighted-regression model. If the model holds, if the regularity conditions hold for approximate normality of parameter estimates, and if each e_j is small relative to $s(d)$, then

$$t = d/s(d)$$ (5.84)

has an approximate standard normal distribution.

Similarly, in the case of ungrouped data, let a constant E_h be associated with observation h, $1 \leqslant h \leqslant N$, and let

$$d = \sum_h E_h(X_h - \hat{P}_1 \cdot_h).$$ (5.85)

Then d has estimated asymptotic variance

$$s^2(d) = \sum_h E_h{}^2\hat{U}_h - \left[\left(\sum_h E_h\hat{U}_h\right)^2 \Big/ \hat{M}\right] - \sum_k \sum_l \hat{f}_k\hat{f}_l S^{kl},$$ (5.86)

where

$$\hat{M} = \sum_h \hat{U}_h$$ (5.87)

and

$$\hat{f}_k = \sum_h E_h(T_{hk} - \hat{\theta}_k).$$ (5.88)

If the model holds, if the regularity conditions are satisfied for approximate normality of parameter estimates, and if each E_h is small relative to $s(d)$, then the distribution $t = d/s(d)$ is approximated well by the standard normal distribution.

5.5   Quantal-Response Models

The logit model is a model of a family of models called quantal-response models. These models have a substantial literature, especially in the area of biological assay and psychology. Finney (1971) provides detailed discussion and an extensive bibliography. The scoring algorithm used with these models goes back to Fisher (1935).

The basis of quantal response models is a known distribution function F with an associated inverse function R such that for $0 < x < 1$, $F(y) = x$ if $y = R(x)$. In the case of logit analysis,

$$F(x) = 1/(1 + e^{-x})$$

and

$$R(x) = \log\left(\frac{x}{1 - x}\right).$$

In probit analysis, the most important alternative to logit analysis,

$$F(x) = \Phi(x),$$

the normal distribution function, and

$$R(x) = \Phi^{-1}(x),$$

the inverse of the normal distribution function.

A general quantal response model is based on the same dependent variables Z_h and independent variables T_{hk} used in (5.1). The probability $P_{i \cdot h}$ that $Z_h = i$ still depends only on the $T_{hk}, 1 \leqslant k \leqslant q$. However instead of (5.1), one assumes that

$$\Omega_h' = R(P_{1 \cdot h}) = \eta + \sum_k \beta_k T_{hk}$$

for some unknown η and $\beta_k, 1 \leqslant k \leqslant q$.

The probit model is slightly less convenient to use in terms of computational difficulty than is the logit model and the two models give very similar results in practice, as noted by Chambers and Cox (1967), among others. Consequently, there are relatively few situations in which probit analysis should be preferred to logit analysis, at least within the social sciences. Thus no use of probit analysis occurs in this book.

A type of regression model is obtained if $R(x) = F(x) = x$ for $0 < x < 1$. Then $R(P_{1 \cdot h}) = P_{1 \cdot h}$ is the expected value of X_h, so that

$$X_h = \eta + \sum_k \beta_k T_{hk} + \varepsilon_h,$$

where each ε_h has mean 0 and variance $P_{1 \cdot h}(1 - P_{1 \cdot h})$. This model is not an ordinary regression model since the X_h do not have constant variances and since the expected value of X_h must lie between 0 and 1. Thus ordinary regression analysis is not the most appropriate method of analysis even in this special quantal response model. Nonetheless, Goodman (1975) has noted that even with ordinary regression analysis, this choice of F and R leads to results very similar to those obtained with logit analysis, especially if the probabilities $P_{1 \cdot h}$ are between 0.25 and 0.75.

Provided that F corresponds to a continuous distribution with density f, the methods described in this chapter for logit analysis are readily applied to general quantal response models.

The analogies to regression analysis still apply, so that computational procedures and asymptotic variance formulas are very similar for the two methods. The principal changes are use of weights

$$U'_{hv} = \frac{[f(\Omega'_{hv})]^2}{P_{1 \cdot hv}(1 - P_{1 \cdot hv})}$$

rather than

$$U_{hv} = P_{1 \cdot hv}(1 - P_{1 \cdot hv})$$

and use of working transforms

$$
\begin{aligned}
Y'_{h0} &= R(P_{1 \cdot h0}), & v &= 0, \\
&= \Omega'_{h(v-1)} + (X_h - P_{1 \cdot hv})/f(\Omega'_{h(v-1)}), & v &\geq 1,
\end{aligned}
$$

rather than

$$
\begin{aligned}
Y'_{hv} &= \log\left(\frac{P_{1 \cdot h0}}{1 - P_{1 \cdot h0}}\right), & v &= 0, \\
&= \Omega_{h(v-1)} + (X_h - P_{1 \cdot hv})/U_{hv}, & v &\geq 1,
\end{aligned}
$$

in computation of maximum-likelihood estimates and use of

$$\hat{U}'_h = \frac{[f(\hat{\Omega}_h)]^2}{\hat{P}_{1 \cdot h}(1 - \hat{P}_{1 \cdot h})}$$

rather than

$$\hat{U}_h = \hat{P}_{1 \cdot h}(1 - \hat{P}_{1 \cdot h})$$

in estimation of asymptotic variances. Corresponding changes occur in expressions such as u_j and u_{jv}. Note that in probit analysis, f is the density function of the standard normal; i.e., $f(x) = \exp(-\frac{1}{2}x^2)/(2\pi)^{1/2}$. As is evident from Garwood (1941), the computational algorithm for quantal-response models is a version of the scoring algorithm, a variant of the Newton–Raphson algorithm used by Fisher (1935). The scoring algorithm reduces to the Newton–Raphson algorithm only in the case of logit analysis.

EXERCISES

5.1   Consider a $2 \times s$ table of frequencies n_{ij}, $1 \leqslant i \leqslant 2$, $1 \leqslant j \leqslant s$. Let the n_{ij} have a multinomial distribution with sample size N and probabilities p_{ij}, $1 \leqslant i \leqslant 2$, $1 \leqslant j \leqslant s$. Show that the log–linear model

$$\log m_{ij} = \log(Np_{ij}) = \lambda + \lambda_i^A + \lambda_j^B,$$

$$\sum_i \lambda_i^A = \sum_j \lambda_j^B = 0,$$

is equivalent to the logit model

$$\omega_j = \log\left(\frac{p_{1 \cdot j}^{A \cdot B}}{p_{2 \cdot j}^{A \cdot B}}\right) = \eta.$$

Show that $\eta = 2\lambda_1{}^A$.

Solution
Note that

$$p_{i \cdot j}^{A \cdot B} = p_{ij}^A / p_j^B$$

and

$$\omega_j = \log\left(\frac{p_{ij}}{p_{2j}}\right) = \log\left(\frac{m_{1j}}{m_{2j}}\right) = \log m_{1j} - \log m_{2j}.$$

If

$$\log m_{ij} = \lambda + \lambda_i^A + \lambda_j^B + \lambda_{ij}^{AB},$$

where

$$\sum_i \lambda_i^A = \sum_j \lambda_j^B = \sum_i \lambda_{ij}^{AB} = \sum_j \lambda_{ij}^{AB} = 0,$$

then

$$\omega_j = \lambda + \lambda_1{}^A + \lambda_j^B + \lambda_{1j}^{AB} - \lambda - \lambda_2{}^A - \lambda_j^B - \lambda_{2j}^{AB}$$
$$= (\lambda_1{}^A - \lambda_2{}^A) + (\lambda_{ij}^{AB} - \lambda_{2j}^{AB})$$
$$= 2(\lambda_1{}^A + \lambda_{1j}^{AB}).$$

Thus ω_j is constant if and only if the λ_{ij}^{AB} are zero.

5.2 Verify the equivalence of (5.2) and (5.3).

Solution

If the log–linear model holds, then

$$
\omega_j = \gamma_j + \sum_{k=1}^{q} \beta_k x_{1jk} - \gamma_j - \sum_{k=1}^{q} \beta_k x_{2jk}
$$

$$
= \sum_{k=1}^{q} \beta_k t_{jk}.
$$

If the logit model holds, then

$$
\begin{aligned}
\log m_{ij} &= \log m_{2j} + \omega_j, & i &= 1, \\
&= \log m_{2j}, & i &= 2.
\end{aligned}
$$

Given this equation, (5.3) follows.

5.3 In Section 5.2, show that if

$$
\omega_{ij} = \alpha_i + \beta_i(j - 2),
$$

then

$$
\begin{aligned}
\eta &= \tfrac{1}{2}(\alpha_1 + \alpha_2), \\
-\eta_2{}^S = \eta_1{}^S &= \tfrac{1}{2}(\alpha_1 - \alpha_2), \\
\eta_j{}^E &= \tfrac{1}{2}(j - 2)(\beta_1 + \beta_2), \\
-\eta_{2j}^{SE} = \eta_{1j}^{SE} &= \tfrac{1}{2}(j - 2)(\beta_1 - \beta_2), \\
\alpha_i &= \eta + \eta_i{}^S, \\
\beta_i &= \eta_3{}^E + \eta_{i3}^{SE}.
\end{aligned}
$$

Solution

Note that

$$
6\eta = \sum_i \sum_j (\eta + \eta_i{}^S + \eta_j{}^E + \eta_{ij}^{SE}) = \sum_i \sum_j [\alpha_i + \beta_i(j - 2)] = 3(\alpha_1 + \alpha_2),
$$

$$
\eta + \eta_i{}^S + \eta_3{}^E + \eta_{i3}^{SE} = \alpha_i + \beta_i,
$$

$$
3(\eta + \eta_i{}^S) = \sum_j (\eta + \eta_i{}^S + \eta_j{}^E + \eta_{ij}^{SE}) = \sum_j [\alpha_i + \beta_i(j - 2)] = 3\alpha_i,
$$

$$
2(\eta + \eta_j{}^E) = \sum_i (\eta + \eta_i{}^S + \eta_j{}^E + \eta_{ij}^{SE}) = \sum_i [\alpha_i + \beta_i(j - 2)]
$$

$$
= (\alpha_1 + \alpha_2) + (\beta_1 + \beta_2)(j - 2).
$$

Thus

$$\alpha_i = \eta + \eta_i^{S},$$
$$\eta = \tfrac{1}{2}(\alpha_1 + \alpha_2),$$
$$\beta_i = \eta_3^{E} + \eta_{i3}^{SE},$$
$$\eta_i^{S} = \alpha_i - \tfrac{1}{2}(\alpha_1 + \alpha_2),$$
$$\eta_j^{E} = \tfrac{1}{2}(\beta_1 + \beta_2)(j - 2),$$

and

$$\eta_{ij}^{SE} = \alpha_i + \beta_i(j - 2) - \eta - \eta_i^{S} - \eta_j^{E}$$
$$= [\beta_i - \tfrac{1}{2}(\beta_1 + \beta_2)](j - 2).$$

The remaining details are based on the observations that

$$
\begin{aligned}
\alpha_i - \tfrac{1}{2}(\alpha_1 + \alpha_2) &= \tfrac{1}{2}(\alpha_1 - \alpha_2), & i &= 1, \\
&= -\tfrac{1}{2}(\alpha_1 - \alpha_2), & i &= 2, \\
\beta_i - \tfrac{1}{2}(\beta_1 + \beta_2) &= \tfrac{1}{2}(\beta_1 - \beta_2), & i &= 1, \\
&= -\tfrac{1}{2}(\beta_1 - \beta_2), & i &= 2.
\end{aligned}
$$

5.4   Verify that in Section 5.4, the Pearson chi-square statistic for ungrouped data can be written

$$\sum_h [X_h(\hat{Q}_h/\hat{P}_h) + (1 - X_h)(\hat{P}_h/\hat{Q}_h)].$$

Verify that

$$S = \sum_h [X_h(Q_h/P_h) + (1 - X_h)(P_h/Q_h)]$$

has mean N and variance

$$\sum_h (1 - 2P_h)^2/U_h.$$

Here $\hat{Q}_h = 1 - \hat{P}_h$ and $Q_h = 1 - P_h$.

Solution

If $X_h = 1$, then

$$(X_h - \hat{P}_h)^2/(\hat{P}_h\hat{Q}_h) = \hat{Q}_h/\hat{P}_h.$$

If $X_h = 0$, then

$$(X_h - \hat{P}_h)^2/(\hat{P}_h\hat{Q}_h)/ = \hat{P}_h/\hat{Q}_h.$$

Thus

$$X_u^2 = \sum_h [X_h(\hat{Q}_h/\hat{P}_h) + (1 - X_h)(\hat{P}_h/\hat{Q}_h)].$$

The summand

$$X_h(Q_h/P_h) + (1 - X_h)(P_h/Q_h)$$

has mean

$$P_h(Q_h/P_h) + Q_h(P_h/Q_h) = 1$$

and variance

$$\left[\frac{Q_h}{P_h} - \frac{P_h}{Q_h}\right]^2 P_h Q_h = \frac{(Q_h{}^2 - P_h{}^2)^2}{P_h Q_h} = \frac{(Q_h + P_h)^2(Q_h - P_h)^2}{P_h Q_h} = \frac{(1 - 2P_h)^2}{P_h Q_h}.$$

Thus

$$\sum_h \left[X_h(Q_h/P_h) + (1 - X_h)(P_h/Q_h)\right]$$

has mean N and variance

$$\sum_h (1 - 2P_h)^2/U_h.$$

Note that this variance may differ considerably from the variance of $2N$ in the chi-square approximation.

5.5 Verify that in Section 5.4,

$$L_u{}^2 = L_{1u}^2 + L_{2u}^2,$$

where $\hat{Q}_h$ and Q_h are defined as in Exercise 5.4,

$$L_{1u}^2 = -2 \sum_h [X_h \log(\hat{P}_h/P_h) + (1 - X_h)\log(\hat{Q}_h/Q_h)],$$

and

$$L_{2u}^2 = -2 \sum_h [X_h \log P_h + (1 - X_h)\log Q_h].$$

Show that L_{2u}^2 has mean

$$-2 \sum_h (P_h \log P_h + Q_h \log Q_h)$$

and variance

$$4 \sum_h \Omega_h{}^2 U_h.$$

Solution

The equation $L_u{}^2 = L_{1u}^2 + L_{2u}^2$ follows since

$$\log \hat{P}_h = \log(\hat{P}_h/P_h) + \log P_h$$

and

$$\log \hat{Q}_h = \log(\hat{Q}_h/Q_h) + \log Q_h.$$

Since X_h has mean P_h, L_{2u}^2 has mean

$$-2\sum_h (P_h \log P_h + Q_h \log Q_h).$$

Since

$$L_{2u}^2 = -2\sum_h \left[X_h \log(P_h/Q_h) + \log Q_h \right]$$

$$= -2\sum_h (\Omega_h X_h + \log Q_h)$$

and X_h has variance U_h, L_{2u}^2 has variance

$$4\sum_h \Omega_h{}^2 U_h.$$

The approximations in Section 5.4 for the mean and variance of $L_2{}^2$ are based on the additional observation that L_{1u}^2 has an approximate $X_q{}^2$ distribution. Thus in large samples, L_{1u}^2 is negligible compared to $L_2{}^2$, provided

$$N^{-1} \sum_h (P_h \log P_h + Q_h \log Q_h)$$

does not approach 0.

5.6 Verify that in Tables 5.4, the model

$$\Omega_{gi} = \alpha_i + \beta_{i1}(T_{gi}) + \beta_{i2}(T_{gi})^2,$$

results in a likelihood-ratio chi-square statistic for ungrouped data of $L_u'^2 = 3{,}339.8$ and a likelihood-ratio chi-square statistic for grouped data of $L_u'^2 = 71.1$.

Solution

Details of the numerical computation are omitted. Calculations closely parallel those for the linear logit model defined in Section 5.3 and those for the general logit model of (5.1) in the case of $q = 2$. For sex i, T_{h1} and T_{h2} are T_{gi} and T_{gi}^2, respectively.

5.7 Use the weights u_{ij2} and the probabilities $p_{1.ij3}$ of Table 5.5 to derive Table 5.7. Note that $\hat{\theta}_1 = 11.258$ and $\hat{\theta}_2 = 11.263$.

5.8 Consider the data in Table 5.8. Let ω_{ij} be the logit for sex i and educational group j. Are these data consistent with the logit model

$$\omega_{ij} = \eta + \beta(j - 2)?$$

Table 5.8

Subjects in 1975 General Social Survey, Cross-Classified
by Sex, Education, and Belief that Most Men Are
Better Suited Emotionally for Politics than Are
Most Women[a,b]

Sex	Education in years	Response	
		Agree	Disagree
Male	$\leqslant 8$	65	46
	9–12	145	153
	$\geqslant 13$	84	133
Female	$\leqslant 8$	83	38
	9–12	255	207
	$\geqslant 13$	76	142

[a] Data tape from the National Opinion Research
Center, 1975 General Social Survey, University of
Chicago, Chicago.

[b] The exact question, as reported in National Opinion
Research Center (1975, page 52) is "Tell me if you agree
or disagree with this statement: Most men are better
suited emotionally for politics than are most women."
There were 59 subjects who were not sure they agree or
disagreed, 2 subjects who did not respond, and 2 other
subjects who did not report their educational level.

If so, find an approximate 95 percent confidence interval for β. What does
this model say about the relationship of education and sex to response to
the question on relative emotional suitability of the two sexes for political
activity?

Solution

Maximum-likelihood estimates of the expected cell counts are shown in
Table 5.9. The L^2 statistic is 7.02 and X^2 is 7.06. Since there are $6 - 2 = 4$
degrees of freedom, the approximate significance level is 0.13 for both
statistics. Thus the model is fairly consistent with the data, although, just
as in Table 5.2, a thorough analysis produces some indication of an interaction between sex and education. Note the adjusted residuals in Table 5.9.
The model implies that only education is associated with response and that
the log odds of a positive rather than a negative response are linear in
education level.

Since $\hat{\beta} = -0.571$ and $s(\hat{\beta}) = 0.0824$, an approximate 95 percent confidence interval for β has bounds

$$-0.571 - 1.96(0.0824) = -0.732$$

Table 5.9

Maximum-Likelihood Estimates of Cell Means
and Adjusted Residuals for Logit Model for Table 5.8[a]

Sex	Education in years	Response	
		Agree	Disagree
Male	$\leqslant 8$	72.6	38.4
		-1.81	1.81
	$9-12$	153.9	144.1
		-1.18	1.18
	$\geqslant 13$	81.7	135.3
		0.42	-0.42
Female	$\leqslant 8$	79.1	41.9
		0.90	-0.90
	$9-12$	238.6	223.4
		1.89	-1.89
	$\geqslant 13$	82.1	135.9
		-1.10	1.10

[a] First line is maximum-likelihood estimate and second
line is adjusted residual.

and

$$-0.571 + 1.96(0.0824) = -0.409.$$

5.9   Consider Table 5.10. Given the subjects response to item A, do sex
and education have any effect on the response to item D?

Solution

For subject h, $1 \leqslant h \leqslant 1391$, let X_h be 1 if the subject agrees with item D
and 0 if the subject disagrees.

Let T_{h1} be 1 if subject h agrees with item A and 0 if the subject disagrees.
Let T_{h2} be 1 if subject h is male, and let T_{h2} be 0 if the subject is female. Let
T_{h3} be the number of years of education of subject h. Consider the two
logit models

$$\Omega_h = \eta + \beta_1 T_{h1}$$

and

$$\Omega_h = \eta + \beta_1 T_{h1} + \beta_2 T_{h2} + \beta_3 T_{h3}.$$

The difference in likelihood ratio chi-squares for these two models is only
1.94. Since there are two degrees of freedom, there is no indication that
given item A, sex and education help predict the response to item D. Of
course, this statement does *not* imply that sex and education are unrelated

Table 5.10

Responses of Subjects in the 1975 General Social Survey to Two Questions on Women's Roles[a,b]

Education in years	Response to item A: Response to item D:	Male Agree		Male Disagree		Female Agree		Female Disagree	
		Agr.	Dis.	Agr.	Dis.	Agr.	Dis.	Agr.	Dis.
0		1	0	0	2	3	1	0	0
1		1	0	0	0	1	0	0	0
2		2	1	0	0	0	0	0	0
3		1	0	0	2	4	0	0	0
4		0	0	0	3	4	0	0	0
5		3	3	4	3	4	2	1	5
6		10	3	0	5	7	1	0	0
7		9	1	0	6	13	1	2	7
8		28	5	5	12	38	5	5	15
9		13	0	3	9	14	0	7	15
10		11	1	5	16	27	3	7	26
11		12	4	9	23	24	4	13	23
12		51	11	38	83	84	14	65	121
13		13	2	12	23	7	0	13	35
14		11	1	11	25	9	1	14	28
15		3	1	5	9	2	0	2	10
16		9	1	8	47	6	0	11	44
17		0	0	2	6	1	0	4	11
18		0	0	1	10	0	0	2	8
19		0	0	4	3	0	0	0	2
20		1	0	1	5	1	1	0	2

[a] Data tape from National Opinion Research Center, 1975 General Social Survey, University of Chicago.

[b] Question A is the question used in Table 3.4 and Table 3.20. Question D is the question used in Table 5.8. Thus A refers to whether women should run their homes and leave men to run the country, and D refers to whether men are more suited emotionally for politics than are women. In all 1391 of 1490 respondents are included in the table.

to item D. As noted in Exercise 5.8, these variables are associated with item D. Residual analyses may also be used to supplement the likelihood-ratio test used here. Such analyses do not appear to provide much evidence for an additional role of sex and education in prediction of response to item D.

References

Anscombe, F. J. (1956). On estimating binomial response relations. *Biometrika* **43**, 461–464.

Armitage, P. (1955). Tests for linear trends in proportions and frequencies. *Biometrics* **11**, 375–386.

Bartlett, M. S. (1935). Contingency table interactions. *J. Roy. Statist. Soc. Suppl.* **1**, 248–252.

Berkson, J. (1944). Application of the logistic function to bio-assay. *J. Amer. Statist. Assoc.* **39**, 357–365.

Berkson, J. (1953). A statistically precise and relatively simple method of estimating the bio-assay with quantal response, based on the logistic function. *J. Amer. Statist. Assoc.* **48**, 565–599.

Berkson, J. (1955). Maximum likelihood and minimum χ^2 estimates of the logistic function. *J. Amer. Statist. Assoc.* **50**, 130–162.

Birch, M. W. (1963). Maximum likelihood in three-way contingency tables. *J. Roy. Statist. Soc. Ser. B* **25**, 220–233.

Birch, M. W. (1964). The detection of partial association, I: the 2×2 case. *J. Roy. Statist. Soc. Ser. B* **26**, 313–324.

Bishop, Y. M. M. (1969). Full contingency tables, logits, and split contingency tables. *Biometrics* **25**, 383–399.

Bishop, Y. M. M. (1971). Effects of collapsing multidimensional contingency tables. *Biometrics* **27**, 545–562.

Bishop, Y. M. M., and Fienberg, S. E. (1969). Incomplete two-dimensional contingency tables. *Biometrics* **22**, 119–128.

Bishop, Y. M. M., Fienberg, S. E., and Holland, P. W. (1975). *Discrete multivariate analysis: theory and practice*. MIT Press, Cambridge, Massachusetts.

Bishop, Y. M. M., and Mosteller, F. (1969). J. P. Bunher, W. H. Forest, Jr., F. Mosteller, and L. D. Vandam (Eds.), *The national halothane study*, Chapter IV-3. U. S. Govt. Printing Office, Washington, D. C.

Bock, R. D. (1970). Estimating multinomial response relations. In R. C. Bose *et al.* (Eds.), *Essays in probability and statistics*, 111–132. Univ. of North Carolina Press, Chapel Hill, North Carolina.

Bock, R. D. (1975). *Multivariate statistical methods in behavioral research*. McGraw–Hill, New York.

Bock, R. D., and Jones, L. V. (1968). *The measurement and prediction of judgement and choice.* Holden–Day, San Francisco, California.

Bock, R. D., and Yates, G. (1973). *MULTIQUAL: log-linear analysis of nominal or ordinal qualitative data by the method of maximum likelihood.* International Education Services, Chicago, Illinois.

Chambers, E. A., and Cox, D. R. (1967). Discrimination between alternative binary response models. *Biometrika* **54**, 573–578.

Cochran, W. G. (1952). The X^2 test of goodness of fit. *Ann. Math. Statist.* **23**, 315–345.

Cochran, W. G. (1954). Some methods for strengthening the common X^2 tests. *Biometrics* **10**, 417–451.

Cochran, W. G. (1955). A test of a linear function of the deviations between observed and expected numbers. *J. Amer. Statist. Assoc.* **50**, 377–397.

Cornfield, J., Kannel, W., and Truett, J. (1967). A multivariate analysis of the risk of coronary heart disease in Framingham. *J. Chron. Dis.* **20**, 511–524.

Cox, D. R. (1970). *Analysis of binary data.* Methuen, London.

Cox, D. R., and Snell, E. J. (1968). A general definition of residuals. *J. Roy. Statist. Soc. Ser. B* **30**, 248–265.

Darroch, J. N. (1962). Interactions in multi-factor contingency tables. *J. Roy. Statist. Soc. Ser B* **24**, 251–263.

Darroch, J. N., and Ratcliff, D. (1972). Generalized iterative scaling for loglinear models. *Ann. Math. Statist.* **43**, 1470–1480.

David, H. A. (1963). *The method of paired comparisons.* Hafner, New York.

Dawson, R. B. (1954). A simplified expression for the variance of the χ^2-function on a contingency table. *Biometrika* **41**, 280.

Deming, W. E., and Stephan, F. F. (1940). On a least squares adjustment of a sampled frequency table when the expected marginal totals are known. *Ann. Math. Statist.* **11**, 427–444.

Dixon, W. J. (1975). *BMDP: biomedical computer programs.* Univ. of California Press, Berkeley.

Dowdall, J. A. (1974). Women's attitudes toward employment and family roles. *Sociological Analysis* **35**, 251–262.

Draper, N. R., and Smith, H. (1966). *Applied regression analysis.* Wiley, New York.

Durkheim, E. (1951[1897]). *Suicide* (J. A. Spaulding and G. Simpson, trans.). Free Press, Glencoe, Illinois.

Dyke, G. V., and Patterson, H. D. (1952). Analysis of factorial arrangements when the data are proportions. *Biometrics* **8**, 1–12.

Fay, R. E., and Goodman, L. A. (1975). *ECTA program: description for users.* Department of Statistics, Univ. of Chicago, Chicago, Illinois.

Feller, W. (1968). *An introduction to probability theory and its applications, I.* (3rd ed.) Wiley, New York.

Fienberg, S. J. (1972). The multiple-recapture census for closed populations and incomplete 2^k contingency tables. *Biometrika* **59**, 591–603.

Finney, D. J. (1964). *Statistical methods in biological assay.* (2nd ed.) Griffin, London.

Finney, D. J. (1971). *Probit Analysis.* (3rd ed.) Cambridge Univ. Press, London and New York.

Fisher, R. A. (1922). On the interpretation of chi square from contingency tables, and the calculation of *P. J. Roy. Statist. Soc.* **85**, 87–94.

Fisher, R. A. (1924). The conditions under which χ^2 measures the discrepancy between observation and hypothesis. *J. Roy. Statist. Soc.* **87**, 442–450.

Fisher, R. A. (1970[1925]). *Statistical methods for research workers.* (14th ed.) Hafner, New York.

Fisher, R. A. (1935). Appendix to Bliss, C. I.: the case of zero survivors. *Ann. Appl. Biol.* **22**, 164–165.

Fisher, R. A., Thornton, H. G., and Mackenzie, W. A. (1922). The accuracy of the plating method of estimating the density of bacterial populations. *Ann. Appl. Biol.* **9**, 325–359.

Fisher, R. A., and Yates, F. (1963[1938]). *Statistical tables for biological, agricultural and medical research.* (6th ed.) Hafner, New York.

Gart, J. J., and Zweifel, J. R. (1967). On the bias of various estimators of the logit and its variance with application to quantal bioassay. *Biometrika* **54**, 181–187.

Garwood, F. (1941). The application of maximum likelihood to dosage-mortality curves. *Biometrika* **32**, 46–58.

Good, I. J. (1958). The interaction algorithm and practical Fourier analysis. *J. Roy. Statist. Soc. Ser. B* **20**, 361–372.

Good, I. J., Gover, T. N., and Mitchell, G. J. (1970). Exact distributions for X^2 and for the likelihood-ratio statistic for the equiprobable multinomial distribution. *J. Amer. Statist. Assoc.* **65**, 267–283.

Goodman, L. A. (1963). On Plackett's test for contingency table interactions. *J. Roy. Statist. Soc. Ser. B* **25**, 179–188.

Goodman, L. A. (1964a). Simultaneous confidence intervals for cross-product ratios in contingency tables. *J. Roy. Statist. Soc. Ser. B* **26**, 86–102.

Goodman, L. A. (1964b). Simple methods for analyzing three-factor interaction in contingency tables. *J. Amer. Statist. Assoc.* **59**, 319–352.

Goodman, L. A. (1965). On the multivariate analysis of three dichotomous variables. *Amer. J. Sociol.* **71**, 290–301.

Goodman, L. A. (1969). On partitioning χ^2 and detecting partial association in three-way contingency tables. *J. Roy. Statist. Soc. Ser. B* **31**, 486–498.

Goodman, L. A. (1970). The multivariate analysis of qualitative data: interactions among multiple classifications. *J. Amer. Statist. Assoc.* **65**, 226–256.

Goodman, L. A. (1971a). The analysis of multidimensional contingency tables: stepwise procedures and direct estimation methods for building models for multiple classification. *Technometrics* **13**, 33–61.

Goodman, L. A. (1971b). Partitioning of chi-square, analysis of marginal contingency tables, and estimation of expected frequencies in multidimensional contingency tables. *J. Amer. Statist. Assoc.* **66**, 339–344.

Goodman, L. A. (1972). A general model for the analysis of surveys. *Amer. J. Sociol.* **77**, 1035–1086.

Goodman, L. A. (1973a). Causal analysis of data from panel studies and other kinds of surveys. *Amer. J. Sociol.* **78**, 1135–1191.

Goodman, L. A. (1973b). The analysis of multidimensional contingency tables when some variables are posterior to others: a modified path analysis approach. *Biometrika* **60**, 179–192.

Goodman, L. A. (1973c). Guided and unguided methods for the selection of models for a set of T multidimensional contingency tables. *J. Amer. Statist. Assoc.* **68**, 165–175.

Goodman, L. A. (1975). The relationship between the modified and more usual regression approaches to the analysis of dichotomous variables. In D. Heise (Ed.), *Sociological methodology, 1976*, 83–110. Jossey–Bass, San Francisco, California.

Goodman, L. A., and Kruskal, W. H. (1954). Measures of association for cross-classifications. *J. Amer. Statist. Assoc.* **49**, 732–764.

Goodman, L. A., and Kruskal, W. H. (1959). Measures of association for cross-classifications, II: further discussion and references. *J. Amer. Statist. Assoc.* **54**, 123–163.

Goodman, L. A., and Kruskal, W. H. (1963). Measures of association for cross-classifications, III: approximate sampling theory. *J. Amer. Statist. Assoc.* **58**, 310–364.

Goodman, L. A., and Kruskal, W. H. (1972). Measures of association for cross-classifications, IV: simplification of asymptotic variances. *J. Amer. Statist. Assoc.* **67**, 415–421.

Grizzle, J. E. (1967). Continuity correction in the χ^2-test for 2×2 tables. *Amer. Statist.* **21**(4), 28–33.

Haberman, S. J. (1972). Log-linear fit for contingency tables. *Appl. Statist.* **21**, 218–225.

Haberman, S. J. (1973a). Log-linear models for frequency data: sufficient statistics and likelihood equations. *Ann. Statist.* **1**, 617–632.

Haberman, S. J. (1973b). The analysis of residuals in cross-classified tables. *Biometrics* **29**, 205–220.

Haberman, S. J. (1973c). *CTAB: analysis of multidimensional contingency tables by log-linear models: user's guide.* International Educational Services, Chicago, Illinois.

Haberman, S. J. (1974). *The analysis of frequency data.* Univ. of Chicago Press, Chicago, Illinois.

Haberman, S. J. (1974b). Log-linear models for frequency tables with ordered classifications. *Biometrics* **30**, 589–600.

Haberman, S. J. (1975). Direct products and linear models for complete factorial tables. *Ann. Statist.* **3**, 314–333.

Haberman, S. J. (1976). Generalized residuals for log-linear models. *Proc. Ninth Inter. Biometrics Conf.* **1**, 104–122.

Haberman, S. J. (1977a). Maximum likelihood estimates in exponential response models. *Ann. Statist.* **5**, 815–841.

Haberman, S. J. (1977b). Log-linear models and frequency tables with small expected cell counts. *Ann. Statist.* **5**, 1148–1169.

Haldane, J. B. S. (1939). The mean and variance of χ^2, when used as a test of homogeneity, when expectations are small. *Biometrika* **31**, 346–355.

Haldane, J. B. S. (1955). The estimation and significance of the logarithm of a ratio of frequencies. *Ann. Hum. Genet.* **20**, 309–311.

Hollingshead, A. B., and Redlich, F. C. (1967[1958]). *Social class and mental illness: a community study.* Wiley, New York.

Ivanov, V. (1962). A general approximation method for solving linear problems. *Sov. Math.* **3**, 476–479.

Kastenbaum, M. A., and Lamphiear, D. E. (1959). Calculation of chi-square to test the no three-factor interaction hypothesis. *Biometrics* **15**, 107–115.

Kullback, S. (1968[1959]). *Information theory and statistics.* Dover, New York.

Lancaster, H. O. (1951). Complex contingency table, treated by the partition of χ^2. *J. R. Statist. Soc. B* **13**, 242–249.

Lancaster, H. O. (1969). *The chi-squared distribution.* Wiley, New York.

Lehmann, E. L. (1959). *Testing statistical hypotheses.* Wiley, New York.

Leslie, P. H. (1951). The calculation of χ^2 for an $r \times c$ contingency table. *Biometrics* **7**, 283–286.

Lewontin, R. C., and Felsenstein, J. (1965). The robustness of homogeneity tests in $2 \times n$ tables. *Biometrics* **21**, 19–33.

Lieberman, G. J., and Owen, D. B. (1961). *Tables of the hypergeometric probability distribution.* Stanford Univ. Press, Stanford, California.

Lynd, R. S., and Lynd, H. M. (1956[1929]). *Middletown: a study in modern american culture.* Harcourt, New York.

Mantel, N. (1963). Chi-square tests with one degree of freedom; extensions of the Mantel–Haenszel procedure. *J. Amer. Statist. Assoc.* **58**, 690–700.

Mantel, N., and Haenszel, W. (1959). Statistical aspects of the analysis of data from retrospective studies. *J. Nat. Cancer Inst.* **22**, 719–748.

Margolin, B. H., and Light, R. J. (1974). An analysis of variance for categorical data, II: small sample comparisons with chi square and other competitors. *J. Amer. Statist. Assoc.* **69**, 755–764.

McFadden, D. (1976). Quantal choice analysis, a survey. *Ann. Econ. Soc. Meas.* **4**, 363–390.

National Center for Health Statistics (1968). *Vital statistics of the United States, 1968*, **2**, Part A. U. S. Govt. Printing Office, Washington, D.C.

National Center for Health Statistics (1969). *Vital statistics of the United States, 1969*, **2**, Part A. U.S. Govt. Printing Office, Washington, D.C.

National Center for Health Statistics (1970). *Vital statistics of the United States, 1970*, **2**, Part A. U.S. Govt. Printing Office, Washington, D.C.

National Opinion Research Center (1972). *Codebook for the spring 1972 general social survey.* National Opinion Research Center, Univ. of Chicago, Chicago, Illinois.

National Opinion Research Center (1973). *Codebook for the spring 1973 general social survey.* National Opinion Research Center, Univ. of Chicago, Chicago, Illinois.

National Opinion Research Center (1974). *Codebook for the spring 1974 general social survey.* National Opinion Research Center, Univ. of Chicago, Chicago, Illinois.

National Opinion Research Center (1975). *Codebook for the spring 1975 general social survey.* National Opinion Research Center, Univ. of Chicago, Chicago, Illinois.

Nelder, J. A., and Wedderburn, R. W. M. (1972). Generalized linear models. *J. Roy. Statist. Soc. Ser. A* **135**, 370–384.

Nerlove, M., and Press, S. J. (1973). *Univariate and multivariate log-linear and logistic models.* Rand Corp., Santa Monica, California.

Neyman, J., and Pearson, E. S. (1928). On the use and interpretation of certain test criteria for purposes of statistical inference. Part II. *Biometrika* **20**, 263–294.

Nicholson, W. L. (1956). On the normal approximation to the hypergeometric distribution. *Ann. Math. Statist.* **27**, 471–483.

Norton, H. W. (1945). Calculation of chi-square for complex contingency tables. *J. Amer. Statist. Assoc.* **40**, 251–258.

Odoroff, C. L. (1970). A comparison of minimum logit chi-square estimation and maximum likelihood estimation in $2 \times 2 \times 2$ and $3 \times 2 \times 2$ contingency tables: tests for interaction. *J. Amer. Statist. Assoc.* **65**, 1617–1631.

Ostrowski, A. M. (1966). *Solution of equations and systems of equations.* Academic Press, New York.

Paykel, E. S., Prusoff, B. A., and Uhlenhuth, E. H. (1971). Scaling life events. *Arch. Gen. Psychiatry* **25**, 340–347.

Pearson, E. S. (1947). The choice of statistical tests illustrated on the interpretation of data classed in a 2×2 table. *Biometrika* **34**, 139–167.

Pearson, K. (1900). On a criterion that a given system of deviations from the probable in the case of a correlated system of variables is such that it can be reasonably supposed to have arisen from random sampling. *Philos. Mag.* **50(5)**, 157–175.

Pearson, K. (1904). Mathematical contributions to the theory of evolution. XIII. On the theory of contingency and its relation to association and normal correlation. *Drapers' Company Research Memoirs, Biometric Series*, **1**.

Pearson, K. (1911). On the probability that two independent distributions of frequency are really samples from the same population. *Biometrika* **8**, 250–254.

Pearson, K. (1916). On the general theory of multiple contingency with special reference to partial contingency. *Biometrika* **11**, 145–158.

Plackett, R. L. (1962). A note on interactions in contingency tables. *J. Roy. Statist. Soc. Ser. B* **24**, 162–166.

Plackett, R. L. (1964). The continuity correction in 2×2 tables. *Biometrika* **51**, 327–337.

Plackett, R. L. (1974). *The analysis of categorical data*. Griffin, London.

Rao, C. R. (1973). *Linear statistical inference and its applications*. (2nd ed.) Wiley, New York.

Rasch, G. (1960). *Probabilistic models for some intelligence and attainment tests*. Nielsen and Lydiche, Copenhagen.

Roy, S. N., and Kastenbaum, M. A. (1956). On the hypothesis of no "interaction" in a multiway contingency table. *Ann. Math. Statist.* **27**, 749–757.

Roy, S. N., and Mitra, S. K. (1956). An introduction to some nonparametric generalizations of analysis of variance and multivariate analysis. *Biometrika* **43**, 361–376.

Stuart, A. (1950). The cumulants of the first n natural numbers. *Biometrika* **37**, 446.

Theil, H. (1970). On the estimation of relationships involving qualitative variables. *Amer. J. Sociol.* **76**, 103–154.

Tukey, J. W. (1962). The future of data analysis. *Ann. Math. Statist.* **33**, 1–67.

Simpson, E. H. (1951). The interpretation of interaction in contingency tables. *J. Roy. Statist. Ser. B* **13**, 238–241.

Uhlenhuth, E. H., Balter, M. B., Lipman, R. S., and Haberman, S. J. (1977). Remembering life events. In J. S. Strauss, H. M. Babigian, and M. Roff (Eds.), *The origin and course of psychopathology: methods of longitudinal research*, 117–134. Plenum, New York.

Uhlenhuth, E. H., Lipman, R. S., Balter, M. B. and Stern, M. (1974). Symptom intensity and life stress in the city. *Arch. Gen. Psychiatry* **31**, 759–764.

Walker, S. H., and Duncan, D. B. (1967). Estimation of the probability of an event as a function of several independent variables. *Biometrika* **54**, 167–179.

Wilks, S. S. (1935). The likelihood test of independence in contingency tables. *Ann. Math. Statist.* **6**, 190–196.

Wilks, S. S. (1938). The large-sample distribution of the likelihood ratio for testing composite hypotheses. *Ann. Math. Statist.* **9**, 60–62.

Woolf, B. (1955). On estimating the relation between blood groups and disease. *Ann. Hum. Genet.* **19**, 251–253.

Yarnold, J. K. (1970). The minimum expectation of X^2 goodness of fit test. *J. Amer. Statist. Assoc.* **65**, 864–886.

Yates, F. (1934). Contingency tables involving small numbers of the X^2 test. *J. Roy. Statist. Soc. Suppl.* **1**, 217–23 .

Yates, F. (1937). The design and analysis of factorial experiments. *Commonwealth Bureau of Soil Science Tech. Comm.*, **35**.

Yates, F. (1955). The use of transformations and maximum-likelihood in the analysis of quantal experiments involving two treatments. *Biometrika* **42**, 382–403.

Yule, G. U. (1900). On the association of attributes in statistics. *Phil. Trans. Ser. A* **194**, 257–319.

Yule, G. U. (1912). On the methods of measuring association between two attributes. *J. Roy. Statist. Soc.* **75**, 579–642.

Example Index

Abortion attitudes, by religion, education, and year
  Source: Data tape from 1972, 1973, 1974 General Social Surveys, National Opinion Research Center, University of Chicago, 262–264, 272–276, 278–284, 290–291
Attitudes toward courts' treatment of criminals, by year of survey
  Source: National Opinion Research Center (1972, p. 29; 1973, p. 57; 1974, p. 55; 1975, p. 58), 120–123
Attitudes toward women staying at home, by sex and education
  Source: Data tapes from 1974 amd 1975 General Social Survey, National Opinion Research Center, University of Chicago, 183–193, 239–240, 247–248, 259–261, 302–303, 310–312, 330–331
Attitudes toward women in politics, by sex and education
  Source: Data tape from 1975 General Social Survey, National Opinion Research Center, University of Chicago, 351–352
Attitudes toward women's roles, by sex and education
  Source: Data tape from 1974, 1975 General Social Surveys, National Opinion

Research Center, University of Chicago, 352–353
Education of husbands and wives in terms of highest attained degrees
  Source: Data tape from 1974 General Social Survey, National Opinion Research Center, University of Chicago, 226–230
Ethnicity and women's role attitudes
  Source: Dowdall (1974), 109, 111–112
Homicides in U.S. in 1970, by race and sex of victim and by type of assault
  Source: National Center for Health Statistics, (1970, pp. 1–183, 6–17), 162–168, 179, 224–226, 236
Homicides: monthly distribution in U.S. in 1970
  Source: National Center for Health Statistics (1970, pp. 1-174–1-175), 82–83
Life stress and memory
  Source: Data file used in Uhlenhuth et al. (1974, 1977), 2–4, 7–8, 23, 79–80
Neurosis and treatment
  Source: Hollingshead and Redlich (1967[1958], p. 260), 154–156
Political views
  Source: National Opinion Research Center (1975, p. 36), 84–85

Psychoses and treatment
    Source: Hollingshead and Redlich
    (1967[1958], p. 288), 112–117, 153–154
Social class, by year of survey
    Source: National Opinion Research
    Center (1972, p. 45; 1973, p. 29; 1974, p.
    29; 1975, p. 41), 155–156
Social class, self-classification by
    Source: National Opinion Research
    Center (1975, p. 70), 23–25
Suicide distribution, by day of week
    Source: Durkheim (1951[1897], p. 118),
    87

Suicide distribution, by region
    Source: National Center for Health
    Statistics (1970, pp. 2-49, 1-65, 6-28),
    53–60
Suicide, monthly distribution in U.S. in
    1968, 1969, 1970
    Source: National Center for Health
    Statistics (1968, p. 1-132; 1969, p.
    1-132; 1970, pp. 1-174, 1-175), 43–52,
    125, 128
Traits desired in parents, by sex of child
    Source: Lynd and Lynd (1956[1929], p.
    524), 93, 95, 99, 106, 151–152

Subject Index

A

Additive logit model, 293–295, 301
  Newton–Raphson algorithm, 295, 299
Additive log-linear model, 102, 133–136, 157
  adjusted residuals, 150
  chi-square statistics, 148
  equivalence to homogeneity model, 141–
    142
  equivalence to independence model,
    139–141
  maximum-likelihood equations, 136–138
  use with Poisson data, 124, 126, 131–132
Additivity condition, 101, 225
Adjusted residuals, 78–79, 82–83
  in additive log-linear model, 130–131,
    150–151
  in equiprobability model, 5–6
  in hierarchical model, 231, 236, 254, 272,
    275
  in homogeneity model, 121–122, 155, 157
  in independence model, 104, 111, 114–
    115, 156, 228–229
  in logit model, 328–330, 342, 351–352
  in log-linear time-trend model, 17–20, 81
  for model of conditional independence,
    188, 242
  in model of constant rates, 46–48
  in quadratic log-linear model, 39
  rankit plot, 20–21
  in symmetry model, 27–28
Analysis of covariance, 215
Analysis of variance, 101, 169, 294
Asymmetry, measure of, 41
Asymptotic covariance matrix, 40, 70, 143
  estimate of, 40, 132, 143, 217, 301, 340
Asymptotic mean, 22, 40, 70–71, 164, 167–
  168, 177, 230
Asymptotic standard deviation, 221
  estimate of, 22, 41, 71, 89, 100, 106, 164,
    212–213, 231–232, 234, 249, 252, 254,
    266–267, 284, 309, 324
  estimate for log cross-product ratio, 116–
    117, 262
  estimate for log odds ratio, 96
  estimate for odds ratio, 96
Asymptotic variance, 22, 57, 70–71, 164,
  168, 212, 221, 230, 242–243
  estimate of, 22, 57, 71, 144, 148, 164, 169,
    218, 220, 295, 342–343, 345
  of estimated log cross-product ratio, 105,
    147, 165–167, 176–177, 180, 222
  of maximum-likelihood estimate in satu-
    rated model, 72–74

B

Bonferroni simultaneous confidence intervals, *see* Simultaneous confidence intervals

C

Chi-square approximation, 75, 149, 188
  in additive log-linear model, 149
  in equiprobability model, 5
  in independence model, 104, 107, 110–111, 115
  in logit model, 325–327, 341, 349
  in log-linear time-trend model, 15
  for model of no three-factor interaction, 223
  in symmetry model, 27.
Chi-square statistic, 46, 50–52, 75, 82, 88, 113, 121, 126, 176, 190, 223–224, 228, 239, 256, 258, 269–270, 308, 325, *see also* Chi-square approximation, Degrees of freedom, Likelihood-ratio chi-square statistic, Pearson chi-square statistic
Closed-form estimate, 254, 259, 275, 286
Coefficient of association, 105
Coefficient of colligation, 105
Column-multinomial model, 134–136, 138, 144, 146, 332
Conditional independence, 161, 185–187, 190, 199–201, 218–219, 221, 226, 243, 245–246, 254, 258–260, 285, 287–288
  adjusted residual, 188, 242
  chi-square statistics, 187, 190
  maximum-likelihood estimates, 187, 190, 194, 200
Conditional probability, 98, 100, 106–107, 184, 199, 248, 263, 276–277, 302
  estimation, 99, 101
Confidence interval, approximate, 71, 82–83, 89, 144, 310
  for cross-product ratio, 106, 117–118, 151–154, 178–180, 189
  for differences in log rates, 57–58
  for log cross-product ratio, 106, 117–118, 152–154, 165–167, 177, 179–180, 189
  in logit model, 324, 351
  in log-linear time-trend model, 22, 80–81

for marginal probability, 97
in quadratic log-linear model, 41–42
Continuity correction, 107–108
Cross-classification, 247, 285
  of three variables, 160, 163, 223, 261
  of two variables, 91–92, 109, 112, 119
Cross-product ratio, 104, 107, 114, 117, 151–152, 178–179, 184–185, 198, 278, 280–281

D

Degrees of freedom, 46, 149, 228
  for additive logit model, 301
  for additive log-linear model, 126, 149
  for comparison of likelihood-ratio chi-square statistics, 18, 133, 257–258, 269, 271, 327, 341, 353
  for hierarchical model, 223–225, 257–258, 286, 289
  for homogeneity model, 121, 123, 155
  for independence model, 104, 110, 114, 155
  for logit model, 308, 325–326, 341, 351
  for log-linear time-trend model, 15–16
  for marginal independence model, 186
  for model of conditional independence, 187, 191
  for model of constant rates, 50, 54, 56–57, 88–89
  for model of no three-factor interaction, 176, 239
  for quadratic log-linear model, 38
  for symmetry model, 26–27
  for variance test, 86–87
Delta (δ) method, 95
Deming–Stephan algorithm, 125, 127, 157, 192, 258, 269, 286, *see also* Iterative proportional fitting
Discriminant analysis, 337

E

Empirical logarithm, 32, 128, 173
Empirical logit, 305, 314, 334
Empirical weight, 305, 314, 334
Equiprobability model, 3–4, 7, 16, 61, 65, 161, 205–206, 254, 259, 287

chi-square test, 4–5
  comparison with log-linear time-trend
    model, 16–17
  residual analysis, 5–7
Exponential decay model, 7–9
Exponential function, basic properties, 9

F

F statistic, 16–17, 50, 56, 75–76, 88–89, 223,
  225, 269
Fisher exact test, 107–108
Fisher variance test, *see* Variance test
Four-way table, 247, 259, 262

G

Generating class, 193, 198–199, 201, 203,
  205–207, 218, 221, 224–225, 227, 231,
  254–255, 258–259, 269–270, 272, 285–
  286

H

Hierarchical model, 160–161, 181, 187,
  197–198, 217, 225, 231–233, 247–249,
  252, 259–270, 268, 271, 293
  adjusted residual, 231, 273, 275
  chi-square statistics, 223–225
  iterative proportional fitting, 217–218, 259
  maximum-likelihood estimation for,
    192–193, 254–255, 270, 285–286
Higher-way table, 247
Homogeneity model, 119–123, 141–142,
  155, 157

I

Independence model, 101–103, 110, 113,
  139, 141–142, 152, 161, 201–203, 205,
  254, 259–260, 287–288
  adjusted residual, 104, 111
  chi-square tests, 103–104, 110–111, 113–
    114, 228
  compared to homogeneity model, 119–120
  maximum-likelihood estimation, 103, 110,
    113–114

Interaction, 103, 160, 170, 192, 199–200,
  203, 233, 265, 268–270, 276, 279, 282–
  284, 309–310
Iterative proportional fitting, 126, 139, 192–
  193, 196, 217–218, 239, 241, 254, 259,
  268–269, 286, *see also* Deming –
  Stephan algorithm

L

Likelihood-ratio chi-square statistic, 86, 149
  in additive logit model, 301
  comparison with likelihood-ratio chi-
    square statistic from a different
    model, 16, 76, 79, 221, 257–258, 260,
    270–271, 326–327, 341–342, 353
  for conditional independence, 187
  in equiprobability model, 5, 79–80
  for hierarchical model, 224–225, 260
  in homogeneity model, 123
  in independence model, 104, 110
  in logit model, 326–327, 341, 350
  in log-linear time-trend model, 15–17,
    79–80
  in marginal independence model, 186
  in model of constant rates, 50–51, 54, 57
  in quadratic log-linear model, 38
  in symmetry model, 26
Linear combination of residuals, analysis
  of, 6
Logarithm, basic properties of, 8
Log cross-product ratio, 101, 105, 107, 115,
  117, 142, 145–147, 151–152, 165, 176,
  178–180, 187, 189, 199, 243, 260–261,
  276, 278–281
Logit, log odds, 292, 318, 321–322
  estimate, 96, 182, 235–236, 277, 279,
    310–311, 324, 330–331
  estimated asymptotic standard deviation,
    96
  in four-way table, 276–277
  in three-way table, 179, 181, 225, 294, 302,
    310–311, 313, 318
  in two-way table, 95, 100, 103
Logit model, 292–293, 295, 303–305, 309–
  313, 316, 331–332, 344–347, 350, 352
  adjusted residual for, 328, 330
  chi-square statistics for, 325–327, 350

normal approximation for maximum-likelihood estimate, 323–324
Newton–Raphson algorithm for, 305, 317, 319
Log-linear model, 25, 60–61, 69, 75, 89, 143, 149–150, 160, 181, 213–295
  applicability to Poisson data, 43, 61
  for conditional independence, 186
  for independence, 101, 110, 113
  relationship to logit model, 292, 313, 316, 332, 346–347
  representation of model of constant rates as, 48, 54
  for two-way table, 133, 136, 144
Log-linear time-trend model, 7–8, 61, 79
  comparison to equiprobability model, 16–17
  large-sample distribution of maximum-likelihood estimate, 22
  maximum-likelihood estimation, 9, 11
  residual analysis, 17, 20–21
Log odds, *see* Logit, log odds

M

Main effects, 102–103, 160
Marginal association, 184–185, 201, 203
Marginal independence, 185, 226
  chi-square test, 185–186
Marginal probability, 94–95, 140–141, 184, 199, 243
  estimation of, 95, 97, 103
Marginal table, marginal total, 185, 192–193, 218–219, 230, 248, 251, 270
Maximum-likelihood equations, 48, 62–63, 85, 136–138, 141, 145, 193, 202, 333
  solution of, 64, 125–126
Maximum-likelihood estimate, 63, 69, 74, 86, 127, 205–207
  in additive logit model, 295, 297–299
  in additive log-linear model, 124–125, 130
  calculation by iterative proportional fitting, 218–219
  with closed-form expression, 254, 269
  for conditional independence, 187, 190, 194, 200
  in hierarchical log-linear model, 270, 285–286
  in homogeneity model, 121, 123, 142

in independence model, 103, 110, 113–114
large-sample properties, 69, 143, 323–324, 339
in logit model, 306, 314, 317, 319, 351–352
in log-linear time-trend model, 9–11
in model of complete independence, 204
in model of constant rates, 45, 49–50, 54–56
in model of no three-factor interaction, 169, 172, 179, 192, 195, 213, 225, 234, 236, 271
in quadratic log-linear model, 30–34
in quantal-response model, 345
in saturated model, 73, 249
in symmetry model, 25
Measure of association, 104–105, 118–119
Model of constant rates, 45–48, 50, 54–56
Model of no three-factor interaction, 161, 181–182, 192–193, 198, 208–209, 213, 225, 233, 239–240, 293–295
  asymptotic variances for, 212, 220, 222, 243–244
  chi-square statistics for, 176, 223, 270
  estimates of log cross-product ratios, 176
  estimation of log odds, 234–235
  iterative proportional fitting, 192, 194, 217–219
  maximum-likelihood estimate, 192, 195, 271
  Newton–Raphson algorithm, 169–171, 236–237
  standardized residual, 230
Multinomial model, multinomial distribution, 61–63, 124, 134–135, 138, 143, 145, 184, 197, 263, 346

N

Newton–Raphson algorithm, 346
  for additive logit model, 295, 297
  for additive log-linear model, 128, 130
  for logit model, 305, 314, 316, 322, 334
  for log-linear model, 65, 68–69, 72, 139
  for log-linear time-trend model, 10–11
  for model of no three-factor interaction, 169–170, 173, 207–208, 210–211, 215–217, 236, 239
  for quadratic log-linear model, 31–32

Normal approximation, 168, 242–243, 342
  for distribution of adjusted residual, 5–7,
    17–18, 48, 78, 104, 111, 121, 130–131,
    150, 188, 328–329, 342–343
  for distribution of estimated log cross-
    product ratio, 165–167
  for distribution of estimated log relative
    risk, 164
  for distribution of maximum-likelihood es-
    timate, 22, 40, 57, 70, 143–144, 322–
    324, 339–340
  for distribution of standardized residual,
    230, 275
  for distribution of standardized value, 28,
    123, 252, 265
Normal equations, 32–33, 128–129

O

Odds, 41, 94, 98, 100, 102, 178–179, 190–
  191, 279, 310
  estimate of, 96, 282, 324
  estimated asymptotic standard deviation
    of, 96
One-way table, 1
Ordinal variable, 302–303
Orthogonal polynomial score, 29

P

Paired comparisons, 293
Parameter estimate, 131, 142–143, 213,
  218–219, 231–232, 249, 253, 261, 266–
  267
Partial association, partial-association
  coefficient, 184–187, 190–192, 196, 201,
  203, 262, 282, 284
Partition of chi-square statistics, 224, 226
Pearson chi-square statistic, 149
  in additive logit model, 301
  for conditional independence, 187
  in equiprobability model, 4, 79–80
  in homogeneity model, 123
  in independence model, 103–104, 110
  in logit model, 326, 341, 348
  in log-linear time-trend model, 25, 79–80
  in marginal independence model, 186
  in model of constant rates, 50–51, 54,
    56–57

  in quadratic log-linear model, 38
  relationship to variance test, 86
  in symmetry model, 26
Poisson distribution, 43–47, 50, 53–54, 62,
  64, 86, 124, 134, 138, 143, 145, 163, 197,
  293
Probit model, 293, 344–345

Q

Quadratic log-linear model, 29–30
Quantal-response model, 344–346

R

R^2 statistic, 17, 50, 56, 76–77, 88–89, 133,
  225
Rankit plot, 20–21
Regression, 10, 13, 17, 21–22, 32, 36, 39,
  65–67, 70, 272, 292, 296, 309, 337, 342,
  344–345, *see also* Weighted least
  squares, weighted regression
Relative risk, 163–165, 168
Residual analysis, 77
Row-multinomial model, 134–138, 143–144,
  146

S

Saturated model, 72, 75, 144, 148, 161, 198,
  218, 232–233, 234–235, 248–249, 252–
  253, 256, 263, 266–267
Scoring algorithm, 346
Selection of models, 223, 272
Simultaneous confidence intervals
  for cross-product ratio, 152–153
  for differences in log rates, 58–60
  for log cross-product ratios, 118, 152–153
Standard deviation, of estimated marginal
  probability, 95–96
Standardized parameter estimate, 231
Standardized residual, 78, 230, 272, 275
Standardized value, 165–167, 169, 181, 232,
  252–253, 265–268, 282–283, 309
  for sample mean in symmetry model,
    28–29

for testing linear trend in proportions, 123
for testing model of no three-factor interaction, 180
Sufficient configuration, 193
Symmetry model, 23–25, 27–29, 84

T

Three-way table, 160, 217, 224, 226, 231, 247
Two-way table, 91, 134, 151, 294

V

Variance test, 86

W

Weighted least squares, weighted regression, 11, 32, 39–40, 65, 69, 128, 143,
173–174, 209, 214–215, 295–297, 305–
306, 308, 314–316, 318, 321–323, 328–
329, 334–335, 339, 342–343, *see also* Regression
Working logarithm, 12, 14, 33, 128, 174, 216
Working logit, 297, 315, 321, 335, 338
Working weight, 321, 335

Y

Yates algorithm, 220, 232–233